Comparative Perspectives on Inquiry-Based Science Education

Stuart Bevins
Sheffield Hallam University, UK

Louise Lehane
St. Angela's College, Ireland

Josephine Booth
Sheffield Hallam University, UK

A volume in the Advances in Educational Technologies and Instructional Design (AETID) Book Series

IGI Global
DISSEMINATOR OF KNOWLEDGE

Published in the United States of America by
IGI Global
Information Science Reference (an imprint of IGI Global)
701 E. Chocolate Avenue
Hershey PA, USA 17033
Tel: 717-533-8845
Fax: 717-533-8661
E-mail: cust@igi-global.com
Web site: http://www.igi-global.com

Library of Congress Cataloging-in-Publication Data

Names: Bevins, Stuart, editor.
Title: Comparative perspectives on inquiry-based science education / Stuart Bevins, Louise Lehane, and Josephine Booth, editors.
Description: Hershey PA : Information Science Reference, [2019]
Identifiers: LCCN 2017060870| ISBN 9781522554394 (hardcover) | ISBN 9781522554400 (ebook)
Subjects: LCSH: Science--Study and teaching (Higher) | Science teachers--Training of. | College teachers--Training of. | Inquiry-based learning. | Comparative education.
Classification: LCC Q181 .C5545 2018 | DDC 510.71/1--dc23 LC record available at https://lccn.loc.gov/2017060870

This book is published in the IGI Global book series Advances in Educational Technologies and Instructional Design (AE-TID) (ISSN: 2326-8905; eISSN: 2326-8913)

British Cataloguing in Publication Data
A Cataloguing in Publication record for this book is available from the British Library.

For electronic access to this publication, please contact: eresources@igi-global.com.

Advances in Educational Technologies and Instructional Design (AETID) Book Series

Lawrence A. Tomei
Robert Morris University, USA

ISSN:2326-8905
EISSN:2326-8913

Mission

Education has undergone, and continues to undergo, immense changes in the way it is enacted and distributed to both child and adult learners. In modern education, the traditional classroom learning experience has evolved to include technological resources and to provide online classroom opportunities to students of all ages regardless of their geographical locations. From distance education, Massive-Open-Online-Courses (MOOCs), and electronic tablets in the classroom, technology is now an integral part of learning and is also affecting the way educators communicate information to students.

The **Advances in Educational Technologies & Instructional Design (AETID) Book Series** explores new research and theories for facilitating learning and improving educational performance utilizing technological processes and resources. The series examines technologies that can be integrated into K-12 classrooms to improve skills and learning abilities in all subjects including STEM education and language learning. Additionally, it studies the emergence of fully online classrooms for young and adult learners alike, and the communication and accountability challenges that can arise. Trending topics that are covered include adaptive learning, game-based learning, virtual school environments, and social media effects. School administrators, educators, academicians, researchers, and students will find this series to be an excellent resource for the effective design and implementation of learning technologies in their classes.

Coverage

- Hybrid Learning
- Curriculum Development
- K-12 Educational Technologies
- Bring-Your-Own-Device
- Instructional Design Models
- Instructional Design
- Web 2.0 and Education
- Game-Based Learning
- Higher Education Technologies
- Virtual School Environments

IGI Global is currently accepting manuscripts for publication within this series. To submit a proposal for a volume in this series, please contact our Acquisition Editors at Acquisitions@igi-global.com or visit: http://www.igi-global.com/publish/.

The Advances in Educational Technologies and Instructional Design (AETID) Book Series (ISSN 2326-8905) is published by IGI Global, 701 E. Chocolate Avenue, Hershey, PA 17033-1240, USA, www.igi-global.com. This series is composed of titles available for purchase individually; each title is edited to be contextually exclusive from any other title within the series. For pricing and ordering information please visit http://www.igi-global.com/book-series/advances-educational-technologies-instructional-design/73678. Postmaster: Send all address changes to above address.

Titles in this Series

For a list of additional titles in this series, please visit: www.igi-global.com/book-series

Performance-Based Assessment in 21st Century Teacher Education
Kim K. Winter (Western Carolina University, USA) Holly H. Pinter (Western Carolina University, USA) and Myra K. Watson (Western Carolina University, USA)
Information Science Reference • copyright 2019 • 295pp • H/C (ISBN: 9781522583530) • US $175.00 (our price)

Modern Technologies for Teaching and Learning in Socio-Humanitarian Disciplines
Larisa Alexandrovna Darinskaia (Saint Petersburg State University, Russia) and Gallina Ivanovna Molodtsova (Saint Petersburg State University, Russia)
Information Science Reference • copyright 2019 • 292pp • H/C (ISBN: 9781522578413) • US $175.00 (our price)

Critical Literacy Initiatives for Civic Engagement
Angela M. Cartwright (Midwestern State University, USA) and Emily K. Reeves (Midwestern State University, USA)
Information Science Reference • copyright 2019 • 242pp • H/C (ISBN: 9781522580829) • US $185.00 (our price)

Digital Games for Minority Student Engagement Emerging Research and Opportunities
Rupanada Misra (Rutgers University, USA) Leo Eyombo (Howard University, USA) and Floyd T. Phillips (New York Institute of Health and Behavior, USA)
Information Science Reference • copyright 2019 • 120pp • H/C (ISBN: 9781522533986) • US $125.00 (our price)

Biometric Authentication in Online Learning Environments
A.V. Senthil Kumar (Hindusthan College of Arts and Science, India)
Information Science Reference • copyright 2019 • 314pp • H/C (ISBN: 9781522577249) • US $195.00 (our price)

Intercultural Foreign Language Teaching and Learning in Higher Education Contexts
Piotr Romanowski (University of Warsaw, Poland) and Ewa Bandura (Jagiellonian University, Poland)
Information Science Reference • copyright 2019 • 323pp • H/C (ISBN: 9781522581284) • US $175.00 (our price)

Multimodal Narratives in Research and Teaching Practices
J. Bernardino Lopes (Universidade de Trás-os-Montes e Alto Douro, Portugal) Maria Clara Viegas (Politécnico do Porto, Portugal) and José Alexandre Pinto (Politécnico do Porto, Portugal)
Information Science Reference • copyright 2019 • 418pp • H/C (ISBN: 9781522585701) • US $185.00 (our price)

Responsible Entrepreneurship Education Emerging Research and Opportunities
Giacomo Marzi (University of Lincoln, UK) and Andrea Caputo (University of Lincoln, UK)
Business Science Reference • copyright 2019 • 198pp • H/C (ISBN: 9781522582892) • US $165.00 (our price)

IGI Global
DISSEMINATOR OF KNOWLEDGE

701 East Chocolate Avenue, Hershey, PA 17033, USA
Tel: 717-533-8845 x100 • Fax: 717-533-8661
E-Mail: cust@igi-global.com • www.igi-global.com

Table of Contents

Detailed Table of Contents

Section 1
Introducing Inquiry-Based Science Education and the Chain Reaction Project

This section begins with an introduction to inquiry-based science education and more specifically the benefits of inquiry to student learning. The second chapter within this section focuses on the Chain Reaction project itself and provides extensive detail on the aims, objectives, and overall structure and implementation of the project.

The National Science Education Standards as well as the 2061 project reaffirm the conviction that inquiry is central to the achievement of scientific literacy. The National Science Education Standards use the term inquiry in two ways: (1) inquiry as content understanding, in which students have opportunities to construct concepts, patterns, and to create meaning about an idea in order to explain what they experience and (2) inquiry in terms of skills and abilities. Under the category of abilities or skills, Bybee includes identifying and posing scientifically oriented questions, forming hypotheses, designing and conducting scientific investigations, formulating and revising scientific explanations, and communicating and defending scientific arguments. It is suggested that many of these abilities and skills are in alignment with those that characterize inquiry-type laboratory work, an activity that puts the student in the center of the learning process. Thus, the following chapter will mainly deal with the inquiry-based science education in the laboratory.

This chapter aims to introduce the policy perspective of the relevant European funding stream, tracing the evolution of the science with and for society call from the first Framework Programme through to Horizon 2020. It also introduces the Pupil Research Initiative (PRI) and how the Chain Reaction project built upon PRI's legacy, using its successes to inform a project that was relevant across Europe. The chapter highlights the key aspects of the chain reaction project, for example the Pupil Research Briefs (PRBs) and Express Yourself Conferences, and how these were adapted, and in the case of the PRBs, expanded upon to be more relevant to different partner contexts and the needs of their education systems and curricula.

Section 2
Comparing the Chain Reaction Experiences Between Countries

This section details the experiences of teacher educators from the different partner countries involved in the Chain Reaction project. While there may be some deviation in terms of structure to each chapter, the focus of all chapters is on the implementation of Chain Reaction in each of country.

This chapter describes aspects of the successes and obstacles in the delivery of chain reaction in school science classrooms in England. It offers an overview of recent education reform in England and moves on to discuss the implementation of the program and provides anecdotal evidence from teachers to support emerging themes from the delivery experience. Issues of "time" restraints and over-burdened "curricular" as barriers to the deployment of inquiry approaches are highlighted by identifying positive outcomes and offering implications for science education across Europe.

This chapter presents a case study of the experience of the Irish Chain Reaction (CR) team, which took place during a time of significant curriculum change in the lower secondary school system. As such, it is hoped that those interested in teacher professional development will find the case of interest while acknowledging the varied cultural, material and structural resources, and limitations that influence the context of any educational change process. The authors have placed a significant focus on describing the Irish context to begin this chapter, initially comparing the old science syllabus with the new "specification," with thought given to the existing modalities of student learning and the nature of teacher professional collaboration and the developments that will be required by the new curriculum. The authors then summarize the plan for CR implementation through a professional learning community (PLC) focused on supporting teacher agency and autonomy in the design of inquiry-based science education (IBSE) classes. Teacher and student reflections of experience are presented.

Ilia State University (ISU) is one of Georgia's leading universities in teacher education and has taken part in a number of EU funded projects aimed at implementing new approaches to science teaching in Georgia, including those with an inquiry-based science education (IBSE) focus. This chapter gives an overview

An inquiry science-based education is commonly followed in a variety of educational contexts around the world and is a key parameter in various national curriculum guidelines. The impetus of this chapter is to record the initial and final reactions of science teachers participating in a series of one-year action research and training program that took place in the University of Crete (UoC) in 2013-2016, identify their perception of the first training course, and explore the impact this data had on the program's redesign for the following training session by the technical board. Teacher reactions and responses regarding what they thought had, and had not, worked well in their classes were taken into account prior to re-designing the training program that the new teachers were going to join the following year. Looking into the general benefits as well as challenges, the authors also examined the overall effect of the UoC IBSE training program to participants as reported by both students and teachers.

Due to the huge development in the communication sector, science education has become very challenging. Inquiry-based science education (IBSE) has been proposed by academic leaders and policymakers as having the potential to increase student involvement and provide opportunities for the development of the key skills desired by employers in the future. Thirty teachers from 15 schools participated in the project and disseminated the knowledge obtained to their colleagues through peer-to-peer knowledge sharing. Most teachers expressed an increase in their understanding of IBSE, and 66% agreed that they received high-quality IBSE/EUPRB briefings. Students showed motivation to learn science and obtained scientific approaches and steps of answering their inquiries in order understand their surroundings, improve their achievement of scientific knowledge, build positive science attitudes, improve their understanding of the nature of science, and develop IBSE skills such as scientific data collection analyses, interpretation, and how to provide evidence to test ideas and develop theories.

The chapter presents how the Chain Reaction project was implemented in the context of science education in Slovakia. The educational reform highlighted inquiry as a leading approach in science education. But teachers were not prepared for implementation of such approaches. There was a lack of resources and

Preface

This book offers comparative perspectives of the experiences of different partners involved in an European Commission funded Framework Programme 7 project, Chain Reaction, which ran for three years between 2013-16. The project, and in turn this book, focuses on the area of inquiry-based science education (IBSE). Defining IBSE is a natural starting point as the book is introduced. It leads us to refer to curriculum and policy which address the concept in more detail. IBSE is viewed as an inspiring way to learn science as it focuses on students' own interests and enables active learning to take place by encouraging them to conduct their own investigations (Rocard, Csermely, Jorde, Lenzen, Walberg-Henriksson, & Hemmo, 2007). In other words, it develops students' scientific skills to ultimately reflect the spirit of professional science.

IBSE has been a key topic of debate and discussion in the science education community globally for many years and has in turn informed policy and practice to varying degrees internationally. The project itself will be detailed more precisely in Chapter 2 however, it is important to note here the key features of the project to inform the readers of this book.

Eleven project members from countries across Europe, plus Jordan, committed to providing practicing science teachers with new understandings of how to support science teaching and learning in the classroom—using an inquiry focus for their teaching. While some countries had existing understanding and knowledge of inquiry approaches, for many of the Chain Reaction partners, getting to grips with a new way of supporting science learning often proved difficult.

Our main objective was to help in developing an inquiry habitus in teachers' practices. The habitus is essentially the way in which a culture is embedded in an individual (Harker, 1984) and how this internalisation becomes the basis for a person's behaviour (Bidet 1979, cited in Harker, 1984). It is the belief or orientation of the teacher towards how their subject should be learned and this ultimately influences their decision-making regarding particular pedagogical approaches.

The book brings together the experiences of the project consortium members to critically reflect on successes and areas for development that emerged from their involvement in the project as well as detailing deployment of an inquiry approach in science classrooms in differing educational contexts.

The experiences of several partner countries are documented in the book:

- England
- Ireland
- Slovakia
- Georgia
- Jordan

- Greece
- Italy
- Germany

Partners report interesting insights into the enabling and disabling factors that they found during the project. The breadth of coverage is true from both a geographical, cultural and political perspective and therefore, offers a true reflection of just how difficult a change process can be. Indeed, when one looks at the literature on teachers' habitus and a changing belief system, it is understandable that change is difficult and takes time for authentic change to happen. Many of these ideas are further explored throughout the book.

As well as the above issues, the book draws on the experiences of a teacher at the forefront of the change process. This teacher, from the Irish context, details his experience of being involved in Chain Reaction and his current practice is situated in the context of student-led approaches in the classroom. He discusses the difficulties of 'working against the grain' with ever present curricular demands and constraints not unique to the Irish context and suggests that the milieu of day-to-day classroom practice can accommodate inquiry approaches.

We summarise the work in a 'reflections and conclusions' chapter which includes three brief reflective accounts from the editors. These accounts are written from a personal perspective and are intended to bring to life the implementation of a large and complex international education project.

The theme of IBSE is both current and relevant in classrooms globally. The increased reference to it within curricular policy documents and the scope of research dedicated to inquiry in the science classroom in the research community highlights strongly how the theme of IBSE is of significant importance. Existing studies advocate how inquiry can facilitate students' motivation and interest in science and understanding of scientific concepts. Potentially, the strongest contribution to knowledge which this current work makes is through the provision of raw accounts of how inquiry in the classroom can actually manifest. A number of funded projects, focused on IBSE, have taken place over recent years particularly in Europe and S.E. Asia. However, through external funding constraints some of these projects have finished abruptly without effective evaluation of the experience and the key learning. This book offers the opportunity to reflect on both the successes and obstacles of a large-scale professional development project to develop teachers' knowledge and understanding of IBSE.

It is envisaged that this book will serve as a useful resource for a variety of stakeholders involved in science education, be it from a reform or practice perspective. For readers looking to become part of a large, complex project like Chain Reaction, this book will offer a lens into the highs and lows of such a project. From a teacher's perspective, the lived experiences of the teacher in chapter 11 which offer some assurances of how difficult it is to change practice while simultaneously documenting his continued drive to achieve effective change.

It can also be suggested that this book would serve pre-service teachers who can potentially see the initial difficulty that practicing teachers can have with employing what can be considered the "gold standard" of how to approach science teaching in the classroom. Projects like Chain Reaction look at constructivist approaches to teaching and learning, in the form of inquiry specifically within our project. The editors' experience is that pre-service teachers often find it difficult to connect such theories to practice. It is hoped that through reading this book, pre-service teachers can begin to see the connection.

Those involved in policy change can also benefit from reading this book as chapters within the book highlight the difficulty in changing teachers' practice. This is something that policy makers need to be cognizant of before introducing changes to the curriculum which may be misaligned with teachers' own beliefs and values about teaching science.

Finally this book will benefit those in the early stages of research, especially those focused on action research, as it outlines the sometimes unpredictable and complex nature of collaboration between academic researchers and classroom teachers.

Chapter 1 discusses inquiry in the science classroom and sets the scene appropriately for following chapters. The chapter characterises inquiry in terms of skills and processes and relates these, particularly, to school laboratory work. The chapter is framed by relevant existing literature and summarises expected benefits of engaging in inquiry approaches. It outlines a variety of international projects focused on inquiry. Finally this chapter extends the discussion to look at argumentation in the science classroom and associates its role in the students' experience of practicing authentic inquiry.

Chapter 2 examines the evolution and aims of the European Commission (EC) Framework Programme 7 Science and Society funding stream, and the policy environment that contributed to this. Additionally, describes the successful project that provided the inspiration for Chain Reaction, before addressing the key aims of the project, its delivery over the three years, the countries involved in the project and critically, the rationale for the project.

Chapter 3 offers critical insight in to the deployment of the Chain Reaction project in England against a backdrop of high stakes testing and teachers' concerns regarding time to use inquiry approaches. It begins by describing the current education climate in England and goes on to discuss the implementation of Chain Reaction. The chapter also highlights some notable successes such as the student research conferences and discusses inquiry in the English education system more broadly.

Chapter 4 describes the experiences and perspectives from the Irish context and highlight a time of curriculum change in Ireland, therefore the experiences of the authors and in turn the project teachers were very much focused on a process of change. The chapter begins by describing the original science curriculum with the consistent message being, that the design of this curriculum led to routines in the classroom focused on product-driven learning. It then details the new science curriculum of which scientific inquiry and the Nature of Science are naturally embedded in its design. The chapter then leads on to describe the project implementation and barriers with detailed vignettes of the teacher workshops that took place in Ireland, together with the in-school practice arising after these workshops. It also presents a summary of the teachers' reflections of being involved in the project with particular reference to the constraints that teachers had. It also provides the students' voices and details summaries of interviews carried out with students who largely held positive views towards inquiry. Finally, the chapter outlines the impact this project has had on practice.

Chapter 5 gives an overview of the project in Georgia, describing first the various shifts in government policy in education since the collapse of the Soviet Union in 1991, including changes to the science curriculum and to teacher education. The chapter then gives a description of the implementation of the project in Georgia; including the challenges encountered and concludes with reflections on the impact of the project from the teachers involved.

Chapter 6 is organised through four sections. The first section describes the Italian secondary school context with section two discussing how the Chain Reaction project was implemented in the Italian educational context. In the third section, the authors briefly explain the main barriers they encountered in the implementation of the Chain Reaction project and how the Chain Reaction activities impacted on local teaching practice. Finally, they reflect on their learning from the project and offer future steps they have implemented.

Chapter 7 presents a summary of the project implementation in Germany. It places a large emphasis on the political association with education and how this informs policy and practice. The authors then detail the project implementation and addresses particular skills that were the focus of students' experiences. For example, they highlight students' communication skills as something which the team in Germany particularly wanted to foster and therefore their project implementation angled their change initiatives towards developing these and other similar skills.

Chapter 8 describes the authors' experiences with a particular focus on how research has informed teacher educators' practice going forward. They do this by specifically capturing teachers' reactions to IBSE and their involvement in the training initiatives at the beginning and at the end of one year. The authors also provide a detailed summary of the Greek Education system. Similarly, to the other chapters, they document their overall project implementation by providing a comprehensive project summary.

Chapter 9 documents the experience of the Chain Reaction project from the Jordanian context with particularly emphasis on cultural and political influences and how these impacted on the experience of all those involved in the project. Similarly, to the other chapters, it then documents the implementation of the project from this context and details the barriers and obstacles they faced during the lifetime of the project.

Chapter 10 situates inquiry in terms of curricular reform in the Slovak context. The chapter describes the Chain Reaction deployment in Slovakia and focuses on professional development issues and the student research conferences as a motivational mechanism for participating students. The chapter concludes with a discussion of project impact with particular emphasis on the university initial teacher education provision.

Chapter 11 presents the experiences of one of the Chain Reaction teachers in designing and delivering his inquiry sessions. It also documents his journey to becoming involved in a national professional development agency to assist with the changing science curriculum in Ireland. This chapter begins by exploring the theme of evolution of practice and details how being involved in a professional learning community supported him in his changing understanding and action of inquiry in the classroom. He details his experience of enacting a pupil research brief (PRB) in Nutrition which he developed himself. Many teachers in the Chain Reaction project used existing PRBs, some of which are showcased at the end of this book. However, this teacher chose to design his own. Part of the sharing of his experience focuses on an interview he conducted with students in his classroom to understand their experiences, this is outlined in the book chapter. He also presents some of the information which a practicing scientist shared with his students during a presentation on what it was like to work as a real scientist. The use of such role models are detailed further through the book. Finally, the teacher in this chapter describes openly his experience of facilitating other teachers to engage in continued professional development at a time of curriculum change in Ireland and shares with the reader how being involved in the Chain Reaction project facilitated working with teachers.

Chapter 12 presents reflections from a project management perspective. It starts with outlining the literature on scientific inquiry, from the perspective of the benefits of inquiry to student learning. The authors examine the experience of teacher educators who are both relatively new and experienced in change initiatives and provide an honest account of whether, having completed the Chain Reaction project, that a change in the practice of the participant teachers occurred and is sustainable. They also offer three reflective accounts focusing on developing experience as a novice teacher educator, managing a large, pan-European project, and teachers' experiences of inquiry. The authors also consider the lessons learned from being involved in such a large-scale project, in terms of what they would do the same and differently next time. They conclude with a brief discussion of what the future may hold for science education in Europe.

Inquiry in the science classroom continues to be regarded as one of the best ways for students to learn and experience scientific thinking and practice. The wealth of existing international literature advocating inquiry approaches as an effective way of learning science is substantial. However, among this body of literature there are relatively few cases reported of inquiry in action. This book is timely as it offers the reader concrete examples of inquiry in the science classroom developed from a large-scale teacher professional development project. Additionally, it includes links to the Chain Reaction materials online, both in English and some of the partner country languages, as well as the Teacher Guidance notes for the project, and the Chain Reaction Toolkit developed by the consortium to help teachers work through the Plants in Space EUPRB. The book's main contribution to knowledge in the field is through the open and honest descriptions, provided by each of the authors, which document techniques that worked, didn't work, and techniques that could work with modification. The authors also highlight teachers' and students' responses to their participation which brings the guided inquiry approach used in the project to life in the day-to-day practice of classroom science.

Over the course of the three years of the Chain Reaction project, across twelve partner countries, 565 science teachers and 4,775 school students, between the ages of 13-17, participated in inquiry-based science in their schools. These are significant numbers and are testament to the dedication of all involved and their commitment to engaging with a guided inquiry approach. Sustainability of the project was an important consideration from the outset of the project and a range of actions have taken place, through each partner, to ensure that maximum sustainability was reached within their educational context. At times, throughout the book, the messages conveyed by the authors do have the contention of the sustainability of inquiry practice at the core of their discussions. While we cannot claim that every participant teacher is using the model continuously and committed to an inquiry approach, each partner is aware that substantial numbers are continuing to develop their knowledge and practice of inquiry in the classroom.

Finally, we acknowledge that this book is overtly pro inquiry and that the editors and contributors are committed to developing and promoting inquiry in the science classroom. However, we concede that there are competing arguments and that inquiry is not always an appropriate approach to take. Therefore, we have taken steps to ensure that the detailed accounts and stories presented in this book are balanced through critical examination and truthful reporting. We do not advocate a 'one size fits all' suggesting that inquiry should be the dominating approach to teaching and learning school science, rather that inquiry should be a part of a teachers' repertoire of classroom methods. Indeed, it can be argued that overreliance on the one approach can adversely affect students' motivation towards learning science. Therefore, through this book, our intention is to offer the Chain Reaction model of inquiry for consideration and potential use as an addition to science teachers' stock of approaches.

REFERENCES

Harker, R. (1984). On reproduction, habitus and education. *British Journal of Sociology of Education*, *5*(2), 117–127. doi:10.1080/0142569840050202

Rocard, M., Csermely, P., Jorde, D., Lenzen, D., Walberg-Henriksson, H., & Hemmo, C. (2007). *Science Education Now: A Renewed Pedagogy for the Future of Europe*. Available: http://ec.europa.eu/research/science-society/document_library/pdf_06/report-rocard-on-science-education_en.pdf

Acknowledgment

The editors would like to acknowledge the help of all the people involved in this book project. Without their support, this book would not have become a reality.

The Chain Reaction project described here and which led to this book received funding from the European Community's Seventh Framework Programme fp7/2007-2013 under grant agreement No 321278. Without this funding the project would not have been possible, and all the authors and editors would like to thank the European Commission for the funding and support received for the duration of the project.

Secondly, we would like to offer our sincere gratitude to the chapter authors who contributed their time and expertise to this book.

Thirdly, the editors wish to acknowledge the valuable contributions of the reviewers regarding the improvement of quality, coherence, and content presentation of chapters.

Finally, the editors would like to express their gratitude to all consortium members and their teachers and students who were involved in the Chain Reaction project. It was a pleasure to work with you on helping to improve science teaching and learning in the classroom.

Stuart Bevins
Sheffield Hallam University, UK

Louise Lehane
St Angela's College, Ireland

Josephine Booth
Sheffield Hallam University, UK

Section 1

Introducing Inquiry-Based Science Education and the Chain Reaction Project

This section begins with an introduction to inquiry-based science education and more specifically the benefits of inquiry to student learning. The second chapter within this section focuses on the Chain Reaction project itself and provides extensive detail on the aims, objectives, and overall structure and implementation of the project.

Chapter 1
Comparative Perspectives on Inquiry-Based Science Education

Rachel Mamlok-Naaman
Weizmann Institute of Science, Israel

ABSTRACT

The National Science Education Standards as well as the 2061 project reaffirm the conviction that inquiry is central to the achievement of scientific literacy. The National Science Education Standards use the term inquiry in two ways: (1) inquiry as content understanding, in which students have opportunities to construct concepts, patterns, and to create meaning about an idea in order to explain what they experience and (2) inquiry in terms of skills and abilities. Under the category of abilities or skills, Bybee includes identifying and posing scientifically oriented questions, forming hypotheses, designing and conducting scientific investigations, formulating and revising scientific explanations, and communicating and defending scientific arguments. It is suggested that many of these abilities and skills are in alignment with those that characterize inquiry-type laboratory work, an activity that puts the student in the center of the learning process. Thus, the following chapter will mainly deal with the inquiry-based science education in the laboratory.

INTRODUCTION AND LITERATURE REVIEW

Science educators have suggested that many benefits accrue from engaging students in science laboratory activities (Dori, Sasson, Kaberman, & Herscovitz, 2004, Hofstein & Lunetta, 1982; Tobin, 1990; Hodson, 1990; Lazarowitz & Tamir, 1994; Lunetta, 1998; Hofstein & Lunetta, 2004; Lunetta, Hofstein, & Clough, 2007). More specifically, they have suggested that, when properly developed, inquiry-centered laboratories have the potential to enhance students' meaningful learning, conceptual understanding, and their understanding of the nature of science. Hofstein and Walberg (1995) suggested that inquiry-type laboratories are central to learning science, since students are involved in the process of conceiving problems and scientific questions, formulating hypotheses, designing experiments, gathering and analyzing data, and drawing conclusions about scientific problems or phenomena.

DOI: 10.4018/978-1-5225-5439-4.ch001

With the reform in science education in the1960s in many countries, the ideal became to engage students with investigations, discoveries, inquiry, and problem solving activities. In other words, the laboratory became the core of the science learning process (Lazarowitz & Tamir, 1994). Many research studies have been conducted to investigate the educational effectiveness of inquiry-based laboratory work in science education in facilitating the attainment of the cognitive, affective, and practical goals. These studies were critically and extensively reviewed in the literature. From these reviews it is clear that in general, although the science laboratory has been given a distinctive role in science education, research has failed to show simplistic relationships between experiences in the laboratory and student learning. Baird (1990) referred to these meta-cognitive skills as "Learning outcomes associated with certain actions taken consciously by the learner during a specific learning episode." (p. 184). Meta-cognition involves elaboration and application of one's learning, which can result in enhanced understanding.

In attempting to develop scientific literacy among students, teachers must create effective learning environments in which students are given opportunities to ask relevant and scientifically sound questions (Penick, Crow & Bonnstetter, 1996). Several studies noted the importance (and value) of questioning skills. Dori and Herscovitz (1999) reported that in fostering 10th grade students' capabilities, posing questions improved their problem-solving ability. Also, Hofstein, Shore and Kipnis (2004) in a previous publication regarding this project, reported that providing students with opportunities to engage in inquiry-type experiments in the chemistry laboratory improved their ability to ask high-level questions, to hypothesize, and to suggest questions for further experimental investigations.

The objectives to be attained in the laboratory depend on a wide range of factors. Among others, these will include the teacher's goals, expectations, subject and pedagogical content knowledge as well as the degree of relevance to the topic, the students' abilities and interests, and many other logistical and economic considerations related to the school settings and facilities. (See Table 1).

In curricular-type projects developed during the 1960s the laboratory was intended to be a place for inquiring, developing, and the testing of theories as well as providing students with the opportunity to 'practice being a scientist'. Many research studies (summarized for example by Mamlok-Naaman, Eilks, Bodner & Hofstein, 2018) were conducted with the goal in mind to explore the effectiveness of the laboratory for attaining the many objectives (both cognitive as well as affective) that had been suggested in the science education literature.

The traditional list of objectives includes:

Table 1. Suggested goals for laboratory activity (Bennett, 2003 pp. 78-79)

Goals for Laboratory Activity
to encourage accurate observation and description
to make scientific phenomena more real
to enhance understanding of scientific ideas
to arouse and maintain interest (particularly in younger pupils)
to promote a scientific method of thought

- Understanding of scientific concepts
- Interest and motivation
- Attitude towards science
- Scientific practical skills and problem solving abilities
- Scientific habits of mind
- Understanding the nature of science (NOS)
- The opportunity to do science

At the end of the twentieth century there was increasing understanding from cognitive sciences that learning is contextualized and that learners construct knowledge by solving genuine, meaningful problems (Wenger, 1998). The school science laboratory can offer students opportunities to have some control of their activities, enhance their perception of sense of *ownership* and *motivation* (Johnstone & Al-Shuaili, 2001). It can be an environment particularly well suited for providing a meaningful context for learning, determining and challenging students' deeply held ideas about natural phenomena, and constructing and reconstructing their ideas. Though a complex process, meaningful learning in the laboratory *can* occur if students are given sufficient time and opportunities to interact, reflect, explain, and modify their ideas. Engaging in *metacognitive* behaviors of this kind enables students to elaborate and to apply their ideas; the process can promote conceptual understanding as well as the development of problem-solving skills. The challenge is to help learners take control of their own learning in the search for understanding while providing opportunities that encourage them to ask questions, suggest hypotheses, and design investigations, "minds-on as well as hands-on" (Gunstone, 1991).

The beginning of the twenty-first century ushered in a new era of reform in science education. Both the content and pedagogy of science learning and teaching are being scrutinized and new standards intended to shape and rejuvenate science education are emerging (National Research Council - NRC, 1996, 2005). The focus was on the fact that inquiry is central to the achievement of scientific literacy. Teaching and learning science by inquiry is an approach that places the student in the center of the learning process. Such an approach poses challenges to both students and their teachers (Krajcik, Mamlok & Hug, 2001). In general, one of the characteristics of current reform is the change in the goals articulated for science teaching and learning namely, that science education should be targeted to all students (attaining scientific literacy for all students) and should be extended beyond the preparation of science oriented students for academic careers in the sciences (van den Akker, 1998). This is in fact a call for also rethink the goals for the learning in and from laboratory work. There are several aspects that characterize current reform. Among these are student's centered learning, learning by the inquiry method, and development of high learning skills such as argumentation, metacognition and asking relevant questions (relevant to the experimental situation). Inquiry in the context of science learning in general and inquiry in the science laboratory in particular are amongst the important components of this reform (Hofstein & Lunetta, 2004; Sere, 2002). It is suggested that many of these abilities and skills are in alignment with those that characterize inquiry-type laboratory work (practical work to include project-based learning), an activity that puts the student in the center of the learning process (see also Hofstein, Mamlok-Naaman, Navon, & Kipnis, 2005; Dori & Sasson 2007). Tobin (1990) based his review on constructivist ideas of providing students with experiences that would enable meaningful learning in the science laboratory. He wrote that:

Laboratory activities appeal as a way of allowing students to learn with understanding and, at the same time, engage in the process of constructing knowledge by doing science (p.405).

To attain this goal he suggested that students should be provided in the laboratory, with opportunities to reflect on findings, clarify understandings and misunderstandings with peers, and consult a range of resources that include teachers, books, websites, and other learning materials. His review reported that such opportunities rarely exist since teachers, in the laboratory, are so often preoccupied with technical and managerial activities. Similarly, Hodson (1993) suggested that although teachers generally professed belief in the value of student-driven, open, practical investigation, in general, their teaching practices in the laboratory failed to support this claim. He also argued that the research literature failed to provide evidence that standard school laboratory activities encouraged knowledge construction. He was critical of the research literature:

Despite the very obvious differences among, for example, practical exercises designed to develop manipulative skills or to measure 'physical constraints', demonstration-type experiments to illustrate certain key-concepts, and inquiries that enable children to conduct their own investigations, there is a tendency for researchers to lump them all together under the same umbrella title of practical work (p. 97).

Tobin wrote that teachers' interpretation of practical activity should be elaborated and made a part of the research design since a laboratory session could be open-ended inquiry in one classroom and more didactic and confirmatory in another teacher's classroom. He claimed, that much practical work is purposeless and often the explicit objectives of the practical work do not coincide with the purpose of practical experiences. In addition – to his opinion, many practical tasks have too many different teaching/learning objectives to focus on during instruction.

EXAMPLES OF INQUIRY-BASED SCIENCE EDUCATION PROJECTS AND STUDIES

Profiles: A European Project in the Framework of FP7

During the last years, the goal of the European Commission's FP7 was to promote inquiry-based science education (IBSE). PROFILES (Professional Reflection-Oriented Focus on Inquiry-based Learning and Education through Science) was one of these European projects in which 22 European institutions participated over 3 years (Bolte, Hofstein, Holbrook, Mamlok-Naaman, Rannikmae, Rauch, & Streller, 2011). The uniqueness of the PROFILES approach was in paying much attention to the raising of the self-efficacy of science teachers, so that they will be able to gain a level of competence and the confidence to teach students in an educationally meaningful and student motivational manner. The project aimed at giving teachers a better appreciation of the changing purpose of teaching science in schools and the value of being involved in networking with their peers. PROFILES recognized the learning needs to be relevant, challenging and rewarding for the students and involves students in the development of educational competences across the learning spectrum through an inquiry teaching-learning approach. The PROFILES' main objectives were to support teachers by providing an inspired, longitudinal, professional development program reflecting a range of stakeholder views and teacher needs. The professional development of science teachers attempts to enable teachers to find ways to enhance student motivation for the learning of science, both in terms of intrinsic motivation (relevance, meaningfulness, importance, as considered by the students) and extrinsic motivation (teacher encouragement, classroom environment and reinforcement of learning). In this way, the PROFILES project attempted to guide teachers to make school science teaching more meaningful in specific cultural settings (Mamlok-Naaman, Hofstein & Penick, 2007). For this, the professional development aims at enhancing the role of the science teacher for inquiry-based science education on a philosophical, curriculum implementation and student motivational/attainment level. The professional development included *school-based interventions by the teacher* to promote inquiry learning for the students and also to engage students in creative, scientific problem-solving and socio-scientific decision-making procedures.

Sere (2002) in France

Sere (2002) in France reported on a long-term project (Lab-Work in Science Education) conducted in seven European nations and wrote that:

The intention [of the study] was to address the problem of the effectiveness of lab-work, which in most countries is recognized as being essential to experimental sciences, but which turns out to be expensive and less effective than wished (p. 624).

The project focused mainly on the effectiveness of lab-work conducted in the context of science learning in the upper secondary schools. Information on practice was gathered through 23 case studies, surveys, and a tool that helps to map and describe the domain of laboratory work. Sere reported that the objectives typically articulated for laboratory work (i.e., understanding theories, concepts, and laws; conducting various experiments; learning processes and approaches; and applying knowledge to new situations) were too numerous and comprehensive for teachers to address successfully in individual laboratory sessions. In response, she suggested that the scope of the objectives for specific laboratory activities should be limited. Science curriculum developers and science teachers should make conscious choices among specific learning objectives for specific lab activities and clearly articulate the specific objectives for their students. Sere's "*targeted lab-work*" project produced a series of recommendations including the need for each laboratory activity to be supported by a particular strategy organized within a coherent long-term program plan with varied-kinds of laboratory work.

Inquiry Chemistry Laboratory (ICL) Program in Israel

A series of 100 inquiry-type experiments were developed and incorporated into the regular chemistry curriculum used in grades 11-12 in the ICL in Israel, most of them dealing with the key concepts taught in chemistry, e.g., acids-bases, stoichiometry, oxidation-reduction, bonding, energy, chemical-equilibrium, and the rate of reactions. The inquiry experiments range from totally 'open-ended' investigation, to those in which the student is asked to conduct only a 'partial inquiry'. Some experiments include design and planning as well as interpretation of results and scientific conclusions, whereas other experiments also include suggested hypotheses and relevant research questions (for more details, see Hofstein, Shore & Kipnis, 2004). The teachers choose approximately 20 experiments, by focusing on certain skills based on their teaching objectives and their students' abilities and interests.

The inquiry experiments are usually performed in small groups, 3-4 students in each group. The number of groups can vary from 4-8, according to the total number of students in the class. During the inquiry stage of the experiment, usually each group of students defines a different inquiry question, conducts a different experiment, and may reach different conclusions. Each group summarizes its inquiry-type experiment in a document fondly called "Hot report", and hands it to the teacher immediately after the lesson, to be assessed.

The students perform the experiments in small groups (3-4), by following the instructions in the laboratory manual. In the first phase (the pre-inquiry phase), the students are asked to conduct the experiment based on specific instructions. The 'inquiry phase' (the second phase) is where the students are involved in more 'open-ended-type' activities such as asking relevant questions, choosing a question for further investigation, planning an experiment, hypothesizing, conducting the experiment (including observations) to accept or reject the hypothesis, and finally analyzing the findings and arriving at conclusions. (Hofstein, Shore & Kipnis, 2004).

Most of the groups are not closely involved with each other's experiments. If teachers want their students to share the knowledge each group has accumulated, and if they want to emphasize and to strengthen certain issues concerning the experiment, they should conduct a summarizing discussion. Knowledge can refer to an inquiry skill and/or chemistry content knowledge. Since the inquiry approach is student-centered, most teachers start the summarizing discussion with students' presentations. For the students, presenting their work requires adequate preparation. The teacher should guide them regarding the way their work should be presented as well as what aspects should be discussed.

Teaching science using the inquiry approach presents challenges both for the teachers and for students (Taitelbaum et al., 2008; Krajcik, Mamlok & Hug, 2001; Lunetta, Hofstein & Clough, 2007). Teaching students by this approach necessitates the involvement of the students in the following activities:

1. Conducting experiments according to the teacher's instructions;
2. Making observations;
3. Raising as many questions as possible;
4. Choosing one question to be analyzed;
5. Constructing a hypothesis for the research question, based on scientific principles;
6. Designing an inquiry experiment to resolve the research question;
7. Making and organising observations once again;
8. Analysing and summarising the inquiry experiment;
9. Presenting the results to the whole class; and
10. Raising more questions.

A laboratory manual was developed to provide the necessary control regarding what students are doing during the laboratory sessions. In addition, the students should prepare "Hot reports" after each lesson. The 'hot report' is the group's product and is prepared in the laboratory during or immediately after the laboratory exercise. The reports were called 'hot reports' since they are conducted during the laboratory exercise and students cannot add additional information afterwards. This ensures that the information is reliable and thus represents the group's activities and deliberations. Figure 1 presents the components of the student's assessment.

Figure 1. The components of the student's assessment

Assessment of practical skills

Continuous Assessment of Students Inquiry Laboratory in Chemistry Observations and "Hot reports"

Social Skills			Conclusions			Inquiry			Observing		Conducting Experiment		
10%		10%	20%			35%			10%		15%		
Interest and curiosity	Cooperation in groups	Communication skills	Criticism and Summary	Conclusions	Presenting results	Planning	Hypothesizing	Questioning	Inquiry stage	Pre-inquiry stage	Handle dexterity	Following instructions	Experiment
													1
													2
													3
													4
													5
													6

Argumentation in the Science Laboratory

The chemistry laboratory may serve as a platform for teaching students how to argument (Karchevich, Hofstein & Mamlok-Naaman, 2013). Several researchers who focused on the issue of argumentation suggested that the inquiry-type laboratory in science education can provide opportunities for students to develop argumentation skills (See a detailed discussion in Hofstein and Kind (2012). However, only few research studies were conducted with the goal in mind of accepting or rejecting this assumption. For example, Rickey and Stacy (2000) found that students who participated in guided inquiry-type laboratories were better at evaluating evidence obtained from their research. Kelly, Druker and Chen (1998) analyzed the discourse in a physics laboratory and found that claims accompanied by justifications are generally given in response to the claims of a colleague in light of the experiment's findings or of the instructions, which may require an explanation or reasoning on the part of the student.

Kind et al. (2011) in the UK investigated the quality of argumentation among 12 to13 year old students in the UK in the context of secondary school physical science program. Their study explored the development of students' argumentation in who undertook three different designs of laboratory-based tasks. The tasks described in their paper involved the students in the following: collecting and making sense of data, collecting data for addressing conflicting hypothesis, and paper-based discussions in the pre-collected data phase about an experiment. Their finding showed that the paper-based task (the 3rd one in the above tasks list) generated larger number of arguments in a unit of time compared to the two other above mentioned tasks. In addition, they found that in order to encourage the development of high-level and authentic argumentation there is a need to change the practice that generally exists in the science laboratories in England. They suggested that more rigorous and longitudinal research is needed in order to explore the potential of the science laboratory as a platform for development of students' ability to argue effectively and in an articulated way.

Katchevich, Hofsrien, and Mamlok-Naaman (2013) focused on the process in which students constructed arguments in the chemistry laboratory while conducting different types of experiments. It was found that, *inquiry-type* experiments have the potential to serve as an effective platform for formulating arguments, owing to the special features of this learning environment. The discourse conducted during inquiry-type experiments was found to be rich in arguments, whereas that during *confirmatory-type* experiments was found to be sparse in arguments. In addition, it was found that the arguments, which were developed during the discourse of an inquiry-type experiment, were generated during the following stages of the inquiry process: hypothesis-building analysis of the results, and drawing appropriate conclusions. On the other hand, confirmatory-type experiment revealed a small number of arguments. In addition, the arguments that were posed in the confirmatory-type experiments were of law-level in their characteristic on detailed analysis of the discourse in the chemistry laboratory, we may conclude the open-ended inquiry experiments, stimulate and encourage the construction of arguments, especially the stages of hypotheses definition, analysis of the results and the conclusions drawing. Some arguments arise by individuals and some by the group. Both types of arguments consist of explanations and scientific evidence, which link the claims to the evidence. Therefore, it is suggested, that mentioned learning environments of open-ended inquiry experiments, is a platform for raising arguments. In this study we wanted to point out the main factors that stimulate raising argumentation in open-ended inquiry experiments, as well as to characterize situations in which argumentation develops a significant discourse.

SUMMARY

In summary, we believe that the inquiry approach in the laboratory, enables the improvement of students' cognitive skills. It is in alignment with findings of an empirical study conducted by Heinz et al. (2017). The study investigated factors that influence the implementation of science inquiry in the education systems of Turkey, Israel, Sweden and the Czech Republic. It was done as part of an EU-funded project called Science-Teacher Education Advanced Methods (2009–2012). The results of their qualitative analysis showed that the four main aspects influence the inquiry-based science education (IBSE): curriculum, assessment, policy and teacher professionalization systems. They claimed that:

While science inquiry is a significant main focus in the US and European science research and education policy, there is a lack of knowledge on the conditions that influence its implementation. In Europe, this can be attributed to the considerable variation that exists between the educational systems in each country. Accordingly, inquiry-based science is implemented to differing degrees in each country's science education system. However, as reforms in education have to be adjusted to local situations and problems, it is necessary to know about the degree, to which IBSE is already implemented in the educational system of a country, especially in the countries' science curricula, its teacher education and teacher professional development (TPD) programs (p. 683).

The focus of this chapter was on the inquiry-type curriculum, on the assessment and on the teachers' professional development. It did not deal neither with cultural aspects, nor with policy influences in different countries. However, it is suggested, that inquiry-type experiments make the students themselves become the active learners, despite some challenges to students and teachers: (1) working in parallel s often causes, (2) the assessment of the students is complicated and time consuming for the teachers, (3) teachers need to undergo special professional development courses in order to implement such a program.

REFERENCES

American Association for the Advancement of Science. (1990). *Science for all Americans, Project 2061*. New York: Oxford University Press.

Baird, J. R. (1990). Metacognition, purposeful inquiry and conceptual change. In E. Hegarty-Hazel (Ed.), *The Student Laboratory and the Science Curriculum*. London: Rutledge.

Bolte, C., Hofstein, A., Holbrook, J., Mamlok-Naaman, R., Rannikmae, M., Rauch, F., & Streller, S. (2011). PROFILES: Professional Reflection-Oriented Focus on Inquiry-based Learning and Education through Science. *Proceedings of the Scientix Conference*.

Bybee, R. (2000). Teaching science as inquiry. In J. Minstrel & E. H. Van Zee (Eds.), *Inquiring into Inquiry Learning and Teaching*. Washington, DC: AAAS.

Dori, Y. D., & Herscovitz, O. (1999). Question-Posing capability as an alternative evaluation method: Analysis of an environmental case study. *Journal of Research in Science Teaching*, *36*(4), 411–430. doi:10.1002/(SICI)1098-2736(199904)36:4<411::AID-TEA2>3.0.CO;2-E

Dori, Y. J., Sasson, I., Kaberman, Z., & Herscovitz, O. (2004). Integrating case-based computerized laboratories into high school chemistry. *Chemical Educator (Boise, Idaho)*, *9*, 4–8. Retrieved from http://chemeducator.org/bibs/0009001/910004yd.htm

Gunstone, R. F. (1991). Reconstructing theory from practical experience. In B. E. Woolnough (Ed.), *Practical science* (pp. 67–7). Milton Keynes, UK: Open University Press.

Heinz, J., Enghag, M., Stuchlikova, I., Cakmakci, G., Peleg, R., & Baram-Tsabari, A. (2017). Impact of initiatives to implement science iquiry: A comparative study of the Turkish, Israeli, Swedish and Czech science education systems. *Cultural Studies of Science Education*, *12*(3), 677–708. doi:10.100711422-015-9704-6

Hodson, D. (1993). Re-thinking old ways: Toward a more critical approach to practical work in school science. *Studies in Science Education*, *22*(1), 85–142. doi:10.1080/03057269308560022

Hofstein, A., & Kind, P. M. (2012). Learning in and from science laboratories. In B. Fraser, K. Tobin, & C. McRobbie (Eds.), *Second international handbook of science education* (pp. 189–209). Dordrecht: Springer. doi:10.1007/978-1-4020-9041-7_15

Hofstein, A., & Lunetta, V. N. (1982). The role of the laboratory in science teaching: Neglected aspects of research. *Review of Educational Research, 52*(2), 201–217. doi:10.3102/00346543052002201

Hofstein, A., & Lunetta, V. N. (2004). The laboratory in science education: Foundations for the twenty-first century. *Science Education, 88*(1), 28–54. doi:10.1002ce.10106

Hofstein, A., Navon, O., Kipnis, M., & Mamlok-Naaman, R. (2005). Developing Students Ability to Ask More and Better Questions Resulting from Inquiry-Type Chemistry Laboratories. *Journal of Research in Science Teaching, 42*(7), 791–806. doi:10.1002/tea.20072

Hofstein, A., Shore, R., & Kipnis, M. (2004). Providing high school chemistry students with opportunities to develop learning skills in an inquiry-type laboratory: A case study. *International Journal of Science Education, 26*(1), 47–62. doi:10.1080/0950069032000070342

Hofstein, A., & Walberg, H. J. (1995). Instructional strategies. In B. J. Fraser & H. J. Walberg (Eds.), Improving science education (pp. 1–20). Chicago: National Society for the Study of Education.

Johnstone, A. H., & Al-Shuaili, A. (2001). Learning in the laboratory; some thoughts from the literature. *University Chemistry Educ, 5*(2), 42–91.

Katchevich, D., Hofstein, A., & Mamlok-Naaman, R. (2013). Argumentation in the chemistry laboratory: Inquiry and confirmatory experiments. *Research in Science Education, 43*(1), 317–345. doi:10.100711165-011-9267-9

Kelly, G. J., Druker, S., & Chen, C. (1998). Students' reasoning about electricity: Combining performance assessment with argumentation analysis. *International Journal of Science Education, 20*(7), 849–871. doi:10.1080/0950069980200707

Kind, P., Wilson, J., Hofstein, A., & Kind, V. (2011). Peer argumentation in school science laboratory – exploring effect of laboratory task feature. *International Journal of Science Education, 33*(18), 2527–2558. doi:10.1080/09500693.2010.550952

Krajcik, J., Mamlok, R., & Hug, B. (2001). Modern content and the enterprise of science: science education in the 20th century. In L. Corno (Ed.), Education Across A Century: The Centennial Volume (pp. 205-238). Chicago: National Society for the Study of Education (NSSE).

Lazarowitz, R., & Tamir, P. (1994). Research on using laboratory instruction in science. In D. L. Gabel (Ed.), Handbook of research on science teaching (pp. 94-127). New York: Macmillan Publishing Company.

Lunetta, V. N. (1998). The school science laboratory: Historical perspectives and context for contemporary teaching. In B. Fraser & K. Tobin (Eds.), *International Handbook of Science Education* (pp. 349–264). Dordrecht: Kluwer Academic Publishers. doi:10.1007/978-94-011-4940-2_16

Lunetta, V. N., Hofstein, A., & Clough, M. P. (2007). Learning and teaching in the school science laboratory: An analysis of research, theory and practice. In S. K. Abell & N. G. Lederman (Eds.), *International handbook of science education* (pp. 393–441). Mahwah, NJ: Lawrence Erlbaum.

Mamlok-Naaman, R., Hofstein, A., & Penick, J. (2007). Involving Teachers in the STS Curricular Process: A Long-Term Intensive Support Framework for Science Teachers. *Journal of Science Teacher Education, 18*(4), 497–524. doi:10.100710972-007-9046-7

NRC, National Research Council. (1996). *National Science Education Standards*. Washington, DC: National Academy Press.

NRC, National Research Council. (2005). *National Science Education Standards*. Retrieved May 29, 2006, from: http://www.nap.edu/readingroom/books/nses/html/index.html

Penick, J. E., Crow, L. W., & Bonnsteter, R. J. (1996). Questions are the answers. *Science Teacher (Normal, Ill.)*, *63*, 26–29.

Rickey, D., & Stacy, A. M. (2000). The role of Metacognition in learning chemistry. *Journal of Research in Science Teaching*, *77*, 915–920.

Sere, G. M. (2002). Towards renewed research questions from outcomes of the European lab-work in science education. *Science Education*, *86*(5), 624–644. doi:10.1002ce.10040

Taitelbaum, D., Mamlok-Naaman, R., Carmeli, M., & Hofstein, A. (2008). Evidence-based Continuous Professional Development (CPD) in the Inquiry Chemistry Laboratory (ICL). *International Journal of Science Education*, *30*(5), 593–617. doi:10.1080/09500690701854840

Tobin, K. G. (1990). Research on science laboratory activities: In pursuit of better questions and answers to improve learning. *School Science and Mathematics*, *90*(5), 403–418. doi:10.1111/j.1949-8594.1990.tb17229.x

van den Akker, J. (1998). The science curriculum: Between ideals and outcomes. In *International handbook of science education* (pp. 421–447). Dordrecht: Kluwer. doi:10.1007/978-94-011-4940-2_25

Wenger, E. (1998). *Communities of practice: Learning, meaning and identity*. Cambridge, UK: Cambridge University Press. doi:10.1017/CBO9780511803932

Chapter 2
Science Education Funding in Europe and the Inspiration for Chain Reaction

Josephine Booth
Sheffield Hallam University, UK

Eleanor E. Byrne
Sheffield Hallam University, UK

ABSTRACT

This chapter aims to introduce the policy perspective of the relevant European funding stream, tracing the evolution of the science with and for society call from the first Framework Programme through to Horizon 2020. It also introduces the Pupil Research Initiative (PRI) and how the Chain Reaction project built upon PRI's legacy, using its successes to inform a project that was relevant across Europe. The chapter highlights the key aspects of the chain reaction project, for example the Pupil Research Briefs (PRBs) and Express Yourself Conferences, and how these were adapted, and in the case of the PRBs, expanded upon to be more relevant to different partner contexts and the needs of their education systems and curricula.

INTRODUCTION

This chapter will introduce the Chain Reaction project, starting with the evolution and purpose of the European Commission (EC) funding stream that financed the project. The second part of the chapter will discuss the rooting of the project concept and design in a successful past project, putting together the consortium that delivered Chain Reaction and an overview of that delivery.

DOI: 10.4018/978-1-5225-5439-4.ch002

A BRIEF HISTORY OF EUROPEAN COMMISSION FUNDING: FROM POLICIES TO PROJECTS

In 2000 the Lisbon European Council set the EU the target of becoming the most competitive and dynamic knowledge-based society in the world, by adopting strategies to offer tailored education and training, promote employability, and to foster social mobility and inclusion.

Economic sustainability is at the core of the European Union. The European Commission (EC) has a series of benchmarks that aim to give an overview of progress in meeting its strategic goal of "smart, sustainable and inclusive growth" (*Europe 2020: A European strategy for smart, sustainable and inclusive growth*, 2010, p8) by 2020 and these run through many of the communications and plans produced by the EC (e.g. *Joint Report of the Council and the Commission on the implementation of the Strategic Framework for European cooperation in education and training*, 2012). In pursuing this aim it is recognised that science, innovation and technology play a key role in the evolution of knowledge based societies and the implications for people of all ages with the need for "new basic skills, including entrepreneurship, science and technology" (*Making a European Area of Lifelong Learning a Reality,* p12). Also the possibility of growing human capital via building networks across Europe for both knowledge and science. 2004's *Innovation Management and the Knowledge-Driven Economy* underlines the importance of innovation and supporting public awareness initiatives to build citizens' trust and confidence to "foster competitiveness in companies and well-being in our societies" (p11). The report also identified the need for structures and schemes to promote innovation in Europe and also to encourage EU cohesion and the spread of good practice via members working together.

As mentioned above, there are a number of reports and communications from the EC and other organisations on the need for citizens to be positive towards science and to have an understanding of science and scientific rationale. This understanding is necessary to be able to engage with the world at large and to provide the skills they need to live in today's knowledge society and to make informed choices. Additionally, as science and technology are recognised as being at the core of socio-economic development (OECD, 2006) it is therefore vital that there are enough people trained and educated to a sufficient level to ensure progress these areas. However it has been found that few young people see science as a subject to pursue. The relative numbers of students taking science subjects at university are falling, with this being seen particularly acutely in relation to girls (OECD, 2006).

It is then unsurprising that the EC has been funding projects to support and advance scientific research and innovation across member countries and beyond since 1984. As *Europe 2020 - A European strategy for smart, sustainable and inclusive growth* noted, greater research capacity in combination with resource efficiency improves competitiveness and helps to create jobs.

Known as Framework Programmes for Research and Technological Development or Framework Programmes and frequently abbreviated to FP, these funding streams set up by the European Commission have had various specific objectives and actions over their consecutive funding timeframes, with FP1-6 lasting five years each, and FP7 in place for seven years. The most recent and ongoing Framework Programme at the time of writing is FP8 - known as Horizon 2020, which covers 2014-2020. The Framework Programmes have evolved over this time, each building on the previous agendas and outcomes, with aims and objectives and funding shifting to reflect changing needs and priorities of society and the economic context in Europe.

However the funding of specific science education projects by the EC stems from policies developed somewhat later, with the launch in 2000 of the Lisbon Strategy, when heads of state and governments across Europe initially emphasised the need to increase the number of people taking up science and technology careers in order to ensure competitiveness and economic growth. For example the working paper *Science, society and the citizen in Europe* produced in November 2000, identified the need for a "science literate" society, able to understand the processes and outcome of scientific research and to understand its benefits, limitations and risks in order to question its use and applications. Additionally, the need was recognised for a more coordinated and broader approach in terms of topic and audience in their approach and also for "a strong European dimension" in initiatives. In order for this to happen it was clear that interest in science would have to increase, in terms of science in school being made more appealing and relevant, and a there being a greater interest in careers in science from young people. Broader challenges highlighted included the lack of women in science and the need to identify and address barriers to participation.

The publication of a *Science and Society Action Plan* in 2001, aimed to support the implementation of a number of the EU's goals; including those around governance and the development of a dynamic knowledge-based economy across Europe. This plan resulted in the "Science and Society" research theme under FP6 which was the first European programme of this type. This aimed to attack "structural weaknesses of European research" (p9) with its actions defined as:

- Bringing research closer to society
- Driving responsible research and application of science and technology
- Stepping up the science and society dialogue and women in science

While overall the aim of the topic was to increase engagement with science, it is in this last action that it targets the fostering the interest of young people in science and careers in science, along with promoting public awareness of science. The evaluation report produced for FP6 notes that €88 million was spent on funding the strand, just 0.5% of the total funding for programme as a whole (*Evaluation of the Sixth Framework Programme for Research and Technological Development 2002-2006*, 2009 p.9). Although it was one of the smaller strands in terms of funding it was one of the most oversubscribed.

It is perhaps because of this that the theme, renamed from Science *and* Society to Science *in* Society was given more substantial funding in FP7, with a budget of €312 million for the period 2007- 2013. The programme was focused on building a framework and societal relationships with the aim of fostering public engagement and creating a sustained two-way dialogue between science and civil society. The Science in Society strand supported 184 projects, encompassing research, coordination and support actions, expert groups, conferences, and seminars (*Ex-post Evaluation of Science in Society in FP7*, 2016) over the seven years of the programme. In terms of expected impact, it aimed to both enable people to be better informed about, and to create a better understanding of, research and innovation, as well as making careers in research and innovation in science more attractive to both genders.

The 2011 Science in Society FP7 call for funding had three "Action Lines":

5.1 First Action Line – A more dynamic governance of the science and society relationship
5.2 Second Action Line – Strengthening potential, broadening horizons
5.3 Third Action Line – Science and society communicate

Under these action lines, areas and activities to be funded were outlined in greater detail, giving topic descriptions, aims and expected impacts. As with previous incarnations of the Framework Programmes, these same areas and activities were opened for application on a yearly basis, with differing actions specified each year. The calls were nested under a series of activities, themes and actions. Thus the strand under which the Chain Reaction project was funded was under *Action Line 5.2 - Strengthening potential, broadening horizons,* under *Activity 5.2.2 - Young People and Science*, with the specific area being *Area 5.2.2.1 Supporting formal and informal science education in schools as well as through science centres and museums and other relevant means*, and the action *SiS.2012.2.2.1-1: Supporting actions on Innovation in the classroom: teacher training on inquiry based teaching methods on a large scale in Europe.*

Writing the Bid for Chain Reaction

Part of the complexity of writing a funding bid for these sorts of streams is the need to integrate all the various layers of priorities and aims. However, the call text was specific in defining the required aims and objectives of potential projects;

This topic will support actions to promote the more widespread use of problem and inquiry based science teaching techniques in primary and/or secondary schools as well as actions to bridge the gap between the science education research community, science teachers and local actors (including providers of informal science education) in order to facilitate the uptake of inquiry-based science teaching. The actions are intended to complement school science curricula and should particularly focus on teacher training activities (pre-service and in-service) and the promotion of European teachers' networks. (Science in Society Work Programme 2012, 2011, p.19).

The project also needed to be based upon a concept that had already proved effective.

The science education strand recommended that prospective bidders look at another EC report when planning and writing their proposal - *Science education NOW: a renewed pedagogy for the future of Europe* which was published in 2007. This report supports the use of Inquired Based Science Education (IBSE) in the classroom, noting that the skills involved in IBSE are important in all areas of life. In contrast to the traditional deductive approach to science teaching, IBSE uses an inductive approach, where the science learning starts with being presented with a problem, which is then answered by the students gaining new knowledge by interpreting the issue, collecting the data in order to find solutions, evaluating their findings and finally communicating their conclusions. This form of teaching emulates, as far as is possible in the classroom, the work of "real" scientists by using the experimental approach and encourages critical thinking and reflection. It is clear that the thinking behind the requirements of the call was very much influenced by the findings of this report, specifically mentioning the IBSE approach and its benefits in the call text. The funding stream was centred on teacher training (a theme in *Science education NOW 2007)* in IBSE with the intention of increasing student interest in Science, Technology, Engineering and Mathematics (STEM) across Europe, thereby enabling active involvement in a knowledge-based society. In addition many of the report's other conclusions are embedded in the call requirements for example, the need for teachers to be part of networks to support their use of the approach (perhaps in combination with the more traditional deductive approach) and the inclusive nature of IBSE in encouraging the participation of girls. Changing science classroom pedagogy and curriculum to positively influence student choices about pursuing science further is a theme that the

OECD (2006) also identified, along with the need for non-stereotyped, accurate information on further study and careers in science.

CHAIN REACTION: THE BACKGROUND

As a response to the suggested shortage of science literate citizens, a consortium of science education specialists from across Europe successfully submitted a proposal to the FP7 Science in Society call to deliver the IBSE focused Chain Reaction project.

Chain Reaction was a Coordination and Support Action, which provided funding for;

- Managing programmes that include coordination and networking activities;
- Dissemination and use of knowledge;
- Studies or expert groups;
- Support for transnational access to major research infrastructures;
- Actions to stimulate the participation of Small and Medium Enterprises, civil society and their networks, rather than being dedicated to pure research.

As already noted, one of the caveats of this Science in Society call was for all proposals to be firmly grounded in an already successfully piloted project - "The actions aimed at here shall already have proven their efficiency and efficacy" (*Science in Society Work Programme* 2012, 2011 p.19). As such, when we set out to design Chain Reaction we first looked to our previous project outputs to guide our thinking. One of the most suitable projects to draw inspiration from was known as the Pupil Researcher Initiative (PRI), a UK programme which ran from 1995 until 2009.

What Was the Pupil Researcher Initiative (PRI)?

The Pupil Researcher Initiative (PRI) was a major UK school science curriculum development initiative devised in 1995 by the Centre for Science Education at Sheffield Hallam University with funding from the Engineering and Physical Sciences Research Council (EPSRC) and the Particle Physics and Astronomy Research Council (PPARC). The overriding aim of PRI was to raise student motivation and achievement in school science through providing exciting innovative and stimulating curriculum materials and activities.

At the heart of PRI was the premise that when Science Technology Engineering and Maths (STEM) subjects are often taught out of context and therefore can seem irrelevant to students' lives, and in turn many students lose or fail to develop an interest. The PPRI set out to tackle some of these issues by exposing students to practicing STEM researchers and their work, as well as introducing activities into the classroom based on real life issues and approaches to their potential solutions. The programme's key aims were to:

- Raise student motivation and achievement in science
- Improve the teaching and learning of science, particularly investigative and ICT skills
- Increase the interaction between schools and the science and engineering research community.

The initiative also worked to raise young people's awareness of the purposes and benefits of science and engineering research and to mobilise researchers to support and enrich school science education. Alongside this was a commitment to forge working partnerships and collaborations with major organisations and initiatives in science education. The plan was to create a culture of celebration of science and engineering achievement within schools and the community.

PRI had three key components:

- Pupil Research Briefs
- Researchers in Residence
- Express Yourself conferences

The Pupil Research Briefs (PRBs) were classroom investigations based on real and current research being carried out in universities and research facilities. The briefs were intended to support the teaching and learning of GCSE and Standard Grade (Scotland) science in the UK. They were designed to offer new ideas, strategies and contexts for teaching and learning experimental and investigative science, while, at the same time, enabling students to gain an insight into the ways that real research scientists and engineers go about their work.

The design of the PRBs attempted to break new ground in curriculum development, drawing heavily on the educational benefits of active student participation in learning and the provision of real and relevant contexts for learning. They were also intended to link and underpin the other two components of PRI, Researchers in Residence and the Express Yourself conferences. Volume 1 of the PRBs was sent to all UK schools in 1996. A couple of years later, this was followed by ten more briefs in Volume 2.

Each PRB posed a different research question or challenge to students thereby supporting student led Inquiry Based Science Education (IBSE).

Researchers in Residence was a project component designed to bring some of the most creative research talents in the UK into secondary schools. The chosen researchers were fiercely passionate about their subjects and their enthusiasm could be seen to ignite a fresh interest in young people. PhD and early career researchers in the STEM subjects volunteered to spend 4-5 days in secondary schools. Their activities ranged from giving classroom support or presentations, arranging visits and attending field trips. Between 1995 and 2006 the Researchers in Residence programme recruited, briefed and placed around 3500 researchers in over 2000 schools working with about 2000 teachers and 300 000 students. The Researchers in Residence component of PRI played a crucial role in the delivery of the aims and goals of PRI. It also expanded the experience of the researchers and was seen as a benefit to all who took part.

Alongside the Researchers in Residence strand of PRI were the ***Express Yourself* conferences**, which were hosted by the researchers and held across the UK annually over the lifetime of the PRI project.

The conferences gave students a forum in which to present the findings of their own investigations. The conferences emulated real research conferences with opportunities for the students to:

- Communicate and share their ideas with other students, teachers and researchers
- Present research papers in seminars chaired by Researchers in Residence
- Present and host displays of their investigations
- Participate in other activities, such as discussing their work with experienced researchers, attending keynote lectures and demonstrations and participating in practical workshop.

How PRI Influenced Chain Reaction

Chain Reaction is a direct descendant of the Pupil Researcher Initiative having drawn on the principles of student focused IBSE and supporting teachers to explore IBSE with their students. As specified in the funding call, the new project had to be based on an existing and proven approach and Chain Reaction used all three of the PRI strands in its design, with the adaptation and updating of 11 of the original PRBs to become the projects Earth and the Universe Pupil Research Briefs (EUPRBs). The Chain Reaction EUPRBs explored a wide range of Earth and the Universe topics ranging from the growth of plants on long distance space flights to calculating the chance of a comet or a large meteorite hitting the Earth.

Although not as extensive as the original Researchers in Residence programme, Chain Reaction sought to "to bridge the gap between the science education research community, science teachers and local actors (including providers of informal science education) in order to facilitate the uptake of inquiry-based science teaching" Science in Society Work Programme (2012 p19) by including role models in the design of the project, to visit schools and support teachers and students in their work on the PRBs. The Express Yourself conferences were also integrated into the project plan, with annual national conferences in each partner country and an international conference. In doing this we sought to bring the core elements of the PRI project to a new audience, combining the PRBs that were most likely to be easily transferable and relevant to curricula of other countries with bringing real scientists into schools and celebrating student achievement.

The EC call to which the project was written focused specifically on a need for 'teacher training activities (pre-service and in-service)' as a means through which basic knowledge could be enhanced, autonomous learning could be encouraged and interdisciplinary/transdisciplinary approaches could be explored. Although the PRI programme did not offer a core continuous professional development (CPD) element, Chain Reaction was able to use the PRB resources and approach as a basis for the development of a comprehensive programme of teacher training which could be delivered by each participant country following an intensive train-the-trainers session at the project kick-off meeting. The project offered partners a great deal of flexibility as to the format of their national training workshops. This was deemed appropriate to ensure CPD was developed with localised requirements and values in mind.

Project-Based and Problem-Based Learning

Having identified the PRI project, and specifically the PRB resources, as the springboard for the Chain Reaction programme, the original proposal also drew rigour from the research on both 'Project-Based Learning and Problem-Based Learning'. According to the literature regarding Project-Based Learning, activities are developed that involve 'the completion of complex tasks that typically result in a realistic product, event, or presentation to an audience' (Barron & Darling-Hammond, 2010 p. 203). Unlike other approaches to learning, Project-Based Learning aims to stimulate understanding through more than just the 'development of content knowledge':

'This approach aims to take learning one step further by enabling students to **transfer** their learning to new kinds of situations and problems and to use knowledge more proficiently in performance situations.' (Barron & Darling-Hammond, 2010 p. 203)

In turn, Problem-Based Learning asks students to work in small groups to undertake investigations which are deemed to be 'realistic and ill-structured' (Barron & Darling-Hammond, 2010 p. 205) as a way to stretch students' investigative skills and solve problems that may have multiple solutions.

Chain Reaction sought to incorporate both Project-Based and Problem-Based Learning by setting multifaceted challenges through the EUPRBs which set to challenge students to think and behave like real scientists. By unpicking the questions asked and by completing the research in teams, the EUPRBs asked students to conduct self-guided investigations that promoted debate and broadened conceptual understanding.

BUILDING OUR CONSORTIUM

As Chain Reaction stemmed from a previously successful UK science education enhancement programme, PRI, the outset of the proposal writing process saw the team at Sheffield Hallam University seeking to bring together a consortium of partners from across Europe. The specific Science in Society Call through which Chain Reaction was funded required a minimum of 10 European partners in any one consortium. Recruitment of partners to *Chain Reaction* initially took place via a network of existing collaborators. This meant that links and relationships between partner institutions were fully established before the project even commenced, and partners were familiar with one another's skill sets and experience. With one additional partner recruited through the EC's CORDIS partner services online the final Chain Reaction consortium consisted of 12 individual partners:

The final Chain Reaction consortium included institutions from both western and eastern European countries and consisted of 10 EC member states and two associated countries (Jordan and Georgia). As a working body the consortium boasted a huge amount of experience in teacher science education, with a variety of backgrounds that complemented each other. Despite a shared philosophy of motivating and

Figure 1. The 12 countries in the chain reaction project

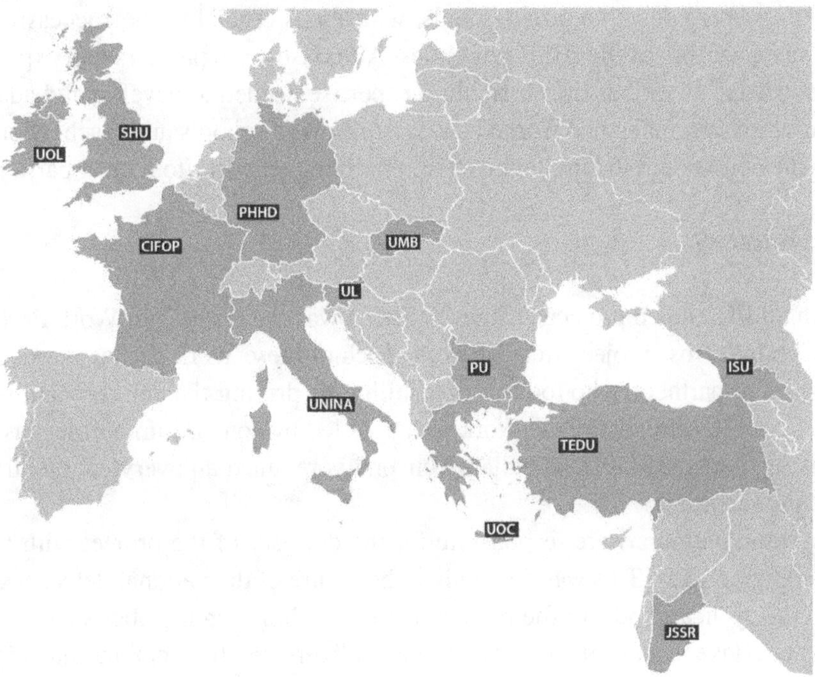

Table 1. The Chain Reaction consortium

SHU	Sheffield Hallam University, UK	UL	Univerza V Ljubljani, Slovenia
UNINA	Universita Degli Studi Di Napoli Federico II, Italy	PHHD	Pedagogische Hochschule Heidelberg, Germany
UMB	Univerzita Mateja Bela V Banskej Bystrici, Slovakia	UOC	Panepistimio Kritis, Crete, Greece
TEDU	Ted Universitesi, Turkey	UOL	University Of Limerick, Ireland
PU	University Of Plovdiv, Bulgaria	JSSR	Jordan Society For Scientific Research, Jordan
CIFOP	Centre International De Formation Pedagogique, France	ISU	Ilia State University, Georgia

inspiring students and teachers in the world of science the individuals involved came from a variety of scientific and educational backgrounds, with a number having taught science in schools. In addition, there was a good gender balance within the consortium, which had a strong female contingent. One of the key strengths to the Chain Reaction project was the strong working relationship that existed at the outset. All consortium members had been actively engaged in the writing of the proposal and were fully committed to its aims and objectives. This created a strong bond between partners that allowed for the consortium to get to work quickly once the funding had been approved. In addition, a number of the consortium members had already been involved in implementing European and IBSE projects. Those partners less familiar with EC projects were able to seek advice and guidance from partners with more experience creating a relaxed, unofficial mentoring programme within the project itself. Chain Reaction offered an alternative framework to other EC project delivery by having all consortium members delivering the exact same content on a cyclical basis. As such partners were able to draw on the understanding and experiences of others while in turn offering support and advice based on their own involvement.

All Chain Reaction consortium members were recruited because of their strong links to teacher education departments; in some cases they were based within a teacher education unit. This enabled effective engagement with teacher education professionals, which was crucial to the successful running of the project and the sustainability of the IBSE principles. All consortium partners had expert scientific and/or education knowledge as well as the curriculum expertise needed to develop and adapt the EUPRBs as necessary. Additionally, many had worked successfully with schools and teachers on previous occasions, allowing the consortium to build upon existing school networks to recruit participating teachers.

Organising Delivery

In common with all EC funded projects, Chain Reaction was organised into Work Packages, which divide the project up into cross project areas of work. Each of these Work packages was then directed by one of the consortium partners, who took responsibility for driving the delivery and reporting on that strand of work. Chain Reaction had nine Work packages led by consortium partners as shown in Table 2. The Coordinating partner, Sheffield Hallam University, retained an overview on all work packages, project finance and reporting.

All 12 consortium partners were responsible for the delivery of the project with a role in each of the individual work packages. This was due both to the nature of the national delivery strands of *Chain Reaction* and to the cyclical model of the project that ran within each member state. Each partner state was required to undertake management, development, delivery and financial monitoring duties relevant to their own national strand.

Table 2. Work packages and work package leads

Work package No.	Work package title	Lead participant
1	Project management and coordination	Sheffield Hallam University, UK
2	Website	University of Crete, Greece
3	Technical Boards (TB) Role	Sheffield Hallam University, UK
4	National EUPRB Development	University Ljubljana, Slovenia
5	Teacher IBSE/EUPRB Briefing	Plovdiv University, Bulgaria
6	In-School Delivery	University of Limerick, Ireland
7	National and international conferences	University of Education, Heidelberg, Germany
8	Impact and dissemination	Jordan Society Of Scientific Research, Amman, Jordan
9	Evaluation	Sheffield Hallam University, UK

DELIVERING CHAIN REACTION

The EUPRB Resources

As the topics and approaches used in the EUPRBs were designed to have a broad appeal, both across abilities and genders, they were developed with a UK audience in mind. Therefore it was necessary for the partner countries other than the UK to adapt the materials both for student use in the classroom and introductory materials for teachers, to their own specific context and curriculum, as well as translating them as necessary. The initial project plan was that this would all take place in the first year of the project. However it soon became evident that time would be required throughout the full life span of the project for updating the materials, based on feedback from teachers and also the partners' experience of delivering CPD. In addition some partners chose to stagger the introduction of EUPRBs to the schools over the period of the three years so that they had the time to get comfortable with the materials and approaches themselves, before introducing them to teachers, as well as giving more time to adapt and translate the materials as appropriate. Other partners presented teachers with a choice of all the EUPRBs each year.

There were a number of ways in which the partners sought to adapt and improve the materials for their context; these ranged from changing the plant species chosen for experiments, altering terminology so that there was a better fit with the curriculum, and adding information to background materials so that they felt more relevant to students, for example providing data on ozone emissions from nearby monitoring stations.

Not all the topics covered in the EUPRBs were suitable for every partner, primarily because they did not fit with their curriculum for the target age group for the project. In addition several partners, sometimes in combination with teachers who were involved in the project, created their own EUPRBs for use in the Chain Reaction project that reflected their own curricula or national contexts more directly. For example; Germany - Solar works © – Construction of a low-budget solar pane), Ireland - Badger Conservation and Culling and Slovenia - Effectiveness of wind turbines. These new resources were shared with the rest of the consortium.

Recruitment of Teachers and Schools

While the project had funding for schools in each country, existing pressures from within schools and the educational systems meant that the UK, France and Ireland experienced schools dropping out in the first year, both before and after teacher briefings had taken place, and the German partner had a school dropout in year 3. Recruitment of schools and teachers also was difficult for some partners over the course of the project, again due to factors both local and national from within the school/education system. For example, a new science curriculum was introduced in England for the year 2015/2016, which meant that teachers were reluctant to commit to any project work outside of their teaching responsibilities and schools were even more unwilling to commit extra curriculum time for projects that were not seen as being core to the curriculum and examination results. In Ireland recruitment was easier in year 2 and 3 as the project had some national recognition and year 1 teachers actively facilitated in the recruitment of year 2 teachers.

In contrast in Georgia, the project was so popular that the partner asked schools to apply, receiving more than 20 applications each year, and undertaking interviews in order to determine which schools would take part in the project. The Bulgarian partner was also able to recruit schools easily and after a project presentation at the National Teachers Conference in Blagoevgrad, there were so many teachers applying to participate in Chain Reaction that the partner was forced to turn schools down. Turkey too was very successful in school recruitment, and along with Italy had extra schools in year 3 of the project, having seven and six schools respectively, rather than the target of five.

Continuing Professional Development (CPD)

Teacher development and support in understanding and implementing IBSE in their classrooms was at the core of Chain Reaction. All partners were committed to training 10 teachers from five schools for each of the three years of the project. Although the teacher educators from each partner were given some guidance as to how this might take place given the experience of the PRI project, it was essentially down to each of them to anticipate how much training would be needed, and the forms that this would take. For some of the teachers, IBSE and the open-ended approach to science teaching that it encourages was very new, while for others it was a familiar concept but perhaps not used due to a lack of confidence or opportunity.

In the UK for example where IBSE is a well-recognised concept, the training was relatively brief, consisting of two formal twilight sessions in year 1 and evolving into one full, more informal day in years 2 and 3, where teachers were encouraged to speak about and share their experiences of IBSE, possible barriers and how the project could be implemented in school. While there was a small amount of practical work, the majority of the discussions were more theoretical, initially based around the nature and process of scientific thinking. Links between the national curriculum and IBSE were explored with particular focus on asking questions and collaboration. Despite the project having funding for schools, it is very difficult to get teachers out of school for the day in England, as schools are reluctant to let them go due to the pressure on teaching time. However once the training had taken place the level of support needed was generally quite low, although support was needed to prepare the students' work for the national and international conferences.

In School Support

Each of the 12 Chain Reaction partners were committed to supporting the teachers throughout their time in planning and implementing the Chain Reaction project in school. As might be expected given the differences in familiarity with IBSE across the consortium, the number of support visits that needed to be made varied between partners and also within partners. For example the Bulgarian partner, reporting that some teachers needed far more support than others and were visited in excess of five times in a relatively short space of time.

In the UK relatively little in school support was needed, as teachers were familiar with an IBSE approach, but were likely to lack opportunity, resources and confidence in its delivery.

Role Models

As already noted, the inclusion of role models, based loosely on the Researchers in Residence part of PRI formed a key part of Chain Reaction, particularly in the aims of bridging the gaps between scientists and schools. All partners recruited and briefed cohorts of role models in every year of the project, generally from inside their institutions in the form of PhD students or early years academics who were keen to support teachers and students in their work on the EUPRBs with school visits and advice.

Express Yourself Conferences

Each partner was required to hold a national student conference at the end of each year of the project. Modelled on the conferences held by the original PRI project, these were called "Express Yourself". For these each school brought teams of students (the number of teams per school was specified by the partner) to present their work and findings to an audience of their peers. Students' work was presented orally and/or in the form of posters, with the idea being that the events were run as though they were genuine academic research conferences, with conference proceedings, invitations, badges and conference bags. Students were coached by role models in school and by their teachers as to how to present their projects. The format of these national events inevitably varied across contexts, but all contained a competitive element for both student presentations and posters. Students' oral presentations were judged by a combination of staff and role models, following a rubric that had been designed with input from the whole consortium.

Partners were also given an "Express Yourself, A young person's guide to presenting their Science", booklet from the PRI project which had been adapted, updated and redesigned by the lead partner. This was then translated and modified for national contexts as necessary. These were printed out and given to schools in order to help students to think about how to structure and plan their presentations. The booklet also contained information about teamwork, getting over nerves and how to put together poster presentations.

As the student conferences were a new concept to the teachers participating in Chain Reaction, partners either discussed the events as part of their in school support visits or held sessions at the partner institutions in order to inform the teachers about the day and what to expect. The majority of teachers were also new to the idea of scientific posters and what they might look like. In the UK and Ireland the release of teachers from school and taking students on trips are very time consuming in terms of gaining parental permission and organising in school cover for teachers, so it was a key part of the partners' role to make sure that the purpose and benefits of the conferences were very clearly outlined. Teachers observed that the conferences provided a great deal of motivation for students in their work on the Chain Reaction projects and in their preparation for the conferences, with work on presentations taking place after school and at lunchtimes.

In the original PRI project the conferences were the end point and culmination of the year's work, however Chain Reaction took this a step further, with the most successful students from the national conferences going forward to represent their country at an International Express Yourself conference.

Each of these conferences included a high profile keynote address from an eminent science professional, interactive workshops and presentations by teams of students. The conferences in years 2 and 3 included an extra inquiry day for students, with teams working in international groups.

Organising and running the national conferences was a new experience for many of the partners, and for some involved far more time than anticipated. However, they were one of the strongest parts of Chain Reaction, introducing students to a range of new skills and experiences (presenting their work to peers, going to a university, for example) and was a rewarding occasion for teachers as they saw their students present the results of their investigations. In total, the 36 Chain Reaction National Express Yourself conferences involved more than 4700 students, with the smaller international events having over a 1000 delegates from the 12 countries in the Chain Reaction project.

FINAL YEAR EXTRA IMPACT TEACHER WORKSHOPS

After two years of Chain Reaction it was clear that while some areas of the project were meeting targets and progressing well it became clear that the aim of creating an international teacher network, with the sharing of practice and experiences, had been less successful than hoped. While in the majority national contexts (particularly where IBSE was a novel concept) teachers were talking about the project and how they had found working with IBSE, opportunities for networking at national conferences was limited by the pressure of time and by the presence of students. At international conferences this was further impacted by language barriers.

As a result of this the lead partner suggested the idea of an international teacher workshop to be held after the final international conference in Bulgaria in 2016, to take place while the students were engaged in the inquiry day. This idea was put to the consortium in October 2015, as it had implications for budget and partner time - particularly for the Bulgarian partners as they would have to host the event. In addition this was not part of the original work in the Chain Reaction contract with the EC and as such required consent from them as the funding body. Despite these issues, the consortium was very keen to proceed with the workshop and the agreement of the EC project officer was obtained.

The workshop had three key aims:

- To enable teacher networking across and within partner countries and to form international links;
- To enable the teachers to share experiences, best practice etc.;
- To further enhance and deepen the learning and experience of IBSE gained as part of Chain Reaction.

The sessions were run by partners from three of the partner institutions, from three different countries (UK, Georgia and Turkey). Two sessions involved the teachers working in international groups to solve problems using an IBSE approach, and the third session was a reflective session on practice and experience of using IBSE and working in the Chain Reaction project. The workshop gave teachers a good opportunity to network, both in their national context and with teachers from other countries, with their participation in the Chain Reaction project and their interest in science teaching and student progress providing common ground. It also allowed all the attendees to see that the challenges faced by science educators in implementing IBSE in the classroom are, to a greater or lesser degree, universal. The workshop also gave teachers the opportunity to exchange ideas and experiences, and to experience being learners for themselves and group working with people with whom they were unfamiliar with. Although the language barrier had been identified before the workshop as being potentially problematic, it did not present barriers to the discussions, with teachers using gestures and drawing to communicate with one another, for example. Teachers who lacked confidence in their English language skills were supported by a member of their country PMB.

Turkey: Teacher Workshop

Chain Reaction had a substantial impact in Turkey, in part because of the novelty of the IBSE approach and enthusiasm with which it was embraced. As a result of this, the lead partner suggested that the project should capitalise on this impact by diverting unspent funding to TEDU in order that they could hold a teacher workshop that would be open to all schools and teachers across Turkey, rather than the limited number that could be fully involved in the project every year. After publicising the event via their Facebook page the Turkish partner received nearly 500 applications for the 75 places at the workshop, the majority of which were from outside of Ankara, despite transport and accommodation not being funded.

The workshop took place at the end of April 2016 and had a number of aims, built on the experience of implementing the project over the course of three years:

- To increase reputation of the Chain Reaction Project across the country
- To disseminate the best practices in the project with maximum number of Turkish teachers
- To discuss problems that experienced Chain Reaction teachers faced with in their classroom practices with inexperienced or new teachers in IBSE implementation
- To share the learning paths of experienced Chain Reaction teachers in terms of how to use questioning effectively, how to promote student-student interaction, and how science festivals could be used as out-of-school opportunity to learn science.
- To develop a network among experienced and inexperienced teachers in IBSE classroom practices
- To provide opportunities for experienced teachers to act as a role model for inexperienced teachers' effective use of PRBs in their classroom practices
- To test the effectiveness of communication among experienced and inexperienced teachers in teacher workshop.

The event combined discussion of IBSE pedagogy across the sciences with Chain Reaction teachers talking about their experiences, the use of hands on "teachers as learners" group work and networking opportunities. An evaluation of the workshop was very positive with teachers responding that they had felt inspired to use IBSE in the classroom and to change their practice.

THE 'WHAT' AND 'HOW' OF CHAIN REACTION ACHIEVEMENTS

During the project design process, Chain Reaction set out seven key objectives. These were:

1. Adapt and disseminate a set of themed inquiry based science resources (EUPRBs) that meet each partner's cultural and curricular needs, to be used for briefing teachers and in the classroom.
2. Promote the use of IBSE in secondary schools across the 12 partners via a programme of teacher development which will engage young people in science through student-led inquiry based research.
3. Help bridge the existing gap between science teachers and the science education community by working collaboratively across schools, teacher training providers and other HE institutions.
4. Promote a European teacher network.
5. Engage in effective dissemination of resources (EUPRBs) and effective practice.
6. Design and deliver effective evaluation and reporting models.
7. Hold 36 national and three international celebration events for teachers and students.

To explore the way in which Chain Reaction met these objectives it is useful to group the longer list above into two main sections:

1. The project content (covering objectives 1, 2, 3 and 7).
2. Dissemination and sustainability (covering objectives 4, 5 and 6).

The Project Content

The Chain Reaction in-school delivery was designed to place the students at the centre of their own scientific investigations. In-school delivery across the consortium had its own character and identity, according to the country and its culture. As such, the approaches employed by the partner countries varied significantly based on localised requirements and pressures, however all approaches to in-school delivery focused on student-led inquiry. The University Ljubljana, for example, had a large focus on having the teachers supported initially by the teacher educators from their institution but then gradually the teacher educators gave the teachers the freedom to complete the inquiry activity alone. Sheffield Hallam University reported that students initially had difficulty with the open nature of the inquiry process but began to understand and enact the practice as they continued to engage in inquiry. The Georgian partner had a focus on socio-scientific relevance, as did the schools in Ireland as they considered inquiry a critical methodology in understanding the importance of socio-scientific issues. The choice of school was interesting with the Bulgarian team, for example, choosing a school for future drivers and mechanics. The message here is that the inquiry approach can be used with all types of schools and students.

While many of the partner schools used existing EUPRBs to begin their investigations, many also went on to design and develop new EUPRBs which were then shared across the consortium. Table 3 illustrates some of the new EUPRBS created as part of the project. As well as sharing these resources across the consortium some were also shared online via the project website.

The use of the adapted and translated EUPRBs, along with the creation of many new EUPRB resources, illustrates the success the project experienced with sharing the IBSE model and providing space and flexibility for partners to be creative and extend the resource portfolio.

Table 3. Examples of the new EUPRBs developed

Partner	New EUPRB developed
UOL	Badger Conservation
UOL	The physics of parachutes
UOL	The effect of drugs on living things
PHHD	Das Sinnesprojekt (The Sense Project)
PHHD	Physikalische und funktionelle Auswertung eines Aufwindkraftmodells (Physical and functional analysis of a model of a solar updraft power plant)
PHHD	Modell eines Aufwindkraftwerkes (Model of a solar updraft power plant)
PHHD	Aufwindkraftwerke – eine alternative regenerative Energiequelle? (Solar updraft power plants – an alternative regenerative energy source?)
PHHD	Bau eines MultiChAcc (Construction of a multi chargeable accumulator)
PHHD	Messung der Hautfeuchtigkeit über den elektrischen Widerstand (Measurment of skin moisture via electrical resistance)
PHHD	Messung der Hautfeuchtigkeit über den elektrischen Widerstand (Measurment of skin moisture via electrical resistance)
PHHD	Handy-Tankstelle – Netzunabhängige Stromquelle für Kleinelektronik (Mobile fuelling station – network-independent power source for small electrical devices)
PHHD	Obsttreibstoff – die Alternative für den Treibstoff der Welt (fruit fuel – the alternative fuel for the world)
PHHD	Myoelektrische Handprothesen (myoelectrical hand prosthesis)
PHHD	Modell zur Nutzung des Aufwindes (Model for usage of updraft power)
PHHD	Düngereffizienz (Fertilizer efficiency)
PHHD	Solar works © (solar works © – Construction of a low-budget solar panel)
PHHD	Green Gym (Green Gym – Usage of exercise machines for power generating)
PHHD	Der beste Schutz vor dem Austrocknen (The best defence against skin dehydration)
PHHD	Energiegewinnung aus Ton (Energy production from sound)
PHHD	Green Lantern (Green Lantern – Light energy analysis of a school)
UL	Smart electric car
UL	Efficiency of water turbines
UL	Effectiveness of wind turbines
PHHD	“Im Aufwind” (Solar chimney power plant)
PHHD	Marmetics (cosmetics)
UL	Smart e-cars
UL	Efficiency of water turbines
UOL	Obesity
UOL	MND
UOL	Ebola
UOL	Forces
UOL	Functions and Expressions in Physics
UOL	Induction Hob

Another marker of success was the 'Express Yourself' conference programme which saw national conferences taking place in all twelve participating countries between February and April of each project year. Every conference had a scientific nature and reflected, as much as possible, professional science conferences with presentations from student teams composed of up to five students per school. Each student team was required to present their findings to their peers, teachers, and a panel of adjudicators made up of practicing science researchers and role models, as well as wider audience members including representatives from local civil society and employers. Role models were selected and invited by each partner team according to their scientific networks and the thematic focus of schools' chosen Chain Reaction project.

The conference agendas all included introductions to the aims of the project, student presentations, poster sessions, a lunch and a final where teams were identified to represent their country at international Express Yourself Conferences. Much like the use of the EUPRBs, each event had its own character and identity, according to the country and its culture.

At the National Express Yourself conferences each year, each partner country chose a team to go forward to the International Express Yourself conference based on set criteria. These were held across Europe as shown by Table 4.

It was decided in Sheffield in 2014, after the first International Express Yourself conference, that future conferences would be accompanied by an additional half day inquiry workshop organized for the students and their teachers. This was trialled in Germany during year 2 and was met with great success. As such it was part of the programme again in Year 3 in Bulgaria. The half day events offered the students an opportunity to work with their peers from across Europe on scientific challenges and puzzles. This extra session meant that in years 2 and 3 the conference was presented over a two-day period.

In Year 3 of the project an exciting teacher networking workshop took place which mirrored the work being undertaken by the students. The teachers were asked to empathise with how the students may feel when faced with a difficult question by undertaking their own scientific inquiry. Much like the students they had to work together to test theories and draw conclusions with their peers from across Europe. A key purpose of this extra activity was to further establish an international teacher network.

Table 5 illustrates the number of attendees at the **National** Express Yourself conferences during Year 1 (2014), Year 2 (2015) and Year 3 (2016). These totals do not include other students participating in the project within school - some schools chose to only bring a representative selection of students to the events based on performance and/or achievement within school. Across the full consortium the National Conferences were attended by a total of 2226 participants and audience members in year 1, 2325 in year 2 and 2245 in year 3. Across the full three years of the project the National Conferences were attended by a total of 6796 participants and audience members.

Table 4. Venues for International Express Yourself conferences

Year 1	Sheffield Hallam University, UK	May 13th 2014
Year 2	University of Heidelberg, Germany	May 11th - 12th 2015
Year 3	Plovdiv Hall 1 of the Hotel "Novotel Plovdiv", Bulgaria	May 9th - 10th 2016

Table 5. Participants at the National Express Yourself Conferences 2014 - 2016

Partner	Total 2014	Total 2015	Total 2016	Total
Students	1387	1722	1666	4775
Teachers	180	183	202	565
Role models	47	46	48	141
Other Participants	612	374	329	1315
Total	2226	2325	2245	6796

Table 6 illustrates the total number of participants attending the three international conferences. The project consortium is proud to have reached so many students and teachers over the course of its three years with 6796 conference attendees at a national level and 713 at our prestigious international events.

Let's turn now to the second marker for the projects achievements, dissemination and sustainability.

Dissemination and Sustainability

At the outset of the project it was anticipated that Chain Reaction would engage directly with 360 teachers. As can be seen in Table 7, the project exceeded these targets by directly engaging with 677 teachers through project delivery and involvement with the conference programme.

The Website

The project had a direct approach to dissemination through face-to-face and peer-to-peer interactions. In addition dissemination with the wider world has been achieved through the main project website, the individual partner sites and social media which offered the forum through which different audiences could be reached.

Table 6. Total No. of participants at the International Conferences (Years 1-3)

	Year 1 (Sheffield Hallam University, UK)	Year 2 (University of Heidelberg, Germany)	Year 3 (University of Plovdiv, Bulgaria)	Total
Students	198	90	77	365
Teachers	36	28	48	112
Role models	4	2	0	6
Other Participants	69	61	100	230
Total	307	181	225	713

Table 7. Total numbers of teachers engaged

Direct teacher engagement including attendance at the national conferences	Direct teacher engagement including attendance at the international conferences	Total
565	112	677

The most important and frequently visited section of the main Chain Reaction website has been the one dedicated to Teacher Resources which features the EUPRBs along with teacher's guidance notes. These have been updated and modified as necessary to bring them in line with the most current research outputs, curriculum developments and local contexts.

An important part of Chain Reaction was networking among educators, IBSE researchers and even aspiring science teachers. During the final year, the Project's website was transitioned to better facilitate this. The forum was built from the ground up to serve as a platform for teachers to interact, exchange information, ask questions and have them answered. It's also a convenient repository for information from our partner teachers aimed at their colleagues in a direct and unmediated fashion.

Along with a main project web interface, twelve individual sites were created for each of the consortium partners. These national websites were created to meet the specific needs of each partner regarding their intended target audience, tone and country specific content. Considerable effort was made to strike the perfect balance between a common uniform branding and individual customisation.

An important advantage of each national website for 10 of the 12 partners was the use of the local language instead of English, which would have presented an additional barrier for non-native English speakers. The national websites directly dealt with the intricacies of each country's educational system and addressed teachers and students in a way that felt natural and familiar. Therefore, dissemination was more efficient across the board. A significant part of our outreach efforts was the use of Web 2.0 technologies, which promote visitor interaction. Several technologies were proposed, but it was agreed that each partner would be free to fine-tune their own strategy to better suit their own needs and target audience. That led to an interesting mix of social media engagement, blogging, video sharing and other web services.

The Chain Reaction Toolkit

A significant *legacy* element of the project is the 'Chain Reaction Toolkit' which can be accessed via the main project website and in the appendices of this book. This resource draws upon the IBSE training provided both *to* and *by* the project's highly training Technical Board. It forms an outline for teachers and educators to understand the message behind Chain Reaction and explore some of the useful hints and tips collated throughout the projects.

As it is a free online resource, teachers can read an introduction to IBSE or choose any of four key areas:

- Translating an EUPRB to the classroom setting
- Inquiry in science education
- Communicating science
- The use of student briefs

These four areas provide a framework for science teachers to understand inquiry in the science classroom and to establish and facilitate an inquiry culture in their own schools. The areas include, sample lesson plans and guidelines, inquiry and communication skills, and thought provoking implications for science education. The toolkit is included in the appendix of this book.

Table 8. Illustrative examples of management at a national level

Turkey	The Turkish partners offered a 3-day pre-training (baseline) workshop which focused on teaching, learning, inquiry and misconceptions. There was a big focus on changing the epistemological orientations of the teachers.
Bulgaria	The Bulgarian partners founded a society for teachers to offer regular meetings and discussions that will encourage reflection and sharing focused on inquiry based science education.
Germany	The German partners chose to focus on facilitating pre-service teachers in developing inquiry-based practices.
Slovenia	The Slovenian partners focused on the use of technology in promoting the Chain Reaction project and used the training workshops to promote a positive alignment between the use of technology and the act of inquiry in the science classroom.
France	In France the focus was on the use of peer-to-peer teaching to support the professional development of in-service teachers. The main aim was to establish reflection between teaching colleagues as a way of developing a culture of inquiry in science education.

Project Sustainability

At the outset of the project in 2013, the consortium discussed a shared vision with regards to sustainability. It was agreed that the project's success rested on each partner possessing a sense of creative and managerial freedom at a national level. By controlling the project within individual national contexts the project itself could ebb and flow with the differing needs experienced from country to country. While the over-arching project management was centralised through the Project Management Board, adaptations were encouraged to ensure the project was fit for purpose and received buy-in from stakeholders. It is this buy-in that ensures the content of the Chain Reaction project continues to be shared across Europe. Table 8 highlights some of the differing approaches taken to the management of Chain Reaction at a national level:

As a three year cyclical project, Chain Reaction was designed in such a way as to offer the twelve European partners opportunity to adapt training, resources and teacher engagement to fit within regional and national contexts. As such the delivery of the project on a national level differed markedly from country to country. To understand further these contextual differences we must begin by exploring the way in which the project was delivered across Europe. The following chapters explore the diversity of experience amongst the different partner countries.

REFERENCES

Barron, B., & Darling-Hammond, L. (2010). *Prospects and Challenges for inquiry based approaches to learning in The Nature of Learning - Using research to inspire practice*. Academic Press.

Dumont, H., Istance, D., & Benavides, F. (2000). *Commission Working Document - Science, society and the citizen in Europe*. Retrieved from https://ec.europa.eu/research/swafs/pdf/pub_public_engagement/ss_en.pdf

Europe 2020 - A European strategy for smart, sustainable and inclusive growth. (2010). Downloaded from: http://ec.europa.eu/eu2020/pdf/COMPLET%20EN%20BARROSO%20%20%20007%20-%20Europe%202020%20-%20EN%20version.pdf

European Commission. (2000). *Commission Working Documents - Science, society and the citizen in Europe*. Downloaded from https://ec.europa.eu/research/swafs/pdf/pub_public_engagement/ss_en.pdf

European Commission. (2001). *Science and Society Action Plan*. Downloaded from: https://ec.europa.eu/research/swafs/pdf/pub_gender_equality/ss_ap_en.pdf

European Commission. (2010). *Communication from the Commission -Making a European Area of Lifelong Learning a Reality*. Downloaded from: http://aei.pitt.edu/42878/1/com2001_0678.pdf

European Commission. (2011). *Work Programme 2012- Capacities - Part 5 Science in Society*. Downloaded from http://ec.europa.eu/research/participants/data/ref/fp7/89382/s-wp-201201_en.pdf

European Commission. (2012). *Joint Report of the Council and the Commission on the implementation of the Strategic Framework for European cooperation in education and training (ET 2020) 'Education and Training in a smart, sustainable and inclusive Europe'*. Downloaded from http://eur-lex.europa.eu/legal-content/EN/TXT/PDF/?uri=CELEX:52012XG0308(01)&from=EN

European Commission Directorate-general for Enterprise. (2004). *Innovation Management and the Knowledge -Driven Economy*. Downloaded from https://publications.europa.eu/en/publication-detail/-/publication/dd46213f-89e1-4c20-ad21-f3adca0b0f7f/language-en/format-PDF/source-85259832

Gago, J. M., Ziman, J., Caro, P., Constantinou, C., Davies, G., Parchmannn, I., . . . Sjøberg, S. (2004). *Europe Needs More Scientists: Report by the High Level Group on Increasing Human Resources for Science and Technology*. Downloaded from: https://www.researchgate.net/publication/259705752_Europe_Needs_More_Scientists_Report_by_the_High_Level_Group_on_Increasing_Human_Resources_for_Science_and_Technology

Organisation for Economic Co-operation and Development Global Science Forum. (2006). *Evolution of Student Interest in Science and Technology Studies Policy report*. Downloaded from: http://www.oecd.org/science/sci-tech/36645825.pdf

Owen, R., Macnaghten, P., & Stilgoe, J. (2012). Responsible research and innovation: From science in society to science for society, with society. *Science & Public Policy*, *39*(6), 751–760. doi:10.1093cipolcs093

Rietschel, E. T., Arnold, E., Čenys, A., Dearing, A., Feller, I., Joussaume, S., . . . Stame, N. (2009). *Evaluation of the Sixth Framework Programmes for Research and Technological Development 2002-2006*. Downloaded from: https://ec.europa.eu/research/evaluations/pdf/archive/other_reports_studies_and_documents/fp6_ex-post_evaluation_expert_group_report.pdf#view=fit&pagemode=none

Rocard, M., Csermely, P., Jorde, D., Lenzen, D., Walberg-Henriksson, H., & Hemmo, V. (2007). *Science education NOW: a renewed pedagogy for the future of Europe*. Downloaded from: http://ec.europa.eu/research/science-society/document_library/pdf_06/report-rocard-on-science-education_en.pdf

Wilkinson, C., Franke, J., & Stroyan, J. (2016). *Ex-post Evaluation of Science in Society in FP7*. Retrieved from http://bookshop.europa.eu/en/ex-post-evaluation-of-science-in-society-in-fp7-pbKI0216492/

Section 2

Comparing the Chain Reaction Experiences Between Countries

This section details the experiences of teacher educators from the different partner countries involved in the Chain Reaction project. While there may be some deviation in terms of structure to each chapter, the focus of all chapters is on the implementation of Chain Reaction in each of country.

Chapter 3
Chain Reaction:
The English Context

Stuart Bevins
Sheffield Hallam University, UK

ABSTRACT

This chapter describes aspects of the successes and obstacles in the delivery of chain reaction in school science classrooms in England. It offers an overview of recent education reform in England and moves on to discuss the implementation of the program and provides anecdotal evidence from teachers to support emerging themes from the delivery experience. Issues of "time" restraints and over-burdened "curricular" as barriers to the deployment of inquiry approaches are highlighted by identifying positive outcomes and offering implications for science education across Europe.

INTRODUCTION

This section begins with a description of recent key policy contexts for science education in England and, prior to devolution of education in 2007, Wales. The description sets the scene for a discussion of the implementation of the Chain Reaction project and its key outcomes in the participating English secondary schools. Three specific elements of the programme are highlighted which represent positive impacts of the programme and issues of concern as well as emphasising barriers to successful deployment of the inquiry approach. The section concludes with a summary of potential further research in the field of inquiry and reflections on our learning.

Science Education in England

Science has a long history in the English school system but prior to 1989 the provision made in different schools and for different school students was somewhat mixed. In the late 1970s, while science was a high-status, attractive subject for many students a number of groups were significantly under-represented with up to 40% of girls doing no science after the age of 14. To regularise education provision across England and Wales the National Curriculum (NC) was introduced in 1989. This mandated that 20%

DOI: 10.4018/978-1-5225-5439-4.ch003

of student's time at ages 14-16 should be devoted to broad and balanced science with a corresponding importance at earlier ages. This was the first time that science became compulsory at primary level with all students from age 5 spending 10% of their time on science.

The 1989 NC signalled the first major effort by government to control provision in schools through curricular, as opposed to financial or assessment means and has formed the basis for many of the subsequent initiatives. The initial curriculum was revised in 1995 and has been clarified and developed by a flurry of government publications and advice since then. Other initiatives that impacted on the changing education system in England and Wales included National Strategies that suggested how the subject should be taught (as opposed to the content to be covered), the creation of a powerful inspection body, the Office for Standards in Education (OFSTED), to oversee teaching practice, a fall in the support offered locally through Local Education Authority (LEA) advisory teachers and a general reduction in the control and coordination of local educational services in favour of a more centrally-driven system leading to, since 2010, one with individual schools operating in isolation or small chains of academies.

As early as 1980, parents were given the right to choose the school their children attended as opposed to being allocated a place by the LEA. To help with these decisions, league tables of schools performance in public examinations were published locally and nationally. The results of OFSTED inspections were also published. This created a competition between schools with examination results as the prime differentiator. This replaced the LEA model based on cooperation between schools and forced schools to actively recruit 'good' students to bolster their results. Anecdotal evidence exists of schools refusing entry to students who did not appear to have good 'chances' in public examinations. The cumulative effect of these changes in the classroom has been to emphasise examination success with a consequent rise in the publication of recall-dominated revision guides as opposed to learning-focused curriculum resources and increased pressure on students and teachers to perform leading to a distinct chilling in the environment for an inquiry-based science experience that projects like Chain Reaction espouse.

However, throughout these changes science retained a privileged position as a 'core' subject, along with English and Mathematics, and remained compulsory for all students from ages 5 to 16. During this time the nature of science was constantly under discussion with arguments about the merits of integrated science (the course organised around topics, often with a socio-scientific element, with the three traditional disciplines covered through these topics) as opposed to coordinated (taught as three separate discipline streams but in the time available for two subjects) or an extended, separate, triple science model (leading to three separate GCSE qualifications) and the significance of inquiry in science. At the same time other subjects were being added to the curriculum and a set of cross-cutting themes were also developed (e.g. health education, citizenship).

An Over-Burdened Curriculum

Perhaps inevitably, by 2005 the curriculum was seen to be significantly overloaded with content and a full review was commissioned to identify tensions and produce suggestions for reform. This produced the 2008 version of the curriculum which claimed to be more flexible and have a reduced content load in science. The key objective of the secondary school curriculum, introduced in September 2008, was to provide a strong framework for exploring STEM subject study and careers and was supported by a programme of study for economic well-being and financial capability. Its aims were to:

- Expand their (school students) horizons for action by challenging stereotyping, discrimination and other cultural and social barriers to choice
- Aim high
- Build a positive and realistic view of needs and capabilities so that they can make effective learning plans, decisions and transitions
- Be aware of changing career opportunities and develop the knowledge and skills to make informed decisions about learning programmes. (DCSF, 2009)

Also important was the key science subject statement which identified work-related factors:

- Discovering how scientific ideas contribute to technological change that affects industry and business, and pupils should experience science in the workplace where possible.

The Labour Government of 2005 – 2010 directed a number of departments to battle a perceived skills gap in STEM fields. The Department for Innovation, Universities and Skills (DIUS) was charged with addressing skills shortages, science in society initiatives and the role and contribution of higher education to the STEM agenda. DIUS' response to the charge was to launch the 'Science So What? So Everything' campaign (2009) which used a range of information including video clips and case studies in an attempt to improve public perception of science and, in particular, the perceptions of young people (see http://sciencesowhat.direct.gov.uk). The STEM Subject Choice and Careers project, introduced by the Department for Children Schools and Families (DCSF, 2006-2011) aimed to bring together elements of careers, science and society, and increase positive attitudes towards science and science education in one initiative that was offered to schools across the UK through a National STEM centre (DfE, 2011 - nationalstemcentre.org.uk).

In their review of the curriculum the Department for Education (DFE, 2011) identified a major issue:

The national curriculum has come to cover more subjects, prescribe more outcomes and take up more school time than originally intended. It is our intention that the national curriculum be slimmed down so that it properly reflects the body of essential knowledge in key subjects and does not absorb the overwhelming majority of teaching time in schools (DfE, 2011).

They also suggested that the curriculum did little to address the perceived need to encourage more students into science careers or even study beyond the compulsory curriculum in in England with most opting out as soon as possible - typically at the age of 16. There were also arguments from educationalists concerned that government interventions had led to the 'de-skilling of teachers' raising the question of whether teachers would be able to make use of increased time and flexibility even if they were offered. The argument included two aspects: the strong emphasis on increasingly rigorous formative assessments with the results published school-by-school to allow parents to avoid schools with apparently 'low' achievement and the removal of teachers as a voice in curriculum development. Both factors tended to emphasise and strengthen centrally-created, results-orientated schemes for 'school improvement' over cooperative, locally-based plans to develop the skills and knowledge of all students in an area. Local schools, and by implication, national results became a political football with politicians at every level pontificating about 'standards' on publication of examination results in the late summer. Michael Gove, the then Secretary of State for Education, exemplified this in his comment to the House of Commons Education Committee by saying that all schools should be 'above average' even though this is a mathematical impossibility.

Academisation and the Unravelling of the National Curriculum

At the same time as changes were made to the NC and the content and nature of the science other initiatives were being undertaken. Prior to 2002 all state-maintained schools were required by law to follow the NC but with the introduction of City Academies in 2002 this constraint was removed. Academies were originally intended to improve provision of state education in areas where the system was seen to be failing students and so were allowed freedom to develop curricula outside the national framework. Since the early 2000's and increasingly since 2010 the number of academies has increased with the majority of state secondary schools now officially recognised as academies and so outside the control of the NC. All are required to take note of the NC and most follow courses that are broadly in line with it in science but increased variation is creeping back into the system. This is further emphasised by another new type of school, called 'free school's, which can be set up by individuals or bodies such as parents, are centrally funded by government and fall outside NC strictures. Faith schools are also allowed to develop their own curricula to reflect their ethos with some schools treating evolution as simply a theory with equal status to intelligent design.

As the government reduced the prescription inherent in the National Curriculum (the 2013 Science orders for students aged 11-14 covered 6 pages whereas the original 1989 document needed over 20 to do the same job) the control of assessment has been strengthened. There has been reduced emphasis on practical work and inquiry skills in the assessment rubrics and a shift to terminal examinations. Table 1 sumarises policy change and initiatives from 1989-2017.

Programme Summary

Chain Reaction was launched into this centrally-controlled, results-dominated educational environment in England in 2013. The project was based around a belief that students would learn to demonstrate research scientists' professional behaviours by engaging in some of the activities typical of those scientists in a supportive school environment. The programme promoted a 'guided inquiry' approach in schools and supported the involvement of science teachers through a structured professional development (PD) component that aimed to inform teachers of how to design and implement guided inquiry through small group projects. Teachers were also given ready-made teaching and learning resources, focused on the Earth and Universe to help guide their student's inquiry projects.

Not surprisingly, the programme in England experienced both successes and disappointments. For example, recruitment of schools went well in years one and two with the optimum sample of ten schools from low socioeconomic areas achieved. Unfortunately only four schools were recruited in year three after a late withdrawal. However, a total of 24 teachers attended the PD sessions, delivered by two members of the Training Board (TB) with input from other members of the programme team, over three years. A total of 578 students participated in the project with 564 attending the three national student conferences and 14 attending the three international conferences.

Table 1. Key policies and initiatives in UK England and Wales since 1989

1989	National Curriculum published specifying skills and content to be taught in all state-maintained schools. Broad and balanced science becomes compulsory for all pupils from 5 to 16.
1991	First school league tables published in the local and national press.
1992	OFSTED created with strong inspection role to replace local authority advisory teachers.
1995	First review of National Curriculum reducing complexity of the document.
2002	First City Academies launched with right to modify NC requirements.
2004	Tomlinson Report, *14-19 Curriculum and Qualifications Reform,* published
2005	White Paper *14-19 Education and Skills* (rejected most of Tomlinson Report's recommendations
2005	White paper *Higher Standards, Better Schools for All,* led to 2006 Education and Inspections Bill
2006	STEM Subject Choice and Careers project
2007	QCA published plans for a more flexible Key Stage 3 curriculum
2008	*Education and Skills Act 2008,* raised school leaving age to 18; Key Stage 3 SATs effectively abolished
2009	*Education for All,* final report of the Nuffield Review of 14-19 education and training
2009	51 Academies opened in September *National Curriculum,* report by Children, Schools and Families Committee
2009	*Apprenticeships, Skills, Children and Learning Act 2009,* created statutory framework for apprenticeships
2009	White Paper *Your child, your schools, our future,* wide-ranging proposals including removal of central government prescription of teaching methods and reduction in the use of private consultants to improve schools
2010	Diplomas scrapped; QCDA scrapped
2011	National STEM Centre and STEM learning network established
2012	Majority of secondary schools achieve academy status.
2013	New National Curriculum published
2015	New GCSE regulations produced increasing emphasis on final examinations as opposed to coursework and, in science, removing the teacher assessment of practical skills.

Implementation

For the purposes of this discussion three elements of the programme have been highlighted for particular focus as they represent important areas of impact within English schools:

- Teacher professional development
- Sustainability
- Student research conferences

Teacher Professional Development—as already mentioned, there is great emphasis on high stakes assessment in science education in England, which tends to mutate classroom inquiries into a process for generating marks in a highly structured assessment regime. Thus, students are often exposed to pre-designed investigations that reflect the requirements of the mark scheme rather than the reality of scientific research activity. This means that teachers attempting to deploy an inquiry often fall into an 'algorithmic pit' with their students achieving little more than completion of the activity and a degree of passive learning (Bevins and Price, 2016). Therefore, it was essential to design professional development sessions that would ensure participating teachers were comfortable not only with the philosophy and aims of the programme but also with delivering inquiry projects in their classrooms. Whether or not inquiry approaches are successful is often at the mercy of the teachers who implement them. If these approaches are misunderstood and incorrectly interpreted as being totally unstructured the benefits are reduced significantly when compared to effective deployment by those whom understand the complexities of successful scaffolding and assessment.

We were also cognisant of desirable characteristics of effective PD for teachers as outlined in existing literature in the field and from our own extensive experience. Therefore, we identified three key characteristics to underpin our model demonstrating that it should:

- Include on and off site school dimensions
- Be collaborative
- Promote development of teachers' classroom practice within a context of enhanced achievement for students

Coolahan (2002) suggests that teacher development is often best promoted within the context of whole school development. Now, at first glance this may support a more top-down approach than we intended but Coolahan also emphasised that this approach need not dismiss the personal and individual needs of teachers and goes on to state that collaboration between teachers within clusters of schools is warranted and supports a more reflective approach. With the above in mind, we opted for a PD package structured through two twilight sessions held at the university combined with follow-up support visits to schools. We engaged the participating teachers in collaborative discussion about inquiry both within the context of Chain Reaction and wider inquiry contexts, while encouraging them to reflect on their existing knowledge and learning.

The first twilight session allowed participating teachers to explore the teaching and learning resources (EUPRBs) and view an appropriate range of equipment in the university teaching laboratories during a three-hour session. This initial session also included discussion that encouraged teachers to reflect on gaps in their knowledge so that these could be addressed. Particular attention was paid to appropriate planning, classroom collaboration and interaction and the time required to deploy a student-led inquiry. Unfortunately, a large proportion of existing research tends to underestimate just how difficult it is for teachers to teach science as inquiry (Anderson, 2000). Nuthall (2005) suggests that teachers often practice within established patterns that arise from 'ritualised routines' within classroom practice. These emerge from the difficulty of managing large numbers of students with diverse needs and learning styles. Therefore, both teachers and students identify structures within the classroom that become fixed to enable comfortable negotiation within them. So, we were mindful that discussion about inquiry which focused on the teachers' understanding of inquiry and how they might successfully deploy such an approach in their own context was important.

The initial twilight session was followed-up by school visits from members of the Technical Board (TB) and wider programme team with each school receiving a minimum of two support visits. These sessions, which typically lasted between one and two hours enabled the participating teachers to discuss their experience of deploying the Chain Reaction approach and highlight problems and successes. For example, during an early visit to a school involved in the first year of the programme a participating teacher highlighted a problem regarding student-group selection. She had initially opted for mixed ability groupings but found that:

large mixed ability groups don't really work, they quickly moved off task and I found it hard to get them back (biology teacher; first year school).

Discussion with the programme team members led the teacher to try a blend of known friendship pairings and mixed-ability groups. She found this approach much more successful and suggested that she would use the approach for future project work. Her success in adopting this later approach resonates with a number of existing studies and several meta-analyses which report that cooperative, small-group project work has significant learning benefits for students (Ginsburg-Block et al., 2006; Johnson and Johnson, 1989).

The final element of our PD model was a two-hour twilight session held at the university towards the end of each programme year. This session had the specific aim of supporting teachers' preparations for the national student conference. Members of the TB and the teachers discussed approaches to developing high quality student presentations, related content and public speaking skills.

Our approach to the PD component of the programme enabled sufficient knowledge transfer to take place between the TB, wider programme members and teachers, about both programme implementation and associated administration procedures, while supporting more collaborative interactions through discussion and debate about specific elements of inquiry-based science education and the Chain Reaction programme. However, the real key to the success of the approach was the follow-up support visits to schools. Teachers found the school visits particularly supportive as this teacher from the second year of the programme stated:

....yeah the twilight session was interesting and got me thinking about inquiry but the extra school visits from Heather and Stuart (programme team members) built my confidence....I felt that if anything went drastically wrong they were there to advise (chemistry teacher; second year).

The involvement of external expertise to support the participating teachers was necessary because the teachers were engaged, in some instances, in substantive new learning which required them to understand new ways of supporting students' learning of content, develop new skills in terms of managing and facilitating inquiry as opposed to more traditional teaching approaches, and to think about their existing practice in new ways. Members of the Chain Reaction team were able to challenge the teachers existing assumptions about inquiry and present the teachers with new possibilities. This is supported by Timperley et al. (2008) who point out:

experts need to know the content of the relevant curricula and what teaching practices make a difference for students. They need to be able to make new knowledge and skills meaningful to teachers and manageable within their practice contexts, to connect theory and practice in ways that teachers find helpful, and to develop teachers' ability to use inquiry and assessment data to inform their teaching (p. 20).

Being able to do this in an authentic environment (the school) enhanced the teacher's confidence in discussing key issues and problems with members of the Chain Reaction team:

its good to be able to gauge how we are developing the project in our school by asking questions during the school visits. It helps me see where I am and where the students are in the project (biology teacher; year two).

Sustainability of the Chain Reaction approach—embedding the programme approach in schools syllabi was an ambitious objective of the Chain Reaction programme. Ambitious because we were mindful of two key obstacles that could very easily prevent achievement of our ambition:

- Existing science education policy
- Traditional school science culture

As previously explained, the education environment in England operating at the time of the Chain Reaction project favoured high stakes assessment outputs which are not conducive to inquiry approaches that rely on project work. Students are trained through pre-designed, recipe-based investigative work in order to comply with mark schemes. Highly structured assessment models leave little room for complex classroom inquiries even if the students' achievements and performance are of the highest quality (Bevins & Price, 2016). Moreover, an overburdened curriculum restricts the time needed to cover required content knowledge which consigns inquiry approaches to a low position on many teachers' priority list of classroom approaches.

However, five of the participating schools, from the total of fourteen, did embed parts of the programme within their schemes of work or as an extra-curricular activity. For example, one school embedded the approach in their syllabi across the whole Year 8 group (ages 13/14). Time was allocated for students to design and develop inquiry projects during the summer term with teachers acting as facilitators to students and, very often, working in collaboration with their students. The students are required to present and discuss their findings to an audience of their peers and teachers at a school science conference towards the end of the school year. A second example concerns the embedding of the approach as a structure for an after-school science club. Prior to his involvement in the Chain Reaction programme the teacher managed an after-school science club, which by his own admission, had little structure and aimed to stimulate members through revision support and occasional demonstrations. After successful involvement in the programme he has taken the project-based model and used it as a scaffold for activities in the science club. Club members can choose any research brief from the range of EUPRBs or identify a topic of their own and engage in a project-based, structured inquiry where they design and deploy their inquiries with facilitation from their teacher.

Students as Researchers: Research Conferences

Perhaps the most successful aspect of the Chain Reaction programme in England was the students' participation in both the national and international 'Express Yourself' student research conferences. As the conferences were designed to reflect the spirit of a professional research conference, we were concerned that many students would find presenting and discussing their research to an audience of their peers and guests such as practising scientists and education researchers a stressful and intimidating element of the programme. Our experience and evaluation data indicates that our concern was unfounded. Indeed, the large majority of the students embraced the experience and surprised us with the quality, detail and delivery of their presentations. A particular element of the approach which may have supported their confidence in engaging in the conferences was that they undertook their research projects in groups of four or five members and presented in the same groups. All of the presentations were well structured and rehearsed with many involving multimedia elements such as video to demonstrate aspects of their work. For example, one particular group of five students, who used the 'Plants in Space'[1] brief to guide their inquiry used video to show key stages of their inquiry while providing a live commentary for the audience.

We suspect that researching and presenting in groups rather than as individuals or pairs had a significantly positive impact on the students' participation. This was achieved through our vision of group work which underpins the programme and rested on the teachers facilitating student ownership of the inquiries:

They're coming to you with nice questions rather than behaviour issues because they're actually thinking about it, which is nice. They're more focused in groups and you can be more of a teacher rather than a crowd controller. It's a much nicer environment (Biology teacher, year 1).

Another key characteristic of student group work we wanted to promote through the Chain Reaction approach was students as 'co-learners' and not simply one student helping another student (Blatchford et al., 2003). To begin with, many of the participating teachers had concerns about their students working in groups based on a belief that students cannot learn effectively from each other and that less able students would simply rely on the more able students to lead the work. These concerns are also reflected in international literature on group work (Lewis and Cowie, 1993; Plummer and Dudley, 1993). However, the teacher's fears were allayed early on as the students displayed levels of engagement and co-operative learning which exceeded their and our greatest expectations:

They adapted really well....... They were comfortable with their learning and that really helped. I do think that really I should perhaps try with some of the other students that I'm not so sure with. Perhaps I need to stretch my comfort a bit and try things like this with them rather than the more traditional teaching (Biology teacher, year 3).

We also have video recordings and evaluation evidence which demonstrate that the students engaged in sharing findings and discussing research projects in the conferences with their peers, both national and international, during poster sessions and conference breaks. The fact that the students took the conferences seriously, and were on task throughout, reflects existing research on the impacts of group work which suggests positive effects on student achievement and attitudes, as well as positive effects on social climate within the classrooms (Johnson & Johnson, 1987; Slavin, 1990). There is also research in the field of developmental psychology that argues the benefits of classroom group work. For example, Pellegrini and Blatchford (2000) suggest that student-student relations can be a successful motivating context as opposed to adult/teacher-student relationships which are less horizontally organised and have a less equitable power structure. Students involved in the Chain Reaction project were offered and accepted ownership of their inquiry projects by their teachers which led to a sharing of power within their groups. We have no evidence to suggest that any of the participating students struggled to participate in their inquiries in any way. Moreover, evaluation data suggests that students placed significant value on being given this kind of autonomy.

In summary, we feel that the combination of a small group approach to inquiry projects and structured professional development for teachers were key aspects of the programme which led some schools to embed elements of the Chain Reaction programme in their science education provision. However, we regret that only a small number of schools were able to embed components of the programme within their schools but suggest that the main reasons for this lie beyond the scope and power of the programme. We now go on to discuss these reasons and identify them as barriers to a successful inquiry approach in schools.

The Consistent Barriers of 'Time and Curriculum'

The main barrier to successful implementation in schools was 'time'. Inevitably, all of the participating teachers experienced challenges as they attempted to juggle the time required for their students' inquiry projects and normal teaching requirements—not only did most of the teachers need to adjust to new or modified approaches to classroom management, teaching approaches and assessment procedures but also wider programme activities such as supporting students to develop their presentations, organising participation in the National Conference and attending PD sessions took valuable time. The foremost concern for the participating teachers was being able to cover mandatory curriculum content needed for examination success and, even though the EUPRBs contained material suited to the science curriculum in England, the teachers stated that the time needed to engage in an inquiry approach through project work can be excessive:

Project work takes too long. If we could fit an inquiry into a single or possibly double lesson then I'd use it more often (physics teacher; third year).

This view is also reflected by some researchers who suggest that models of inquiry often involve extensive practical work (Kirschner et al., 2006) which they view as a questionable use of time. While we would argue that this need not be the case we do recognise that the research briefs used in Chain Reaction predominantly involve some level of practical work. We make no apologies for this but acknowledge that investment in extra time is required to effectively engage in this type of project-based inquiry.

The issue of time was also suggested as a reason which prevented most of the schools from embedding the Chain Reaction approach into their curricula or schemes of work even though five did manage to embed some aspects. Anecdotal evidence from teachers representing the nine schools which did not embed, at least, some components of the approach indicates that senior management within those schools were fearful of the time required to engage in inquiry-based project work and felt that coverage of subject content would suffer if they were to accede to including the approach in future school planning:

Our SMT (senior management team) think that I'll take too much time doing practical work and stuff. They're scared that I won't get through the content the kids need for the exam (biology teacher; third year).

This clearly has significant implications for the inclusion of an inquiry approach in a science teachers' repertoire of classroom methods. The strong indications are that within the current environment in England, focused on high stakes assessment, it is very difficult for teachers to deploy inquiry as a classroom learning approach. This, in turn, has strong implications for student autonomy and inquiry skills development necessary for further science study and careers. The professional science community has demonstrated, over recent years, that new graduates do not possess the required practical skills when entering employment (Roberts, 2002; Hofstein et al., 2011). A lack of commitment to inquiry approaches from schools will not solve this problem.

We suspect that, while issues of time and management of inquiry approaches are understandable and reasonably defensible, a key factor in not using project-based inquiries is a lack of confidence among many of the participating teachers. Interviews with the teachers show that the five schools who have embedded elements of the Chain Reaction approach demonstrate that they are comfortable with, and value, inquiry approaches. Furthermore, of these ten teachers, six have a career background which demonstrates experience of working in science laboratories or similar industries and all have teaching experience of a minimum of five years. We are mindful of Carter's (2008) view that teachers often lack experience in using inquiry approaches and therefore, are not as skilled as in other classroom methods which would result in a lack of confidence to try something unfamiliar. Also, as mentioned earlier, we note Nuthall's (2005) comments regarding teacher's adherence to fixed patterns and routines in the classroom in order to make sense of the day-to-day melee of teaching. Nevertheless, we will now go on to report some key positive impacts from the programme in England.

POSITIVE IMPACTS

Those schools who have embedded the inquiry approach either in their syllabus or in after-school science clubs have indicated strong impact on students' motivation and achievement. For example, one particular school have designated an area in the science department for displaying students' work through project-based inquiries to celebrate students' engagement and successes after receiving requests from the students to publish their work beyond whole school presentations.

Of particular note is that members of our Initial Teacher Education (ITE) team, who were involved in the programme as Technical Board (TB) members, have embedded sessions in the Bachelor of Arts undergraduate degree teaching programme that reflect the Chain Reaction programme and its focus on inquiry. Clearly this has implications for future ITE students and their careers as teachers in English schools. In the light of our view that inadequate attention to inquiry based approaches in ITE science we welcome this and will monitor future impact on ITE students.

However, probably the main positive impact of the programme has been the students' participation in the national and international student research conferences. This is particularly pleasing as the main objective of the programme was to promote a guided inquiry approach to science teachers while developing their knowledge and skills. The fact that the students engaged in their projects as enthusiastically and skillfully as they did is, we suggest, testament to the power of an inquiry approach.

CONCLUSION

We put forward a claim that the Chain Reaction programme was successful in engaging school students in inquiry projects which promoted small group inquiries and aimed to develop skills such as designing an investigation, data gathering and analysis, and presenting findings. Based on evaluation evidence, we also suggest that the participating teachers benefitted from involvement in terms of developing a greater understanding and experience of inquiry as an approach to science education. However, while some schools have demonstrated a commitment to embedding components of the programme in their schools many have not. The key reason for this appears to be the perceived amount of time required to design and deploy a classroom inquiry. While the large majority of the participating teachers admit to a belief that inquiry can be a creative and effective method for teaching and learning science, they add that a requirement to cover curriculum content prevents them from using inquiry. This perception of inquiry as a lengthy and complex approach to learning science presents a challenge to advocates of inquiry. Science education researchers would do well to focus their attention on teachers' perceptions of inquiry and, where possible, present models of successful implementation by science teachers for replication. Additionally, existing literature suggests that inquiry approaches rely heavily on the knowledge and skills of the teacher, as indeed, does good teaching generally (Barron and Darling-Hammond, 2010). Therefore, where inquiry is poorly understood by teachers there is clearly less chance of teacher engagement with the approach. Research that investigates teachers' understanding of inquiry approaches would contribute much to the field.

One way in which teachers' understanding of inquiry can be supported is through professional development activity designed to guide their implementation and assessment of inquiry projects. This was a particularly successful element of the programme in England with the teachers approving the approach and stating that more professional development opportunities should be made available to support their learning and understanding of inquiry.

Of particular note in the English context was the significant success of the student conferences. Having the opportunity to present their inquiry projects, from design through to analysis and findings, to an audience of their peers and members of the public inspired rather than deterred participating students'. The conferences provided an additional goal for the students, beyond the identified learning goals of their projects, and provided a 'check' for their learning as they reflected upon their projects and findings in order to present brief summaries.

This chapter was introduced with a review of developments that suggested teachers were operating as disempowered delivers of a curriculum designed elsewhere and chasing rigorous but unrepresentative measures of student achievement. To a large extent our view has not changed and although a new curriculum has been introduced in England there is no evidence that it will change science education for the better in any significant way. However, our belief and faith in inquiry approaches to teaching and learning science has grown as a result of the Chain Reaction programme through both teacher and student participation and successes. Of particular note, and a source of some pride to us and the teachers involved in the project, is the way that students, from a variety of backgrounds and ability ranges, have taken to the role of research scientists, when given the chance, and the professionalism, commitment and flair they demonstrated when presenting their research to their peers. Our learning from the programme has significantly influenced our continuing study of inquiry approaches and endorsed our advocacy of this type of approach to classroom science education.

REFERENCES

Anderson, R. D. (2002). Reforming science teaching: What research says about inquiry. *Journal of Science Teacher Education, 13*(1), 1–12. doi:10.1023/A:1015171124982

Barron, B., & Darling-Hammond, L. (2010). Prospects and challenges for inquiry-based approaches to learning. *The nature of learning: Using research to inspire practice*, 199-225.

Bevins, S. C., & Price, G. (2016). Reconceptualising inquiry in science education. *International Journal of Science Education, 38*(1), 17–29. doi:10.1080/09500693.2015.1124300

Blatchford, P., Kutnick, P., Baines, E., & Galton, M. (2003). Toward a social pedagogy of classroom group work. *International Journal of Educational Research, 39*(1), 153–172. doi:10.1016/S0883-0355(03)00078-8

Coolahan, J. (2002). *Teacher education and the teaching career in an era of lifelong learning*. Academic Press.

Department for Children, Schools and Families. (2009). *National Curriculum*. QCDA.

Department for Education. (2011). *Maths and science education; the supply of high achievers at A level.* Research report DFE – RR079.

Department for Education. (2006-2011). *Stem subject choice and careers: STEM Choices*. DfE.

Department for Innovation, Universities and Skills. (2009). *Science so what? So everything.* Available online Retrieved on June 29th 2011 at: *English Secondary school students' perceptions of school science and science and engineering*. Author.

Ginsburg-Block, M. D., Rohrbeck, C. A., & Fantuzzo, J. W. (2006). A meta-analytic review of social, self-concept, and behavioral outcomes of peer-assisted learning. *Journal of Educational Psychology, 98*(4), 732–749. doi:10.1037/0022-0663.98.4.732

Hofstein, A., Eilks, I., & Bybee, R. (2011). Societal issues and their importance for contemporary science education—A pedagogical justification and the state-of-the-art in Israel, Germany, and the USA. *International Journal of Science and Mathematics Education, 9*(6), 1459–1483. doi:10.100710763-010-9273-9

Johnson, D. W., & Johnson, R. T. (1989). *Cooperation and competition: Theory and research.* Edina, MN: Interaction Book Company.

Lewis, J., & Cowie, H. (1993). Cooperative group work: Promises and limitations. A study of teachers' values. *Education Section Review-British Psychological Society, 17*, 77–77.

Nuthall, G. (2005). The cultural myths and realities of classroom teaching and learning: A personal journey. *Teachers College Record, 107*(5), 895–934. doi:10.1111/j.1467-9620.2005.00498.x

Plummer, G., & Dudley, P. (1993). *Assessing Children Learning Collaboratively*. Chelmsford, UK: Essex Development Advisory Service.

Roberts, G. G. (2002). *SET for success: the supply of people with science, technology, engineering and mathematics skills: The report of Sir Gareth Roberts' review*. Academic Press.

Timperley, H., Wilson, A., Barrar, H., & Fung, I. (2008). Teacher professional learning and development. Academic Press.

ENDNOTE

1 Plants in Space is one of the EUPRB teaching and learning materials which focuses on the students taking the role of research scientists working in a university department of Bioastronomy who are tasked with undertaking preliminary work to design a life sustaining unit for a manned flight to Mars.

Chapter 4
Chain Reaction:
The Irish Context

John O'Reilly
University of Limerick, Ireland

Liam Guilfoyle
University of Oxford, UK

Louise Lehane
St Angela's College, Ireland

ABSTRACT

This chapter presents a case study of the experience of the Irish Chain Reaction (CR) team, which took place during a time of significant curriculum change in the lower secondary school system. As such, it is hoped that those interested in teacher professional development will find the case of interest while acknowledging the varied cultural, material and structural resources, and limitations that influence the context of any educational change process. The authors have placed a significant focus on describing the Irish context to begin this chapter, initially comparing the old science syllabus with the new "specification," with thought given to the existing modalities of student learning and the nature of teacher professional collaboration and the developments that will be required by the new curriculum. The authors then summarize the plan for CR implementation through a professional learning community (PLC) focused on supporting teacher agency and autonomy in the design of inquiry-based science education (IBSE) classes. Teacher and student reflections of experience are presented.

INTRODUCTION

This chapter presents a case study of the experience of the Irish Chain Reaction (CR) team, which (uniquely within the project) took place during a time of significant curriculum change in the lower secondary school system. As such, it is hoped that those interested in teacher professional development will find the case of interest while acknowledging the varied cultural, material and structural resources, and limitations that influence the context of any educational change process. We have placed a sig-

DOI: 10.4018/978-1-5225-5439-4.ch004

nificant focus on describing the Irish context to begin this chapter, initially comparing the old science syllabus with the new "specification", with thought given to the existing modalities of student learning and the nature of teacher professional collaboration and the developments that will be required by the new curriculum. We then summarise the plan for CR implementation through a professional learning community (PLC) focused on supporting teacher agency and autonomy in the design of Inquiry-Based Science Education (IBSE) classes drawing on their experience of the prompts provided by the Pupil Research Briefs (PRBs). In essence, this represents the epistemology developed by the Irish CR team and is informed by a brief review of the literature pertaining to IBSE professional development (PD).

In reviewing the project implementation, the focus is on teacher sense-making within the PLC, considering initial workshops that tried to establish a culture of cooperation and the sharing of existing understandings of IBSE and further workshops that reviewed ongoing work. Core experiences that the teachers found valuable are explored alongside obstacles and issues that arose. This is contextualised in relation to new, emerging models of IBSE that include psychological and power dimensions in the classroom. We conclude with a consideration of the impact of CR on the Irish context, to include our own learning as teacher educators and the likely long-term influence of CR on the new Irish curriculum. Reflections are offered on the aspirations we held in relation to developing teacher agency and autonomy and the practical realisation of same, comparing our experience with lessons emerging from recent reviews of the curriculum change process in Scotland.

BACKGROUND

Setting the Scene: Previous Irish Curriculum

The curriculum in Irish post-primary schools is divided into two cycles, the junior and senior cycles. The previous junior cycle general science syllabus (2003-2016) is transitionally being removed from schools as part of a wider curriculum reform. This is in part driven by PISA reports which show that Irish students' science proficiency was considerably lower than the OECD average with one in nine students scoring low on the assessment scale (Perkins et al., 2012). The educational culture in Ireland emphasised the memorisation of facts (Cullinane & Liston, 2015) due in part to the assessment structure of the varying syllabi.

In the science syllabus, mandatory experiments were completed as a cookbook approach and seen as a "tick the box" exercise for students to complete. Resources and time constraints often meant that students did not complete these experiments and simply wrote information into the laboratory notebook directed by the teacher (Broggy, Boyle & Forde, 2012). Two prescribed "investigations" were submitted for external assessment, although students had the option to complete their own investigation, a chance for them to engage in open inquiry. However, the Chief Examiners Report (2010) suggests that only 0.6% and 0.7% of Lower and Higher Level students respectively chose this option. While this is a small cohort of seemingly adventurous students, most used investigations from previous years for their answers and the average mark received was 59.5% for Lower Level students (Chief Examiners Report, 2010). Some teachers reported that students preferred to be "spoon fed" and did not like opportunities for independent learning (Cheevers et al., 2006). All of these factors contributed to a culture where genuine IBSE experiences do not take place.

The third component of the assessment was the completion of a summative examination (worth 65%), intended to test students understanding of the content, the investigative approach, and the application of science process skills. However, it can be argued that the structure tested students' memory rather than their conceptual understanding of the content or scientific processes (NCCA, 2013). Senior cycle syllabi have similar assessment issues driving students to simply learn to the test, in this case for a high-stakes exam that determines university entrance. All of these reasons, perhaps, provide evidence for why Irish teachers tend to support constructivist beliefs to a lesser extent than other OECD countries, and report stronger preferences for structuring practices and weaker preferences for student-oriented practices relative to these other countries (Shiel, Perkins & Gilleece, 2009).

This, in turn, leads to routines in the classroom focused on product-driven learning. Indeed Nuthall (2005) refers to the "ritualised routines" (p.895) of teaching where challenging learning is avoided, as the collective (student and teacher) focus is to arrive at the answer in the most expedient manner possible. The use of scientific inquiry activities focuses on the process of learning science, rather than the product, by doing it in such a manner that it resembles the practice of science in the real-world context. It has been well documented that Irish science curricula are content heavy and this overload and the perceived pressure to prepare for high stakes summative examinations may act as barriers to allow teachers to explore with constructivist practices (Cosgrove, Shiel, Oldham & Sofron, 2004) such as scientific inquiry (Crawford, 2007). Therefore, while IBSE may be abstractly mentioned in the old science syllabus, for example, genuine opportunities to engage in such practices were not always considered feasible. The factors considered above result in power relations in the classroom where a "didactic contract" (Brousseau, 1998, cited in Donnelly, O'Reilly & McGarr, 2014) defines student roles as compliant and where Irish teachers may believe establishing student ownership in any significant way is too challenging.

The issue, therefore, remains in relation to how IBSE is viewed within curricular policies. The old science syllabus in the Irish context viewed inquiry as the completion of recipe-like, mandatory experiments and the conflation of experiments with investigations. This is compounded by the isolation of key moments within the syllabus where "inquiry" should take place, i.e. "investigations". However, this can lead to students seeing the cognitive learning of science through inquiry as being separate to the procedural and personal dimensions of inquiry. Forcing students and teachers to engage in prescriptive experiments, does not allow a genuine, organic, and authentic approach to learning science to occur.

The cultural focus on the product rather that the process seen within the Irish science curricula, teachers tended to work in isolation, a sort of balkanisation of practice. This can be further illustrated by the findings of the OECD "Teaching and Learning International Study" report (summarised by Shiel et al., 2009) which shows that comparatively Irish teachers do not generally take part in professional collaboration with colleagues and merely exchange resources without dialogue of practice/methodological approaches. It can be argued that the previous junior cycle curriculum did not call for a focus on professional collaboration. This, however, is changing through the new junior cycle science specification, which will now be discussed.

New Irish Curriculum

In September 2016, the junior cycle science specification was introduced to 1st-year students entering post-primary education. The term "specification" was introduced to communicate the intention to move away from what was perceived to be the excessive prescription of content within closed learning objectives and the mandatory experiments of the old syllabus that supported a "teach to the test" mentality.

The new specification has open learning outcomes that require teachers to make sense of the associated learning and, in principle, offer flexibility in the specific content that might be explored and how. This should include experimentation which is no longer prescribed and where students have the opportunity to design their own investigations.

The science specification is made up of four contextual strands each with a number of learning outcomes. These learning outcomes were designed to be open in nature with interlinks between concepts found in each of the strands. The strands are Biological World, Chemical World, Physical World and, lastly, Earth and Space. Coincidentally the latter is closely connected with the theme of the EUPRBs. However, elements of the EUPRBs can be found throughout the contextual strands, for example, the learning outcome linked to photosynthesis in Biological World is mirrored in the EUPRB entitled "Plants in Space." The strands are unified by a Nature of Science strand, which permeates all of the contextual strands.

There is a strong focus on scientific inquiry where students are active participants in their classroom to include, for example, the opportunity for students to design, implement, and evaluate experimental investigations, facilitated by the teacher. For example, two of the stated learning outcomes of the new Junior Cycle Science Specification are:

NOS2: Students should be able to recognise questions that are appropriate for scientific investigation, pose testable hypotheses, and evaluate and compare strategies for investigating hypotheses

NOS3: Students should be able to design, plan and conduct investigations; explain how reliability, accuracy, precision, fairness, safety, ethics, and selection of suitable equipment have been considered

The assessment associated with the new science specification is the completion of two classroom-based assessments (marked locally with feedback), a written assessment, and a summative state examination (marked externally). The classroom-based assessments are related to the nature of science learning outcomes and are intended to act as assessment for learning opportunities.

Figure 1. New science specification strands (NCCA, 2016)

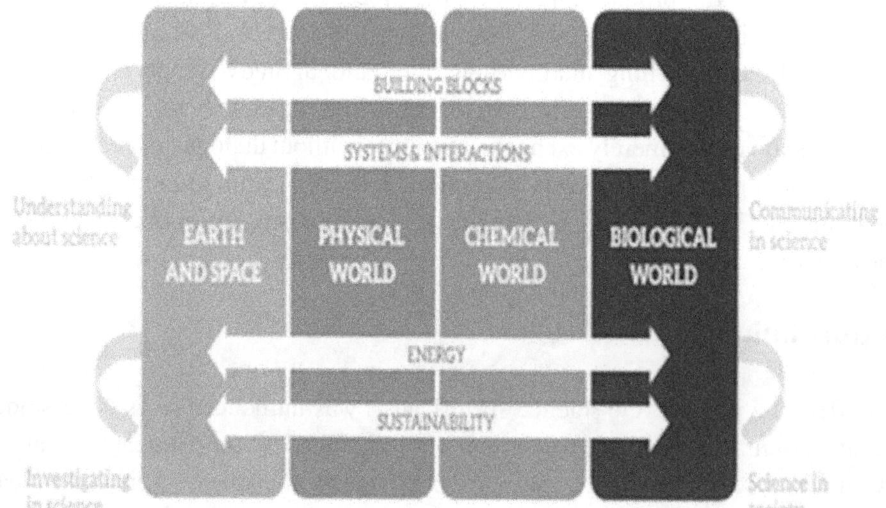

The transition to the new Junior Cycle Framework has been met with contestation due to industrial unrest around the assessment attached to the new curriculum. While this has mostly related to the issue of teachers assessing their own students, concerns have also been expressed about the open nature of the new learning outcomes. This has meant that CR took place at a time when there was serious unrest in schools. Initially, one teacher union did not allow its teachers to engage in professional development relating to the new specification or to engage in the classroom-based assessments (McGuire, 2016). Such tensions have meant that the transition into the new specification and the engagement in professional development modelled around the new specification have been limited. As such, there has been a great deal of confusion in relation to the proposed changes to teaching, learning, and assessment (King, 2017).

PROJECT SUMMARY

This section will consider the epistemological approach adopted by the Irish CR team. From the outset, our intention was to invite schools and teachers to participate in the formation of a PLC focused on collectively enhancing approaches to IBSE. We wanted to stress that while the PRBs were a significant resource to challenge and prompt thinking about IBSE, they were not a technical prescription to be copied in their classrooms and that the core learning associated with CR would be only be achieved by drawing on the collective input of all members of the PLC. As such, the PRBs represented an opportunity to "disturb practice" and try something new, both in terms of content and pedagogy. We considered the formation of our Technical Board (TB) a critical factor in securing buy-in during a time of significant industrial unrest where teachers could not be seen to be engaging with PD that was associated with the new science specification. Our TB included well-known and respected teachers in our region who were also engaged in our Professional Development Service for Teachers (PDST) and the Junior Cycle for Teachers (JCT) science team. The PDST is tasked with general and ongoing PD while the JCT is explicitly focused on PD for the new science specification. This allowed us to both learn from the experience of these bodies but also to contribute to their thinking as outlined below. The teachers on our TB also brought with them a network of colleagues to invite to participate in CR and helped to develop trust and credibility in the work of CR. We maintained contact with our national curriculum agency (NCCA) through the Project Management Board (PMB) who gave feedback on the PD process throughout and incorporated emerging experiences of the CR teachers in the work of the science development group tasked with designing the new science specification.

In summary, we hoped to secure teacher investment in CR through making the PD relevant to future changes in the science specification, through energising participants as resources for each other, and by providing scaffolding during the PD process that introduced new, practical ideas in relation to IBSE. It was hoped that this would be effective in developing Pedagogical Content Knowledge (PCK) which Van Driel and Berry (2012) note requires that PD should be closely aligned with professional practice while providing opportunities to enact novel instructional approaches followed by individual and collective reflection on experience. In their review of the literature pertaining to Inquiry PD, Capps, Crawford, and Constas (2012) argue that the desired outcomes of teacher PD include enhanced teacher knowledge, changes in beliefs and practice, and enhanced student achievement. We hoped that CR would support enhanced teacher knowledge in content terms through the provision of the PRBs focused on earth and space, a new strand in the junior cycle science specification that does not neatly align with the disciplinary backgrounds of Irish science teachers. We aimed to enhance pedagogical knowledge in the modelling, experience of, and reflection on cooperative approaches to learning that are coherent with the Nature of Science. Together this knowledge would be used flexibly and creatively by teachers to design their own PRBs. CR aimed to provoke changes in beliefs and practices through the practical application of enhanced teacher knowledge and subsequent reflection within the PLC in a content area and with pedagogical approaches that are closely aligned with the changes in our national curriculum. Ultimately, by influencing beliefs and practice as a PLC, we aspired to develop teacher professional agency to take the initiative in designing and delivering IBSE classes.

We were conscious of the inertia associated with changes in beliefs which Biesta, Priestley, and Robinson (2015) associate with issues relating to conceptions of the role of the teacher, the nature of students and the purpose of education with respect to teacher agency in Scottish curriculum reform. They argue that overall limitations in teacher agency are most strongly driven by a lack of collective understanding of the purpose of education that derives from vague understandings of top-down reform, leading to an instrumental focus that does not consider broader purposes. Within CR, this might be considered as just doing "inquiry" without considering the broader purposes of science education so we attempted to broaden discussion at all workshops to move beyond mere technical consideration of content or pedagogy. Biesta et al. (2015) highlighted that, in relation to students, positive relationships were seen as key but that teachers saw their role as directive and active in fixing student deficits. In defence of these relationships, teachers might protect their students from reform that they perceived as too challenging or beyond student ability, which was perceived as fixed. We were conscious of this potential hegemonic assumption in the work of CR and, as will be illustrated below, the PD process was successful in challenging and changing some of these assumptions about student ability. Finally, in relation to the role of teachers Biesta et al (ibid., 2015) noted that teachers made use of the language of policy but were sceptical of taking on a truly facilitative role and acting in the role of agents of change in curriculum design. Indeed, there was a nervousness about being "required" to be autonomous in practice. This is clearly at odds with a "bottom-up" approach that supports ownership of change processes, so within CR we attempted to avoid the perception that the TB were the voice of authority or that there was a single "right" way to do IBSE. This is consistent with a view of IBSE where student ownership of learning is necessary to allow for knowledge construction in science including posing questions, planning investigations, and critiquing and debating findings (Anderson, 2002; NRC, 1996; Osborne, Erduran, & Simon, 2004). However, this also requires ownership of the process on the part of teachers who must allow students to challenge and contest ideas, including their own, something students are traditionally reluctant to do (Donnelly et al., 2014). This professional partnership was conceived of as a space where contested views of practice could be discussed in an open and trusting manner and where disagreement was an inevitable part of the sense-making process, something recently highlighted by Bissaker (2014).

Throughout the work of CR, the PMB were in frequent contact with policy makers (NCCA) and PD agencies (JCT) to reflect on the ongoing work and to contribute our emerging understandings to support the development of the new science specification.

Project Implementation and Barriers

Overview of Implementation

Each year of the project began with an initial two-day workshop in the host institution, the University of Limerick. Here, teachers became familiar with the Technical Board and the other teachers in their new PLC, learned about the CR project, and began to develop their collective understanding of IBSE. Given the typically limited experiences that Irish teachers have of working as curriculum makers in professional collaboration, reviewed earlier in this chapter, some time was given to the discussion of the nature, purpose, and collective expectations of the PLC. The structure of these workshops allowed for open and frank discussions of any concerns that teachers had regarding inquiry, and the potential barriers to success. Teachers took part in a lesson study, to see inquiry being modelled and to experience inquiry as a student. A debrief took place after the lesson study to allow participants to reflect, as teachers and learners, on the experience they had engaged in. Before participants left, an online learning community was set up to keep the PLC in contact and an action plan put in place for the coming months that included plans for ongoing support, a school visit from a role model practicing scientist, a second (mid-term) teacher workshop, and national/international conference dates.

It was important that this workshop also be an immediately valuable process for teachers to engage in. To provide additional value to the participating teachers the process of engaging Chain Reaction was mapped to the School Self-Evaluation (SSE) process which post-primary teachers in Ireland are mandated to engage with. SSE is a collaborative, reflective process of internal school review that involves gathering and analysing evidence, drawing and reporting conclusions, and implementing an improvement plan (DES, 2016). The TB outlined how the process of engaging in Chain Reaction during the year would clearly map onto the SSE, but equally how a PLC meeting could also be linked to the SSE process. Teachers were invited to express their expectations about what participating in the CR PLC might or should involve. There was concern that expectations should be realistic in relation to the existing crowded curriculum and that practical pedagogies should be engaged with to allow for experiential learning and strong student self-regulation, where mistakes were welcomed and shared in a cooperative, not competitive environment. In relation to their own professional learning, teachers expressed the hope that they could fit the work into their existing practice and that they could be open to sharing experiences in an honest and vulnerable manner that honoured their efforts. Teachers were uncertain of their own knowledge of IBSE and questioned whether they were "doing it right" or, indeed, if "results" would follow from IBSE. Concern was expressed about potential loss of control in classroom management and teachers were unsure as to how they would manage noise levels during student group work. Critically, it was noted that there was "no recipe to follow" and that many of their aspirations were at odds with typical classroom routines, which are heavily focused on textbook use. In this, they hoped for the support of the CR team, their principals, and the rest of the PLC in being open to new ideas and feedback on developing practice.

A PLC refers to a community in which teachers can share their individual practices with the aim of searching for "good practice" based on the outcome of the collective inquiry (Wong, 2010). This involved "disturbing practices" as teachers reflected and analysed their own teaching. The steps of SSE provide a systematic process to facilitate the reflection on the descriptions and explanations of teaching. During the professional development training, the CR team helped to scaffold the teachers' progression through these three steps (gathering, reporting, and developing a plan) in developing a framework to evaluate their teaching and learning. Many of the key principles of SSE are integral in the CR community of professional development: collaboration, inclusiveness, reflection, evidence-based decisions, flexibility, leadership, improving teaching and learning, and communication.

Many PRBs that were produced as part of the CR project focused on topics of Earth and Space, which were new additions to the Junior Cycle Framework science specification. The CR team carefully reviewed and linked these PRBs with the new specification whilst being considerate of the industrial relations, which were ongoing with teaching unions in Ireland surrounding the new Junior Cycle. The CR team always sought to motivate participating teachers with what could be intrinsically valued such as the opportunities for professional learning, collaboration, and an exploration of their subject area, as well as what could be of extrinsic value, such as gaining new methodologies, resources like PRBs for teaching the new specification, and data for the SSE process.

Initial Workshop: Making Thinking About IBSE Visible

Working in groups, teachers discussed what they thought inquiry is and what it looks like; from the perspective of the teacher and the student. Teachers reflected on the elements of inquiry that they were already implementing in their classrooms, as well as features of inquiry that they would like to implement in future teaching. This gathered evidence of existing practice and allowed the teachers to reflect on improvements and developments that they could make to their own professional practice.

The PLC initially discussed inquiry as including the following: demonstrations and experiments; linking science to the news and current affairs; asking questions; classroom discussions; pupil debate; pupil group work; inviting visitors into the classroom; field trips; and highlighting inter-disciplinary links. Then, in pairs, teachers discussed the questions "How do you 'do' scientific inquiry in your teaching?" and "What does the inquiring student look like?" in order to create a shared understanding of what inquiry was understood to mean. These questions were selected by the TB to place a focus on action (teacher and student) which would be followed by analysis to determine categories in responses. Teachers placed their thoughts on post-it notes that were then placed on the board within a diagram to facilitate comparison. These became the talking points for discussion and debate. This process allowed for the co-construction of shared meaning for IBSE between all teachers. A primary model was created through mediated discussion (Figure 2.). This co-constructed understanding of IBSE and what is important for IBSE is outlined below.

How Do You "Do" Scientific Inquiry in Your Teaching?

- **Developing Reasoning:** Teachers discussed how they repeatedly questioned "why?" in order to push students beyond basic descriptive responses to more reasoned or evidenced claims. The long-term goal being the creation of a habit-of-mind for rationalisation and reasoning.
- **Encouraging Motivation:** Teachers used 'hooks' to draw students into a problem and develop motivation by presenting something that is almost certain to spark student curiosity. Teachers spoke about the need for excitement and a desire in students to learn more. Demonstrations, videos, or even sets of data could be used for this purpose. Ethical or socio-scientific issues were often stated by teachers to be extremely motivating factors for their students.
- **Fostering Responsibility:** There was broad acknowledgement that students had a responsibility both for their own learning and to the broader society who is funding their education so they might contribute and solve the problems of tomorrow. The need for students to fully participate was reiterated and it was pointed out that there is often a desire for an easy solution or shortcut to the answer but that no learning can occur without effort.

Figure 2. Co-constructed model of understanding of IBSE (Year 3)

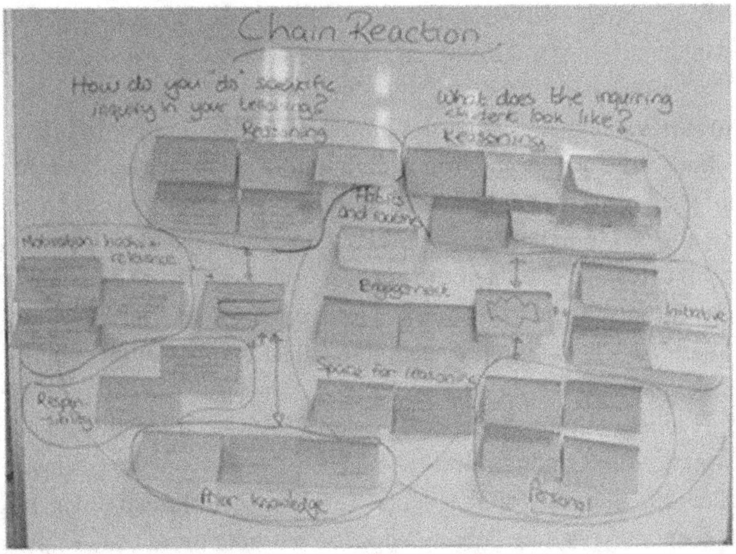

- **Tapping into Prior Knowledge:** Teachers recognised that for any inquiry to happen, it was necessary to orientate students, prepare the scaffold upon which students would continue learning, and weed out any misconceptions that might cause hindrance. A number of different techniques were used by teachers to elicit the prior knowledge of students, but most teachers agreed that this was a necessary step regardless of its form.

What Does the Inquiring Student Look Like?

- **Reasoning:** Just as teachers would attempt to develop reasoning in their students, they felt an inquiring student would engage in and be able to show thinking and reasoning. An inquiring student would be able to make connections with prior knowledge, providing reasoning for their answers, and ask questions. Teachers offered praise for students who might *'try to catch you out'*, suggesting that they see this is the sign of an inquiring student.
- **Initiative:** A student who looks for other solutions, even if looking for an easier route, were thought to be examples of inquiring students. These may often appear to be the louder, more confident, outspoken students but teachers also spoke about the quieter student whose voice could be the most provocative if allowed the space and environment to express themselves. The willingness to put forward new ideas, questions, and solutions was seen as central to the inquiring student.
- **Interactive Collaboration:** This component of the inquiring student was, for teachers, a way of encapsulating the interactive and collaborative nature of scientific investigation. The willingness to work productively with others was seen as an important feature of the inquiring student, but teachers also noted the necessity to build this confidence in students to make contributions. Here, good student-teacher relationships were thought to be essential.

Overlap: Inquiring Students and Teachers

In Figure 2 above, the diagram shows overlap between concepts that apply to both the student and teacher. These were collated under 'Habits and Routines', 'Engagement', and 'Space for Reasoning':

- **Habits and Routines:** Habits and routines are relating to both 'of-mind' and 'of-practice'. These habits and routines are for the classroom community as a whole, both students and teachers. Teachers would need to facilitate the creation of habits and routines for students. These might be habits of questioning and rationalising, or the routines of independent investigation and the associated classroom logistics. Teachers felt they needed to break the habit of stepping in with a student who seems to be struggling and create a habit of providing a little more time, or some guiding questions. Similarly, teachers felt they needed to adjust their practice to allow some wait-time for a student to think when answering questions.
- **Engagement:** Teachers felt that successful inquiry in the classroom depended on wholehearted engagement from both the students and the teacher. Active, engaged, questioning, and receptive students need not always be the loud students. Teachers thought that it was important to recognise that quieter students may well be equally engaged but take more time to process, or need an opportunity to speak their mind.
- **Space for Reasoning:** In order to get the reasoning that teachers desire of students, and to encourage quieter students to contribute, teachers felt it was important to create an environment that was safe and comfortable. This means students have the time to think, but also the safety to *'question the system'* without fear of reprimand, and the teacher should also be able to have the space to say *'I don't know the answer to that question',* essentially letting go of the concept of the teacher as 'all knowing' and instead acknowledge students for asking challenging and insightful questions. This process of acknowledgement was thought to be a route to building confidence in students and to creating this culture of questioning.

Initial Workshop: Lesson Study Experience

A significant part of the first workshop was devoted to a modified form of lesson study where the participating teachers, rather than secondary school students, acted as the "class" and two of the technical board members taught one of the PRBs, namely "Green Light". Since CR was encouraging teachers to consider disturbing practice in their own classrooms it was considered important to provide a concrete experience of an IBSE lesson which could be potentially team-taught, in a double period (80 minutes) and with limited resources (pen and paper). Not only would teachers likely experience a novel teaching methodology (jigsaw) through the lesson study but also an approach to IBSE that was classroom rather than laboratory-based. The latter was considered important in challenging the conflation of experiments and practical work with inquiry, a significant issue in Irish schools as described above. It was also hoped that this experience would demonstrate that the CR team could "walk the walk" in IBSE teaching and be open to feedback following the lesson. As such it could be considered that the lesson study approach adopted was coherent with the 3D model of inquiry (Bevins & Price, 2016) in the professional development context to include content (new science and teaching methodology), skills (managing a jigsaw classroom, modelling IBSE) and psychological (reflection and openness to feedback). The CR team stressed that this experience was not presented as a "perfect" model of an IBSE classroom to be copy/

pasted into their own classroom but rather as a concrete experience to facilitate reflective dialogue. Immediately following the PRB, class teachers were invited to critique, question, and offer feedback on the learning experience. The ensuing discussion was framed by the earlier co-constructed model of IBSE; evaluating the lesson, from both a student and teacher's perspective, in light of the co-constructed idea about what IBSE should be.

We as teacher educators wished to use this approach for several reasons linked to the establishment of credibility attached to engaging in a PLC with experienced teachers. It was important to create a space where reflective dialogue was encouraged, not only for personal fulfilment but also for our own professional development. To that end, after each workshop we summarised the collective PLC experiences to inform future engagements with teachers. A disconnection often persists between academic knowledge and the practical dimensions of teaching (das Graças C. de Arruda Nascimento & dos Reis, 2017) with teachers often reporting the unrealistic messages being shared by teacher educators removed from classroom practice.

Lesson Study Using the Green Light PRB

The "Green Light" PRB places students in the role of scientists tasked to support the development of an advertising campaign to encourage people to switch from using traditional, filament light bulbs towards more efficient compact fluorescent equivalents. The teacher introduced the task and formed the class into groups of five where one student takes on the role of chair to lead the work of the group. Printed materials were provided to the chair who led the group through a description of the work involved. Students must become experts in one area relating to the overall task namely i) how filament bulbs work and problems associated with them ii) how compact fluorescent bulbs work and their efficiency iii) public opinion in relation to these light bulbs iv) calculating the cost difference between using the different bulbs in a typical household. Once students were assigned to one of the four tasks, they were directed to an area of the classroom with those from other groups focused on the same task. There they engaged with print material (a single sheet) that provided some key information and asked questions about the area in which they had to become experts. Students discussed this and took their own notes before returning to their original (jigsaw) groups. Here the chair led them in discussion to share their information with the rest of the group and together they designed a poster drawing on all expert information in answer to the brief to persuade the public to buy compact fluorescent lightbulbs. The class finished with a presentation of the posters and discussion of how the work they conducted might be similar to that of a scientist ("scientific inquiry") and reflective dialogue ("teaching thoughts"), which was captured on the board for discussion, Figure 3.

Teachers highlighted the pedagogical affordances of the jigsaw methodology in providing a very structured approach for differentiation (by task and difficulty) in a mixed ability classroom, the accountability of every student to contribute to group success thus avoiding passengers, the truly facilitative work of the teacher who acted primarily in a "quality assurance" mode, the development of communication and thinking skills, and creativity in the production of the poster. They noted student independence, the space to make mistakes through "productive failure", and the opportunity and time for teachers to stand back and make decisions carefully in relation to praise and feedback. In discussing the IBSE components of the class, teachers focused on the fact that students had significant responsibility and that they made the majority of the decisions relating to critiquing information, data analysis, and representing and communicating information while working as a team. In relation to the ethical responsibility, it was noted that

Figure 3. Teacher reflections on "Green Light" PRB experience

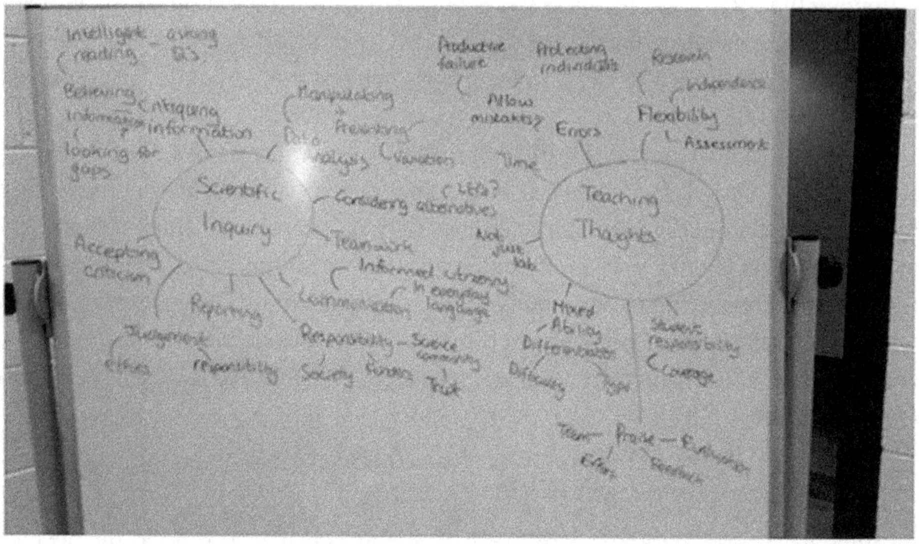

they chose not to include information on the posters that the compact fluorescent light bulbs included mercury, which prompted a discussion and consideration of the risks involved. This led to consideration of the influence of market forces on the work of scientists and associated responsibility. Teachers also noted that they could not conclude that compact fluorescent lightbulbs were the environmentally favourable choice, as other light sources had not been considered such as LEDs. This prompted discussion about the potential to build on this class to an extended experimental investigation where students could test various light bulbs in relation to energy use, light output, heat generation and so on. It was recognised that many of these ideas and future directions for work could be generated by the students, giving them significant ownership over the work.

The CR team teachers also shared thoughts about their experience of the class, reflecting on the "odd" experience of standing back from the students as they worked, feeling somewhat superfluous to requirements! In one class a student had made a particularly insightful comment during the expert group discussion and one of the CR teachers was about to jump into the discussion to highlight and extend upon this thinking. The team teacher, however, suggested that the students should be left undisturbed as to intervene would be to highlight the "correct" answer and stifle further discussion. At the conclusion of each of the initial workshops, there was a very positive attitude to the experience and a sense that a good start had been made to the realisation of the expectations of the PLC.

Post-Workshop Implementation

Following the workshop, teachers began implementing IBSE using the resources available to them through CR or other self-created pupil research briefs. Some of these classes were observed by the CR technical board, allowing for reflection and support for on-going improvement. All teachers came back to the University of Limerick, 4-5 months after their initial workshop, for a second day-long workshop to check in and reflect on the process as a PLC. Following this, the teachers began working with their students to represent and communicate their research findings using posters, which were displayed at the Irish National "Express Yourself" Conference. A judge- and peer-selected team from one school, and their teacher, then progressed to the International "Express Yourself" Conference where an end-of-project teacher workshop also took place. Each of these components is discussed in further detail below, followed by a discussion of the themes of the teacher and student experience of IBSE as they emerged through these activities:

- ***Support through school visits and the online community*** – As one of the aims of the project was to create a Professional Learning Community (PLC), teachers and project coordinators were on hand to offer any advice, comment, or indeed provide help to the teachers. This was largely done interactively through online platforms (Google Community Year 1, Edmodo in Year 2 and 3). Members of the TB also visited schools at the request of the teachers to help them develop the plan, provide feedback, and answer any questions they had. Some of these school visits were recorded with permission. There were some tensions inherent in the PD process in that, while we aspired to support teacher autonomy, it proved very difficult to secure teacher engagement with the online supports in uploading ideas or developing PRBs for feedback such that collaboration between the workshops was very limited.
- ***Mid-term support workshop*** – The workshop provided an opportunity to reflect on practice as a community, to share experiences of what went well and not so well, to ask questions, and to learn some additional methodologies to facilitate inquiry. At the request of teachers, the TB members provided some new learning about the 5E model of scientific inquiry, cooperative learning, concept mapping, and questioning methods to support IBSE. Some of the time was also devoted to re-examining the worries and concerns that were expressed in the first workshop in light of the experiences teachers now had with IBSE. The final part of the workshops involved preparing for the National "Express Yourself" conference. As a Professional Learning Community, it was important to get teachers input on how they would like the national conference day to operate. Teachers were also given a folder with a number of resources such a sample PRBs, journal articles, and information on alternative teaching strategies that could be used in line with inquiry. This folder would also serve the school in their completion of the SSE.
- ***Role models*** – Each school was visited by a role model who was a practising scientist. It was important for students to see how science is done outside of the classroom, and to gain an appreciation of how science really operates in the field. The role models also discussed with students the work they did during the inquiry-based activity and answered any questions that the students had. The project coordinators tried to recruit role models whose work reflected the topic on which students carried out the inquiry-based lesson. For example, one school carried out a PRB on badger conservation. The role model who visited this school was a conservationist who has featured on both radio and television programmes on the topic. Another school that had carried out the "Collision Course" PRB had an astrophysicist visit their class.
- ***National Express Yourself Conference*** – The posters that students created to communicate their inquiry were displayed at the national conference. Role models, teachers, and other students browsed these displays, asked questions about the investigations, and voted for a 'winner' to represent Ireland at the International Conference (see next section). Students also engaged in a mini-PRB which was facilitated by a member of the TB. Both students and teachers had opportunity to share their thoughts and feelings about their engagement in CR in both public and private ways. Surveys were used for private responses, while graffiti boards and on-stage interviews were used

to share reflections amongst the whole group. Finally, plaques were awarded to each school, specifically to the teachers, in order to publicly acknowledge their commitment and engagement with CR.

- ***International Express Yourself Conference*** – The group of students selected from the national conference, and their teacher, attended the international conference. Here, students presented their research with oral presentations to an international audience and again engaged in inquiry activities. This was an opportunity for both students and teachers to see the work of other students and ultimately furthering the goal of dissemination of the IBSE approach, methods, and tools. In addition, in communicating their research to a body of their peers, students mirrored the work of practising scientists, highlighting a more holistic understanding of the nature of science.
- ***End-of-project teacher workshop*** – One of the main events at the International Conference was a workshop for the teachers who travelled with students. This was a unique opportunity for teachers to widen their network of colleagues, and to contrast their practice across different national contexts. Mixed groups of teachers engaged in collaborative inquiry activities which put them, once more, in the role of a student. Teachers were facilitated in using the opportunity to think and talk about IBSE with their international peers.

Through all of these activities, the TB reflected on responses from students and teachers. It was important to consider how teachers and students engaged and interacted with IBSE in the classroom. For this reason, classes were recorded when a TB member would visit the school. This would allow the TB members to reflect in a much deeper way on the implementation of CR. Similarly, discussion and activities between the PLC were recorded to capture teachers' ongoing thoughts and feelings about their engagement with the IBSE approach. Finally, interviews and surveys were conducted at numerous stages throughout the three years of the project which helped to provide the TB with an overall sense of the teacher and student experience. Some of the major points of reflection that emerged are noted here.

Teachers Reflections

The teachers involved in CR very clearly valued the **space** to think and talk about their practice. This message resonated throughout where teachers would talk about needing time, not just for classroom work in a seemingly overloaded curriculum, but for professional conversation, collaboration, and development. After the project, one teacher stated:

It was great to have the space and freedom to construct my own understanding of the nature of inquiry with my students and the teachers involved in the project. The space to experience productive failure, to try new ideas and to test my own assumptions with new classroom practices and reflection was a refreshing escape from the everyday confines of the science syllabus.

The theme of **vulnerability** emerged at numerous points throughout the project. The vulnerability that came with trying something new was high. This was particularly felt as teachers perceived a necessary loss of control when becoming more student-centred. Additionally, given Ireland's somewhat isolated teacher culture, teachers were not used to or comfortable with exposing their practice for others to see. Bringing another practitioner or observer into the classroom became a source of anxiety for some. Teachers also wondered what other teachers in their school, or even the parents of their students, might think of their new practice. They were particularly conscious that colleagues, management, parents, or even students would perceive IBSE as time wasting and not sufficiently focused on content that might feature in an examination. For the teacher to engage wholeheartedly in IBSE they would need to overcome the thoughts that their practice would be negatively criticised by those around them and to challenge their own vulnerabilities.

Teachers were often concerned that mixed ability classrooms would not be a suitable environment for inquiry. Their **expectations** of students were often that there would need to be a lot of support and guidance in order to get, at least some, students through the process. Where teachers felt that some students might actually be able to achieve with a little more time, it was this very time that was seen as an unavailable luxury and so increased teacher-control was seen as necessary. Speaking of the coursework projects in the old curriculum, one teacher said:

All science teachers know from doing the projects how awkward this is going to be, honestly... Those projects drive me batty, because we don't have time to let the kids experiment. We just don't have time!! They can do a bit of it, but you have to guide people along. And mixed ability classes...that is really hard

What some teachers in the PLC came to realise and share was that their students, when given their opportunity, surpassed the teacher's expectations. On reflection, one teacher questioned herself about the expectations she had for her pupils, asserting that quite often these expectations are too low. The PLC began to consider that this might become a self-fulfilling prophecy and that setting higher expectations of student achievement in inquiry would be a more beneficial stance. In each year, this change in what teachers expected of students was evident as students consistently surpassed the teachers' expectations in some way.

Teachers' worries and concerns at the beginning of the project were that their classroom practice was restricted by the context of the current curriculum, the lack of adequate time, space or resources, as well as an exam-focused educational culture.

If the exam doesn't assess it, then what is the point? Even if it helps in college or later in life, the primary objective is to get into college which is done via examinations

Teacher **agency**, which results "from the interplay of individual efforts, available resources, and contextual and structural factors" (Biesta & Tedder, 2007, p. 137), is necessarily impacted by the perception of restriction, accountability, and lack of time or resources. Teachers needed space and time for professional collaboration to be provided, as it was through CR, and support in the face of feeling vulnerable, in order to motivate their individual efforts. They needed the classroom resources to reduce the preparation time, and the pedagogical resources for guiding their practice of IBSE. CR also provided these. What CR could not address was the particular school and curricular context where the teacher's practice was situated. These factors limited the agency that teachers could exercise which will be considered further in the conclusion of this chapter.

Despite the vocalised understanding and appreciation of IBSE from both teachers and students, the project was not without its barriers. Teacher dropout was an issue within the Irish context, particularly in year 1. Professional dialogue between members of the project technical board discussed possible reasons for this and it was suggested by one member, an experienced teacher, that teachers wanted to be given off the shelf PRBs because of factors like time constraints and weak self-efficacy towards IBSE given the lack of experience in curriculum making. It can also be argued that teacher scepticism could serve as a possible reason for the disengagement of teachers. Teacher-attitudes towards IBSE is, therefore, a crucial determinant of whether IBSE professional development organically translates into classroom practice.

Another constraint identified by some teachers was the support they received from the management structures of their schools, particularly with reference to supporting the teachers' attendance at the professional development workshops. Teachers reported the difficulty they had with being able to free up time to engage in workshops and tasks related to Chain Reaction. Harland & Kinder (1997) concede that management support is critical for effective professional development. While the creation of a PLC is significant for supporting teachers, additionally support from management and indeed other colleagues is needed back in the school context.

Student Reflections

Examining the various sources of data, from surveys, to interviews, to graffiti boards, it is very evident that the majority of students enjoyed the process of engaging in inquiry and the methods of learning that they experienced during the CR project. In interviews, they stated that they *"enjoyed working together as a team"*, *"the way that you could teach yourself"*, and that *"it makes you more of an independent learner"*. Some students felt science *"was easier to understand cause [sic] we were going away and putting it into our own words"* and that they *"were able to work like a real scientist"*. On graffiti boards at the national conference event, students continue to provide positive comments relating to collaborative work and independent research, but also added that having control and choice over what they would learn was important to students (Figure 4.) What is interesting, however, is that students seemed to focus on the process (learning in groups as fun, taking responsibility…) rather than the product of learning.

A barrier that was highlighted through this project was the disconnection between the perceived student appreciation and development of IBSE practice and the final product presented at both the National and International "Express Yourself" conferences. The students appeared to have developed some understanding of IBSE practice, however, their ability to articulate the science was lacking at times. Therefore, it can be argued that the students placed greater focus on the process rather than the product. Interestingly, the latter is normally the case in a traditional classroom. This has emphasised a certain tension between the IBSE experience of the students and the conceptual learning that should take place with such practice and the balanced focus on the process and the product. It may also highlight the need to develop students' communicative skills and their ability to use scientific language. All should be viewed as critical to an authentic learning experience. In relation to Bevins and Price (2016) IBSE model, the first dimension of inquiry (scientific knowledge) appeared to be weak in students' learning outcomes. As Bevins and Price (2016) suggest, the inquirer (students) need to be able to navigate within and manage the interactions between all three dimensions for a meaningful productive inquiry experience to take place. The authors consider that issues relating to limited teacher agency and student focus on process and skills emerging from CR are very significant in the context of curriculum reform in Ireland, which we will return to in the conclusion.

In terms of feedback of the inquiry process, students mentioned several benefits such as the ability to work in teams and the autonomy that they were given. However, there were occasions when the use of teamwork did prove a barrier between students. Personal conflict between students and the dominance of certain group members was often identified in the classroom observations. Interestingly, one of the Irish in-service teachers experienced similar tensions during the International workshop between partner

Figure 4. Graffiti board (National Conference Year 3)

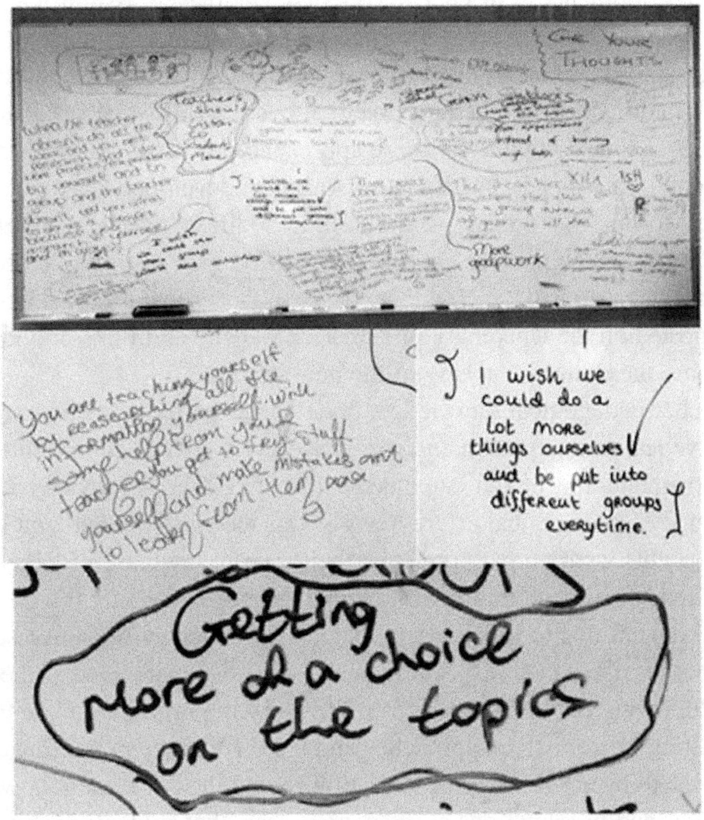

countries. He stated that *"like with any group, there were strong personalities. Trying to manage that, can be difficult"*.

The power relations between teachers and students was also seen as a barrier. Donnelly et al. (2014) identified power relations within an inquiry classroom. Evidence of these power relations exists when looking at some of the exchanges between teachers and students in the classroom observations. For example, in Year 1, one of the teachers Mary (pseudonym) mentioned how she would *"guide you along the right path in case we think you are completely going off left field"*. This perhaps signifies a 'guess what's in the teacher's head' mentality, in other words, a teacher's cognitive economy (Lemke 1990), which needs to be overcome if authentic inquiry is to take place.

IMPACT

In describing the impact of CR we must return to the central goal, namely to support teacher agency in designing and implementing IBSE to include influences on teacher beliefs and assumptions which translated into improvements in student learning. We will also consider the influence of CR on the national curriculum reform process and the work of the JCT tasked with supporting science teacher PD in relation to the new specification.

Examples of teacher learning are presented in the previous section with the core impact summarised here. In relation to developments in teacher knowledge, CR played a significant role in establishing existing conceptions of IBSE and then building on this in relation to implementing IBSE beyond the laboratory, in the classroom and as an overall habit of mind. It would not be fair to claim that these developments became embedded in teaching practice, but CR did some productive work in opening the participants' minds to the practicality of these approaches in existing classrooms and in preparation for future development in the science specification. Teachers developed an expanded view of scientific inquiry and were able to link this directly to the JCF. In addition, teachers expressed appreciation for the earth and space PRBs which provide content that will be very useful to address the new strand in the science specification, an area in which they expressed concern about the depth of their knowledge. The PRBs are now hosted on the University of Limerick website and links provided to the JCT so that other schools might use them in the teaching of the new strand.

In relation to teacher pedagogical knowledge, none of our participants had previous experience of the jigsaw cooperative learning approach and this proved very useful in providing a concrete scaffold for the practical implementation of (an expanded view of) IBSE. As reviewed above teachers were able to identify a range of pedagogical affordances of this teaching method and alignment with IBSE and many of their students went on to experience their own jigsaw-based PRBs. This also assisted in challenging expectations of student ability with many teachers expressing the realisation that they had limited student achievement due to low expectations. A few teachers remained sceptical of the utility of cooperative approaches like jigsaw, especially due to what they perceived as the poor content knowledge demonstrated following the PRB experience. This may in part derive from the ritualised routines that characterise existing transmissive approaches and the difficulties in shifting classroom culture for students and teacher alike over a brief implementation period. It was evident however to the TB, from classroom observations and presentations at the conferences, that students highlighted the importance of the process of inquiry over the comprehension of scientific content.

CR provided an opportunity for teachers to collaborate on lesson design and share understandings of practice supporting their sense-making of the change process. This will be critical for the future success of the JCF reform. While teachers appreciated the opportunity to share ideas, discuss existing practice and recognise common challenges and work together in pairs at their schools they were less willing or inclined to share their PRB designs with other members of the network. It is likely that this was influenced by the general business of practice where CR was an additional task to add to existing workloads. But it is of note that while teachers appreciated professional dialog during the workshops, there was very little engagement between schools on the IBSE design process despite the provision of online networking facilities and the encouragement of the TB. As such the influence of CR on teacher agency was mixed as will be reviewed in the conclusion below.

CR provided an effective space for engagement with policy makers through the NCCA and PD agencies through the JCT. An education officer with the NCCA who was responsible for leading the development group for the new science specification joined the CR PMB at the beginning of the project. It was recognised that the learning approaches advocated by CR were very coherent with the new science specification and that the work produced by teachers and students provided the NCCA with concrete examples of Irish teachers' and students' capacity to adapt to new modalities of teaching and learning within the realistic constraints of the classroom environment. The development of teacher knowledge through the PLC, building on existing conceptions of IBSE and positioning teachers as agents of change who acted as resources for each other was also aligned with the new JCF. This concrete data was of particular use during the period of industrial unrest as CR acted as a useful intermediary to allow new approaches to be tested in the classroom without reference to the JCF. This work also led to the invitation of the PI of CR to join the development group where the PD experience and CR approaches were directly relayed to the development group as food for thought in the specification design. This was influential in extending conceptions of IBSE away from a simple "scientific method" towards the representation of the Nature of Science in the specification, providing concrete examples of pedagogical approaches and considering the broader purposes of science education.

This learning also found purchase with national PD agencies as several members of the TB found roles with the JCT and PDST. In addition to the above JCT members have integrated CR PD experiences including lesson study into the national PD structure. The JCT PD methodology has also adopted a PLC-like approach that brings the science teaching staff from multiple schools together during day long PD sessions that emphasise the teachers as knowledge generators in the face of experiences that "disturb practice". The positioning of teachers as the key resources of change that honours their professional judgement is a key feature of the work of the JCT that follows from CR and indicates a marked change from historical "top-down" PD.

CONCLUSION

Brookfield (1995, p.1) cautions that "One of the hardest things teachers have to learn is that the sincerity of their intentions does not guarantee the purity of their practice". Our intention within CR was to support teacher autonomy in a PLC to develop approaches to IBSE in the context of curriculum reform, a new science specification and changes in teacher professionalism. What follows are our reflections as teacher educators on the TB and include some thought about the lessons from CR for the future of science curriculum development in Ireland.

As described above, teachers seemed to value the focus on autonomy significantly during the workshop sessions and were able to work openly to identify emerging professional learning. In this though there was a tension between our positioning of teachers as resources for each other in the PLC and knowledge generators and their desire to have some "off the shelf" exemplars they could apply in their classrooms. We considered the PRBs and pedagogical experiences provided by the lesson study to be adequate in this regard but there was a consistent demand for more and concern that teachers were not "doing" IBSE right. In addition, the work developed by students, while demonstrating an increased awareness of scientific processes was disappointing in terms of scientific content understanding. This may reflect several factors to include a struggle in the nature of facilitation of IBSE classrooms where classes may have been more open than guided and teachers were unsure when to intervene and act in a directive or quality assurance manner. The lack of time to implement EUPRBs was also almost certainly an issue. As such there may be parallels here with the experience of curriculum reform in Scotland described by Biesta et al. (2016) where teachers felt that they were "forced" to be autonomous in implementing changes that they did not fully understand. This desire for prescription is in tension with a simultaneous desire of teachers to have more of a say in how they do their work. In any case, the false dichotomy between scientific processes and content knowledge must be avoided with a focus on the simultaneous development of both.

If Irish teachers are to have autonomy that is driven by their own volition and not "forced", they must be provided with time and collaborative structures to collectively make sense of and plan for the implementation of the open learning outcomes of the new specification, especially in relation to the nature of science outcomes with respect to IBSE. In terms of competence, the desire to extend professional capability is dependent on the factors described above in relation to autonomy and the capacity of teachers to work together in the development of scientific and pedagogical knowledge to be openly shared with colleagues, ideally as a form of action research nested within the SSE process. Our experience on CR suggests that Irish teachers certainly have the openness to engage in this process but that some work will have to be done to sustain this as an ongoing activity considered to be a normal, habitual part of professional practice. Finally, the most significant people that teachers will have to relate to are in fact each other, not surprisingly as Fullan (2007, p.74) reminds us that "teachers are convinced by what other teachers do and think". The JCT are again doing effective work in this regard, encouraging teachers to learn from each other as they disturb practice and challenge assumptions while trusting their professional judgement.

This approach can potentially lead to the development of "Professional Capital" (PC) which Hargreaves and Fullan (2012, p.1) claim "…relates to one's own or group worth, particularly concerning assets that can be leveraged to accomplish desired goals". PC includes human capital (knowledge and skills of the individual) and social capital (exists between teachers providing access to human capital of others and positive interactions) that can develop effective decisional capital (making good, discretionary judgements). It is perhaps the capacity of students and teachers alike to exercise decision-making that will determine the potential of the system to overcome the ritualised routines that dominate current practice, but research indicates that this will not be easily achieved (Donnelly et al., 2014, p.2043):

In an experiment, I don't like to be inventive. I follow instructions

CR has provided significant impetus to curriculum and professional development agencies and the teacher educators involved in Ireland to support the ongoing change process.

REFERENCES

Anderson, R. (2002). Reforming science teaching: What research says about inquiry. *Journal of Science Teacher Education, 13*(1), 1–12. doi:10.1023/A:1015171124982

Bevins, S., & Price, G. (2016). Reconceptualising inquiry in science education. *International Journal of Science Education, 38*(1), 17–29. doi:10.1080/09500693.2015.1124300

Biesta, G., Pristley, M., & Robinson, S. (2015). The role of beliefs in teacher agency. *Teachers and Teaching, 21*(6), 624–640. doi:10.1080/13540602.2015.1044325

Bissaker, K. (2014). Transforming STEM education in an innovative Australian school: The role of teachers' and academics' professional partnerships. *Theory into Practice, 53*(1), 55–63. doi:10.1080/00405841.2014.862124

Broggy, J., Boyle, R., & Forde, J. (2012). Current Views on the Practical Assessment of Science at Junior Cycle in Ireland. *Proceedings at the Science and Mathematics Education Conference.*

Brookfield, S. (1995). *Becoming a critically reflective teacher.* San Francisco: Jossey-Bass.

Campbell, T., Abd-Hamid, N., & Chapman, H. (2010). Development of instruments to assess teacher and student perceptions of inquiry experiences in science classrooms. *Journal of Science Teacher Education, 21*(1), 13–30. doi:10.100710972-009-9151-x

Capps, D. K., Crawford, B. A., & Constas, M. A. (2012). A review of empirical literature on inquiry professional development: Alignment with best practices and a critique of the findings. *Journal of Science Teacher Education, 23*(3), 291–318. doi:10.100710972-012-9275-2

Cheevers, C., Eivers, E., & Shiel, G. (2006). *Implementing the revised Junior Certificate Science Syllabus: What teachers said*, Available online http://www.education.ie/en/Publications/Inspection-Reports-Publications/EvaluationReports-Guidelines/insp_jcssreview_pdf.pdf

Cosgrove, J., Shiel, G., Oldham, E., & Sofroniou, N. (2004). A survey of mathematics teachers in Ireland. *Irish Journal of Education, 35*, 20–44.

Crawford, B. A. (2007). Learning to teach science and inquiry in the rough and tumble of practice. *Journal of Research in Science Teaching, 44*(4), 613–642. doi:10.1002/tea.20157

Cullinane, A., & Liston, M. (2016). Review of the Leaving Certificate biology examination papers (1999–2008) using Bloom's taxonomy – an investigation of the cognitive demands of the examination. *Irish Educational Studies, 35*(3), 249–267. doi:10.1080/03323315.2016.1192480

das Graças, C., de Arruda Nascimento, M., & dos Reis, R.F. (2017). Teacher education: Perceptions of teachers who are entering the municipal education system in Rio de Janeiro. *Educação e Pesquisa, 43*(1), 49–64.

Department of Education & Science. (2003). *Junior Certificate Syllabus.* Dublin: Stationary Office.

Department of Education & Science. (2016). *School Self-Evaluation Guidelines.* Available online https://www.education.ie/en/Publications/Inspection-Reports-Publications/Evaluation-Reports-Guidelines/School-Self-Evaluation-Guidelines-2016-2020-Post-Primary.pdf

Donnelly, D., McGarr, O., & O'Reilly, J. (2014). "Just be quiet and listen to exactly what he's saying": Conceptualizing power relations in inquiry-oriented classrooms. *International Journal of Science Education, 36*(12), 2029–2054. doi:10.1080/09500693.2014.889867

Fullan, M. (2007). *The New Meaning of Educational Change.* Routledge.

Hargreaves, A. (1989). *Curriculum and Assessment Reform.* Milton Keynes, UK: Open University Press.

Hargreaves, A., & Fullan, M. (2012). *Professional capital: Transforming teaching in every school.* Teachers College Press.

Harland, J., & Kinder, K. (1997). Teachers' continuing professional development: Framing a model of outcomes. *Journal of In-service Education*, *23*(1), 71–84. doi:10.1080/13674589700200005

Hayes, S. (2011). *A Critical Examination and Evaluation of the Place of Science in the Irish Transition Year* (Unpublished thesis). University of Limerick.

King, D. (2017). *Junior Cycle Reform – Why Context Matters: A context-centric analysis of curriculum reform in lower Irish secondary education* (500th ed.). University of Sheffield. Available http://etheses.whiterose.ac.uk/18148/1/David%20King%20UoS%20Full%20Thesis%20FINAL%20Sept%2017%20Junior%20Cycle%20Reform%20Why%20Context%20Matters.pdf

Lemke, J. (1990). *Talking science: Language, learning, and values*. Norwood, NJ: Ablex.

McGuire, P. (2016, February 22). Where now for junior cycle reform? *The Irish Times*. Retrieved from https://www.irishtimes.com/news/education/where-now-for-junior-cycle-reform-1.2539986

National Council for Curriculum and Assessment. (2006) *Junior Certificate Science, Draft Guidelines for Teachers*. Available online http://www.curriculumonline.ie/uploadedfiles/JC/JCScienceTGlin.pdf

National Research Council. (1996). *The National Science Education Standards*. Washington, DC: The National Academy Press.

NCCA. (2012). *Towards a Framework for the Junior Cycle: Innovation and Identity*. Dublin: NCCA.

NCCA. (2013). *Background Paper and Brief for the Review of Junior Cycle Science*. Dublin: NCCA.

NCCA. (2016). *Junior Cycle Science Specification*. Available online https://www.curriculumonline.ie/getmedia/153bc83f-9848-49f0-ad87 0a0d6b9b596c/Specification-for-Jr-Cycle-Science-EV_20160126-(1).pdf

Nuthall, G. (2005). The cultural myths and realities of classroom teaching and learning: A personal journey. *Teachers College Record*, *107*(5), 895–934. doi:10.1111/j.1467-9620.2005.00498.x

Osborne, J., Erduran, S., & Simon, S. (2004). Enhancing the quality of argumentation in school science. *Journal of Research in Science Teaching*, *41*(10), 994–1020. doi:10.1002/tea.20035

Perkins, R., Shiel, G., Merriman, B., Cosgrove, J., & Moran, G. (2012). *Learning for life: the achievement levels of 15 year olds in Ireland on mathematics, reading literacy and science in PISA 2012*. Education Research Centre.

Shiels, G., Perkins, R., & Gilleece, L. (2009). *OECD Teaching and Learning International Study (TALIS): Summary Report for Ireland*. Education Research Centre.

State Exam Commission. (2010). *Junior Certificate Examination Science, Ordinary Level Chief Examiner's Report. Higher Level Chief Examiner's Report*. Athlone: SEC.

Van Driel, J. H., & Berry, A. (2012). Teacher professional development focusing on pedagogical content knowledge. *Educational Researcher*, *41*(1), 26–28. doi:10.3102/0013189X11431010

Van Kampen, P. (2015). *Summary of the Junior Cycle Science Specification Process*. Available online http://www.juniorcycle.ie/Curriculum/Subjects/Science/Science

Varley, J. P., Murphy, C., & Veale, O. (2013). At the crossroads: The impact of new Irish science curricula on first year post-primary students. *Research in Science Education*, *43*(1), 275–298. doi:10.100711165-011-9252-3

Wong, J. L. N. (2010). What makes a professional learning community possible? *Asia Pacific Education Review*, *11*(2), 131–139. doi:10.100712564-010-9080-6

Chapter 5
Implementation of the Chain Reaction Project in Georgia

Marika Kapanadze
Ilia State University, Georgia

ABSTRACT

Ilia State University (ISU) is one of Georgia's leading universities in teacher education and has taken part in a number of EU funded projects aimed at implementing new approaches to science teaching in Georgia, including those with an inquiry-based science education (IBSE) focus. This chapter gives an overview of the Chain Reaction project in Georgia, including discussion of the teacher professional development scheme, the perceptions of teachers following participation and implementation of the project, and the impact of Chain Reaction on students' motivation and teachers' professional development.

INTRODUCTION

There have been several major educational reforms implemented in Georgia since the collapse of the Soviet Union in 1991. Since 2004, new versions of national curricula for both primary and secondary schools have been piloted and put in place. Modern science education reforms emphasise the importance of students gaining experience in inquiry. This means that teachers must cultivate the knowledge, skills, and thought processes necessary to mentor their students through authentic investigations (Windschitl, 2003). The new science curriculum in Georgia has an emphasis on inquiry-based and student-centred teaching and learning activities (Kapanadze et al., 2010; Slovinsky, 2012). An IBSC (Inquiry-based Science Curricula) supports students performing science activities rather than merely memorising facts and theories. The curriculum also aims to motivate students; students work to find solutions to real problems, the resolution of which requires them to apply ideas they have learned previously (Edelson, 2001; Singer et al., 2000). Despite several new versions of textbooks being published as a result of the new Georgian National Curricula, most teachers are still not sufficiently prepared to implement student-active learning methods and do not possess the skills necessary to apply the learning processes required by the new curricula. Additionally, school laboratories are rarely properly equipped.

DOI: 10.4018/978-1-5225-5439-4.ch005

As a result of this, in-service teachers' education has been identified as a key point for supporting the reforms, with new standards for teacher education being developed and approved by the Ministry of Education and Sciences of Georgia (TPDC, 2012). The new standards identify specific competencies required from science teachers to enable them to effectively achieve the desired outcomes defined in the national curricula. The Teachers Professional Development Centre (TPDC) under the Ministry of Education and Sciences of Georgia, which is officially responsible for in-service teacher training in the country, organizes centralised short-term training. These professional development courses are based primarily on traditional approaches to science education, and many feel that teachers do not get any practical benefit from them. The traditional science education approach focuses mainly on factual knowledge (what we know), while inquiry-based science education aims at asking deeper questions (how we know) and providing understanding of scientific processes and the nature of science itself (Bartholomew et al., 2004). Teachers should be made aware that by engaging students in inquiry they become active learners able to obtain their own knowledge. Through inquiry, scientists gather information and propose hypothetical predictions and explanations while accessing and building on prior experiences and knowledge (Shope & McComas, 2015). Opportunities to engage students in inquiry claim to improve educational achievement in the sciences while also cultivating positive attitudes toward science, which in turn may increase participation in science careers (Kanter & Konstantopoulos, 2010).

An analysis of the current TPDC science training modules reveals that few of them are learner-centred and include sessions that require substantial interaction between trainers and teachers. The module "Planning of the learning process and teaching and learning strategies" is one example which is based on modern teaching and learning approaches (TPDC, 2012). Most of the modules, however, remain teacher-centred and need improvement. For example the physics module, in which teachers are trained to solve physics tasks, relies on lectures to deliver information and the teachers participating in the training are exposed to predominantly traditional, outdated methods of teaching and learning which increase the likelihood of them reproducing this in their own classrooms.

While this reform process is on-going, the low quality of science teaching in schools and the impact of that lack of quality on students' motivation and success remains an issue, which is often discussed and reported on during education conferences and meetings in Georgia (e.g. Chinchaladze & Burjanadze, 2016). As a result it might be that new teacher education programs are required in Georgia in order to improve the quality of science teaching in schools.

The first most important project in science education implemented in Georgia was the Tempus project SALiS (2010 – 2012), coordinated by ISU. First workshops and trainings about student-centred and IBSE approach were organised at ISU with the help of EU universities (Kapanadze & Eilks, 2014). In 2013, the science education research centre SALiS was established at the university.

Following the first *Chain Reaction* project consortium meeting at Sheffield Hallam University in the UK, the project team in Georgia prepared general information about the project and a brief description of the activities involved. This information was announced in Georgia and disseminated via social media and published on the ISU web page. From the outset, the project received a great deal of interest from Georgian schools; despite the heavy reliance on traditional teaching methods described above, many teachers in Georgia are open to innovation and eager to obtain professional development and knowledge about modern teaching methods. In each project year the number of applications for participation in *Chain Reaction* increased; in 2013-2014, we received 16 applications for participation from schools all over Georgia, in 2014-2015, 21 applications, and for the last year, 2015-2016, 26 applications.

As a result of the enthusiasm by teachers for the project (and in direct contrast to other contexts where recruitment proved difficult) the ISU team had to create a selection procedure with criteria for participating schools and teachers; for example, participating teachers need to be open minded, ready to work with us, and eager to implement project ideas in their schools and regions. The main idea behind the selection was to aid the spread of the principles of IBSE and *Chain Reaction* throughout the country by covering different regions across Georgia. A key expectation was that participating teachers and students would later engage other schools in their respective regions with ideas about inquiry-based learning. However, one of the most important preconditions was the support of a given school's administration; if teachers appeared to suspect that the administration would not support the principles of the project, the school was not selected. Nonetheless, the selection of schools was difficult and time-consuming, as all of the applicants were qualified and motivated.

Table 1 presents the list of participant schools and the modules implemented during the three project years.

In order to be able to start the project in schools we first had to translate the project materials - Pupils Research Briefs (PRB) into Georgian (most teachers in Georgia do not speak English) and adapt them to suit our particular national needs, both in terms of curriculum and local context. For the project's first year we were able to translate two PRBs, but from the second year onwards each teacher received six translated PRBs). English versions of all *Chain Reaction* materials were published on the project web page. Consequently, teachers were able to enlist the help of their students or English teachers at schools and implement other PRB as well.

After discussing all of the available options, the selection of the PRB to be used in classroom was made with both students and teachers jointly reaching decisions about the inquiry module they wanted to use. In some cases, the decision was made as a result of a specific school context and awareness of local issues.

Table 1. Participating schools and implemented PRBs

Year	Schools		Modules
2013 – 2014	1	State School N1, Telavi	Green Light Plants in Space
	2	State School N12, Gori	
	3	School "Kartuli Skola", Kutaisi	
	4	School "Mtsignobartukhutsesi", Tbilisi	
	5	Buckswood International School, Tbilisi	
2014 – 2015	1	State School N1, Bolnisi	Green Light Plants in Space Feed the World Green Heating Out of Site, Out of Mind PHEPPS
	2	State School N33, Kutaisi	
	3	State School N42, Tbilisi	
	4	Newton Free School, Tbilisi	
	5	Aia-Gess Gymnasium, Tbilisi	
2015 – 2016	1	State School N2, Batumi	Green Light Feed the World Green Heating Out of Site, Out of Mind
	2	State School N3, Borjomi	
	3	State School N4, Telavi	
	4	State School N12, Rustavi	
	5	Bakurtsikhe State School, Gurjaani	

PROJECT IMPLEMENTATION

Teacher Professional Development in Chain Reaction

Before starting the project in schools all participating teachers were trained at ISU, using translated and adapted Chain Reaction materials, with the PD program being developed by the ISU project team, taking into consideration the professional needs of Georgian science teachers.

The professional development (PD) program devoted to implementing IBSE in Georgian science classrooms consisted of at least 50 hours of training (seminars, lectures, and workshops) over a period of several months. Each training day consisted of practical work, problem solving and inquiry activities for the teachers, as well as seminars and discussions about IBSE.

As discussed previously, while Georgia's current national science curricula is based on inquiry, very few Georgian teachers are experienced in developing students these skills. For the participating teachers, it was important not only to introduce and adapt the PRB, but also to provide a theoretical background for IBSE and a discussion of its importance for student success. At the beginning of the PD program, one session was devoted to theoretical issues of inquiry-based learning. Teachers worked in groups and discussed the IBSE approach in contrast with other teaching strategies.

The PRBs were based on the needs of the curricula in England, as they were created by the Centre for Science Education at Sheffield Hallam University in the UK. As a result, the introduction session included discussion about adapting these modules to Georgia's national science curricula. Georgian teachers also developed their own ideas regarding the PRBs. These ideas were discussed among the teachers during the workshops, in schools, and were also developed by teachers with their students for presentation at national conferences. One of the examples of teacher-student innovation was to study LED lamps in the *Green Light* PRB alongside filament lamps and CFLs.

During the PD program, the participating teachers worked on different activities, all of which focused on issues related to everyday life with a student-centred emphasis on problem solving that relies on scientific inquiry. The sessions were oriented toward enhancing specific inquiry skills such as experimenting and investigating alongside wider skills, such as creating and cooperating. Principles such as creative thinking, problem solving, and designing learning environments that emphasise socio-scientific issues and concerns were introduced by the project staff and engaged with by the participating teachers. The main part of each stage of the PD courses involved discussions among the teachers and between the teachers and project staff as illustrated in Figure 1. These discussions were referred to frequently when teachers were interviewed as being of significance in terms of their learning and in sharing new and developing new ideas. These sessions enabled teachers to feel empowered in terms of formulating approaches, innovating and being motivated about improving their own classroom practice.

After several PD sessions, participating teachers went on to work with their students in the classroom, supported by the project staff throughout this process of project implementation – for example, by organising Skype conferences with the participating schools and in-person visits. Visits were planned according to the needs of the schools - and took place twice during while the PRBs were ongoing. These visits were very important for the teachers, providing them with support from the ISU team not only at the university but also directly in the classroom and allowing discussion about the project activities.

The types and content of discussions, experiments, and questions varied greatly from one school to another during visits due to the topic and content of the project. However, the motivation of students to produce something important and creative was present in all participating schools. We observed numerous

Figure 1. Model of chain reaction PD program at Ilia State University

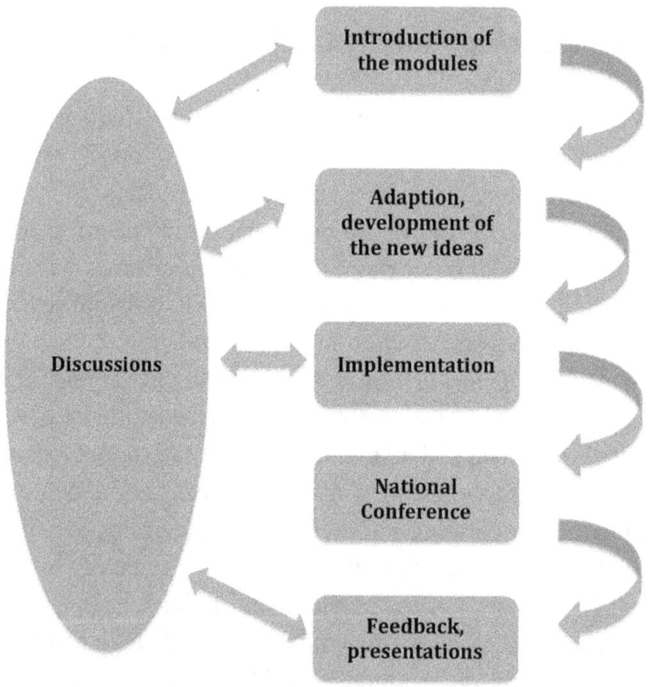

groups of students with innovative ideas and plans, and clear indication of the impacts of their findings in helping them to see the importance and relevance of science to society, which may have the potential to lead some students to choose to study science and continue on to careers in science related fields.

Students' Visits in Different Laboratories

During the project period, students visited different laboratories to conduct inquiries, arranged by their teachers, as the facilities at schools are generally poor. These visits were very important for the students so they could see how scientific work is carried out in real laboratories.

Students from Buchswood International School, State School N12 Rustavi and State School N42, Tbilisi visited the SALiS Laboratory at Ilia State University several times, carrying out measurements of photometric properties of different lamps with PASCO sensors during the studies for *Green Light*. Students from School ''Kartuli Skola'' and State School N 33 from Kutaisi visited the science laboratory at Akaki Tsereteli State University in Kutaisi, again carrying out measurements for studying properties of different types of lamps for *Green Light*.

Students from Telavi State School N4 investigated soil structure and its properties, visiting several different laboratories to carry out their research; at Telavi State University they studied the content of nitrates and calcium in a soil and plants, and at Ivane Javakhishvili State University in Tbilisi, they identified heavy metals in infiltrates with the atom adsorption method, and at the SALiS laboratory at ISU, Tbilisi they did investigations of soil structure and its chemical properties.

Bacteriological analyses of landfill soil samples were done by the students of Borjomi State School N3 at the laboratory of the Borjomi Mineral Water Factory, having obtained special permission for conducting studies there. Students from Bakurtsikhe State School, Gurjaani did chemical analyses of infiltrates and different types of soils for the *Out of Site Out of Mind* PRB at the Nonorganic Chemical Institute laboratory in Tbilisi and at Gurjaani sanitary laboratory.

This cooperation between students and scientists was a key part of the project in Georgia, helping to close the existing gap between schools and universities and motivating students to study science at a deeper level. The students had the opportunity to engage directly in scientific work, to gain insight into scientific research methods, and to become more familiar with modern technology and the laboratory environment.

National Conferences

Chain Reaction national conferences took place for each of the three years of the project, enabling students to present and discuss their work with their peers, as well as interacting with scientists and consider with them different scientific ideas and possibilities.

National conferences were attended by more than 120 students and their teachers each year, along with science experts from the Ministry of Education and Sciences of Georgia, the Teacher Professional Development Centre of Georgia and students and professors from various universities. Each conference consisted of two parts, with students both presenting their project results in front of an audience and also displaying the models they had constructed during the project. Both parts of each conference were interactive and engaging not only for the participating students and teachers but also for the conference guests and jury.

National conferences were a significant event, not only for the Georgian schools but also for ISU itself. Conferences took place in the ISU conference hall, giving the students the experience of both visiting and presenting in a university. A live video stream was organised during the conferences, so that parents, friends and other participating countries could follow the event. Each conference session was similar to a real scientific event, with a conference chair and discussions after each presentation. Students asked questions that reflected their knowledge about the development of the inquiry modules, for example, concerning the procedure of inquiries and results of the studies. Prizes were given out for a variety of categories e.g. best model, creativity approach, best project, etc. and all participating students and teachers received certificates.

IMPACT

Impacts and Outcomes for Students

The project had some far-reaching impacts upon students in Georgia, allowing them to influence their surroundings, local policy and to develop projects that had a wider visible impact. For example the *Plants in Space* PRB led one group from State School N12 from Gori to develop their own system for growing plants using aquaponics based on their work, which enabled them to launch an agricultural business for the school. Batumi State School N2 chose the *Feed the World* PRB, and developed their own system for growing potatoes in Adjara, which was later presented to an agricultural institute for further inquiry and development. Rustavi State School N12 and State School N33 from Kutaisi constructed their own solar panel systems for heating the school building's water supply and installing a new water heating system for a swimming pool respectively, based on the *Green Heating* PRB.

Students from State School N42 from Tbilisi built water turbines for use in their village to generate electricity from the village's small river, based on the *PHEPPS* PRB. The students tested various turbines of various sizes, shapes, and number of blades, and constructed the most effective model.

Out of Site, Out of Mind PRB addressed issues of litter recycling and landfill construction, a problematic issue in Georgia. Awareness of the region's ageing landfill system led to students from School N4 from Telavi performing experiments, visiting different laboratories, and creating their own landfill model, which they presented to the Mayor of Telavi, who committed to address the issue and to start new landfill construction works based on the investigation results of the students' project. School students and teachers also discussed activities such as building a green garden in the city centre and preparing sorting baskets for litter with the city government. Also as a result of this PRB and awareness around this problem, State School N3 from Borjomi suggested building a landfill liner from clay sourced from the region, planning to test local resources for creating new materials to protect the environment from infiltrates.

The *Green Light* PRB a module about energy-efficient lamps was of particular interest for Georgian students and teachers as Light Emitting Diode (LED) lamps are unusual in Georgia. Students planned and conducted experiments into the energy efficiency of three types of lamps: incandescent, fluorescent, and LED, with varying levels of teacher support. Two schools from the first project year, Buckswood International School and ''Mtsignobartukhutsesi'', discussed the results of their respective studies with their school administrations, leading to the decision to replace all ordinary bulbs in school buildings with LED lamps; enabling the students to understand the importance of science for their everyday lives, and the obvious and significant impact of their work.

The activities described above gave the students a real insight into the importance and potential impact of science and scientific research, both in their local area and more widely, allowing them to positively influence policy at a school and local level, and to see the practical benefits of the work they undertook.

Impact on Teachers

The final stage of the PD occurred after the National Conference and involved final discussions with the teachers. Teachers summarised their work and commented on the importance of the project for their students and themselves, frequently emphasising the significance of participating in *Chain Reaction*. They reported about students' increased motivation towards learning science and development of inquiry skills. They discussed also about the improvement of their own classroom practices and increased knowledge about IBSE. Project staff provided feedback about the inquiry modules completed as part of the project, and gave recommendations for further development in classroom practice and on taking the project work forward.

The impact of the project implementation on the participating teachers and students, as already noted above, was significant around the country. To gain more insight relevant to the results of the project, interviews were carried out with the participating teachers (at the end of the project) with questions around the main strands of the project (i.e. PD, PRBs, in classroom work, conferences), and about teachers' personal views and challenges, allowing teachers to share their own ideas and perceptions of the project. Teachers identified their lack of experience in IBSE prior to *Chain Reaction* and described how the project had a significant impact on their professional development including changing their views on teaching methods. Teachers also stressed the importance of the project for their students; helping them to understand the importance of scientific knowledge for everyday life and the links between different subjects. They also mentioned that students developed inquiry and knowledge transfer skills.

One physics teacher noted that before her involvement in *Chain Reaction*, IBSE was only about the development of teachers' ideas by the students, stating *''after project PD courses and implementation of project activities, I know what it means and how my students can conduct research activities during my lessons''*. Having been part of the project her students were able to plan and conduct inquiry independently: "*When my students plan their activities themselves, they make a list of what they need for inquiry and perform all activities independently, this is one step forward for them and for me as a teacher. They are able to discover laws in physics! We should change our teaching approach and give students more freedom.*" Stressing the importance of students' involvement in the learning process, she stated: "*When I see that not all the students are active I have to change my plan and try to involve the whole class.*" She told us that participation in *Chain Reaction* changed her understanding of inquiry. This was echoed by others including a biology teacher who commented that before *Chain Reaction* she seldom used inquiry activities, and her prior knowledge about IBSE was very different from that she had gained during the project.

Several teachers noted that students gained both scientific knowledge and an interest in science from an inquiry approach: "*They learned how to conduct inquiry, analyse data, and present study results in front of a big auditorium. I've seen that for my students, a more effective way to gain knowledge is practical work and inquiry*". A second teacher stated: "*Conducting and implementing Chain Reaction was the best way for my students to get knowledge. Neither study books nor theory-based lessons are as effective for the students for gaining an interest in science as Chain Reaction.*"

Several teachers considered that participation in *Chain Reaction* was a valued teaching experience, with a chemistry teacher observing that "*My students have done very interesting inquiries at school; they have presented their results in front of a wider audience. They gained experience in conducting inquiry. This is my first project in IBSE. It was hard work, but I'd like to take part in other similar projects. Chain Reaction is my best teaching experience.*" The experience that the project gave to students was also noted by other teachers, with one noting that alongside her learning to plan inquiry activities and conduct such activities with her students, for them it was the first time they considered themselves to be scientists; "*This was the main motivation for them to learn science subjects*". Another chemistry teacher from Kutaisi State School N33 stressed that participation in *Chain Reaction* enhanced her students' motivation to learn science and made them proud of their work "*After the Chain Reaction conference my students were very proud of their work and PRB they presented there.*"

On the topic of low cost experiments for implementation of IBSE, a physics teacher from Aia Gess Gymnasium from Tbilisi stated; "*Infrastructure at schools is less a problem then teachers' views and thinking about teaching approaches. I am able to conduct inquiry with low cost technology; I got these ideas from the Chain Reaction workshops. Now I know how to motivate my students.*" A biology teacher from the same school stressed the importance of sharing experiences with other teachers: "*For implementation of IBL (Inquiry-based Learning), teacher experience is very important. We learned through our practice, and we learned from other teachers in Chain Reaction.*"

It is clear from these responses that for teachers in Georgia the *Chain Reaction* project changed their understanding about IBSE, with teachers stressing in addition the importance of the project in teachers cooperating with one another, for building the scientific knowledge and development of inquiry skills of students and highlighting that participation in the project was a valued teaching experience.

Sharing Experiences and Practice

Participating teachers were very keen to share their experience in *Chain Reaction* with other teachers and as a result presented their work around IBSE implementation in their classrooms at different national and international conferences, often independently of the project. These presentations shared the knowledge and experience gained from the project and the importance of inquiry-based projects for students' assessment. They stressed the importance of project-based learning for students' motivation and development of inquiry skills.

A physics teacher from the first project year from Buckswood International School Tbilisi presented at the teachers national conference in Tbilisi in 2016. Her presentation concerned the development and implementation of project-based learning at her school during and after *Chain Reaction*. There was great interest towards the project from the conference participants. She answered their questions about the inquiry modules, about her work with the students and presented a booklet with the results that was published for the *Chain Reaction* national conference in 2014.

Additionally, participating teachers shared their experiences of the implementation of IBSE at a conference on educational technologies in Georgia in November 2015, at the third regional conference in Batumi in 2016, at the regional science teacher conference in Telavi in February 2017 and at the first national science teachers conference, organized by the SALiS center at Ilia State University in April, 2016. Teachers presented examples of best practice and shared information with the others about the project. As noted previously, the *Out of Sight, Out of Mind* PRB, regarding recycling and landfill construction has particular relevance in Georgia and there was an in-depth discussion about the implementation of this PRB and about the development of the idea of a modern landfill during these conferences.

Students' Presentations

During the project implementation period participating students further developed inquiry modules and presented their studies at different national and international conferences beyond those that took place as part of the *Chain Reaction* project. Project results were presented at the national conference organized by the Millennium Challenge Georgia, 2015, at the Georgian Technical University conference for young scientists in Tbilisi, 2015, at the international student conference in September 2015, in Milan, Italy, at the students' third national conference Chemistry tomorrow – 2016 in Georgia and at the international conference "Chemistry without Borders" in January 2017 in Israel.

One of the participating students said: "*We presented Chain Reaction modules and our study there. Students and teachers from different countries attended the conference. We answered questions and there was a very interesting discussion following our presentation about green energy.*" One biology teacher of one of the participating schools told us: "*After presenting the results at different conferences my students don't want to terminate their research so they ask me to continue our work on inquiry projects. They are very motivated and have a great interest in science.*"

Students from the Newton Free School Tbilisi participated in the National Olympiad in science subjects, presenting *Chain Reaction* project inquiry activities and its results. They were awarded the first prize and visited the CERN research centre in Switzerland with the *Chain Reaction* teachers.

From these students' work and presentations it is clear that they have an interest in science and scientific inquiry, and are continuing to work on their research projects.

Barriers/Challenges

Georgian teachers were clearly enthusiastic about *Chain Reaction* and developed project activities with great interest, despite many of the participating teachers not previously being familiar with IBSE. The participants undertook significant efforts to learn about IBSE and collaborate with the project staff and other participating teachers to gain knowledge about further developing the PRBs and implementing project activities in the classroom.

Both project staff and participating teachers, however, encountered obstacles during the development and implementation of the project activities. For example, due to the national curriculum, science subjects are usually taught for just two hours per week, with teachers being expected to cover the content of the curriculum, and many claiming the classroom hours are not sufficient to cover all activities in sufficient depth. As a result, all participating teachers stressed the importance of time in implementing *Chain Reaction*. The teachers told us that conducting inquiry activities needs a lot of time and they required double hours for science classes, as students do not have the opportunity to complete work, they leave the classroom with unanswered questions, meaning that the lessons are unsuccessful. As a result of these time constraints, many *Chain Reaction* teachers worked on the project activities with their students after regular school hours, sometimes late into the evening.

Another challenge for the project implementation in the classroom was the number of students per class; this is particularly significant in state schools. In Georgia, the number of students per class varies from 25 to 35 in state schools and from 10 to 15 at private schools. Teachers noted that each student should be able to work and do experiments, after which they should compare their results and discuss them. A clear message from participating teachers was that the number of students in each classroom is an important factor when conducting inquiry activities, as large groups carrying out experiments can be difficult.

The lack of equipment in school laboratories was another barrier for the project. All participating schools purchased science equipment and materials for the inquiry activities using project funding. These funds provided significant support for all participating state schools as well as for private schools.

Teachers also mentioned exams as an obstacle. Georgian students are expected to pass exams at the end of 11th grade for their school certificate, and students who intend to study at university must pass national examinations. These examinations are not based on national curriculum demands and require students to develop and recall only theoretical knowledge. Teachers claim that preparation for these examinations takes a lot of classroom time and is a reason why many 11th-grade students are unable to conduct inquiry.

The lack of professional literature available in the Georgian language was also mentioned as a barrier for teachers; there are no books available for teachers about modern science teaching methods and approaches. During the *Chain Reaction* project, participating teachers received all project materials and support from project staff in Georgian language as their sole sources of information.

CONCLUSION AND RECOMMENDATIONS

On the basis of data analyses performed after the implementation of *Chain Reaction* PD courses, the Georgian science teachers reported the following:

- Increased enthusiasm about science teaching;
- Increased knowledge of inquiry-based science education (IBSE);
- Improved classroom practice;
- More positive attitudes about professional development;
- Changing thinking about science teaching and learning; and
- Collaborative working with other colleagues.

The project's implementation also impacted Georgian students. In interviews, participant teachers reported the following about their students:

- Increased motivation for science and science learning;
- Increased enjoyment of science;
- Development of group-working skills; and
- Development of inquiry-based learning skills.

From the feedback provided by the teachers, it is clear that *Chain Reaction* in Georgia has been a success, with students experiencing increased motivation in science and teachers gaining knowledge about IBSE. Teachers were able to collaborate to develop existing modules or implement new ones, gaining more familiarity with inquiry projects and being able to develop their own approaches.

The professional development program that took place in Georgia as part of *Chain Reaction* successfully promoted the development of teachers' competences in IBSE and supported the implementation of inquiry-based learning in science classes. ISU has recommended that the Teacher Professional Development Centre in Georgia, and also other organisations and universities implement the *Chain Reaction PD program* as a model for teacher professional development program on IBSE in Georgia.

As a result of the experience of *Chain Reaction* in Georgia it can be concluded that educational reform and innovation within the school system must be combined with teacher involvement and improvement of the practice of continuous professional development. Such further development may aid in enhancing the motivation of more students and embedding IBSE in science teaching in Georgia.

REFERENCES

Bartholomew, H., Osborne, J., & Radcliffe, M. (2004). Teaching students "Ideas-about-science": Five dimensions of effective practice. *Science Education*, *88*(5), 655–682. doi:10.1002ce.10136

Chinchaladze, I., & Burjanadze, T. (2016). *Experiments and Students Motivation*. Retrieved 8.05.17 from http://salis.iliauni.edu.ge/wp-content/uploads/2016/04/journal1.pdf

Edelson, D. C. (2001). Learning-for-use: A framework for the design of technology-supported inquiry activities. *Journal of Research in Science Teaching*, *38*(3), 355–385. doi:10.1002/1098-2736(200103)38:3<355::AID-TEA1010>3.0.CO;2-M

Kanter, D. E., & Konstantopoulos, S. (2010). *The impact of a project-based science curriculum on minority student achievement, attitudes, and careers: The effects of teacher content and pedagogical content knowledge and inquiry-based practices*. Retrieved on 24.04.2017 from http://onlinelibrary.wiley.com/doi/10.1002/sce.20391/full

Kapanadze, M., & Eilks, I. (2014). Supporting Reform in Science Education in Central and Eastern Europe - Reflections and Perspectives from the Project TEMPUS-SALiS. *Eurasia Journal of Mathematics, Science & Technology Education, 10*(1), 47-58.

Kapanadze, M., Janashia, S., & Eilks, I. (2010). From science education in the soviet time, via national reform initiatives, towards an international network to support inquiry-based science education – The case of Georgia and the project SALiS. In I. Eilks & B. Ralle (Eds.), Contemporary science education (pp. 237-242). Aachen: Shaker.

Reaction, C. (2013). Retrieved 24/04/2017 from http://www.chreact.eu

SALiS. (2013). *Science Education Research centre*. Retrieved from https://salis.iliauni.edu.ge/

Shope, F., & McComas, W. (2015). Modeling scientific inquiry to guide students in the practices of science: The ED3U teaching model of conceptual change in action. In Inquiry-Based Learning for Science, Technology, Engineering, and Math (STEM) Programs: A Conceptual and Practical Resource for Educators. Academic Press. doi:10.1108/S2055-364120150000004013

Singer, J., Marx, R. W., Krajcik, J. S., & Chambers, J. C. (2000). Constructing extended inquiry projects: Curriculum materials for science education reform. *Educational Psychologist*, *35*(3), 165–178. doi:10.1207/S15326985EP3503_3

Slovinsky, E. (2012). SALiS and educational policy in Georgia. *Proceedings of the Student Active learning in Science Final Conference*.

TPDC. (2012). Retrieved 24/04/2017 http://www.tpdc.ge/uploads/pdf_documents/sabunebismetyvelo.pdf (in Georgian)

Windschitl, M. (2003). Inquiry projects in science teacher education: What can Investigative experiences reveal about teacher thinking and eventual classroom practice? *Science Education*, *87*(1), 112–143. doi:10.1002ce.10044

Chapter 6
How to Relate Research on Students' Views and Teacher Education About Inquiry With Dissemination Activities

Italo Testa
University Federico II, Italy

Silvia Galano
https://orcid.org/0000-0002-3510-8658
University Federico II, Italy

Alessandro Zappia
Liceo Nitti, Italy

Giuliana Capasso
University Federico II, Italy

Luigi Antonio Smaldone
Caserta Planetarium, Italy

ABSTRACT

Inquiry-based science education (IBSE) has been acknowledged as central in many curriculum reforms since mid-nineties. In this chapter, the authors aim to investigate the way IBSE was implemented by Italian science secondary school teachers in the context of the Chain Reaction Project. Thirty teachers attended a professional development course aimed at familiarizing them with IBSE principles. Then, the teachers implemented in their classrooms a teaching-learning sequence (TLS) for at least 5 hours. About 900 students were involved in the activities. Effectiveness of the TLS was measured by administering a pre- and post-implementation questionnaire about Nature of Science. Findings suggest that TLS inquiry aspects have been only partially adopted by the teachers. Moreover, they found that students that obtained the lowest score in the post-test questionnaire were taught by the teachers that mostly transformed the proposed TLSs. The data suggest the need to develop specific training courses to support teachers in the implementation of new methodologies as IBSE.

DOI: 10.4018/978-1-5225-5439-4.ch006

INTRODUCTION

Inquiry Based Science Education (IBSE) has been acknowledged as central to many curriculum reforms since the mid-nineties. The implementation of inquiry practices and their effectiveness have been increasingly debated since the 2000s also in Italy, due to some recent reforms that try to innovate the teaching practice. However Italian school science curricula are still, in many respects, strongly content-laden and teachers have a limited knowledge of methods and approaches related to IBSE. The aim of this paper is to describe how the activities proposed in the framework of the Chain Reaction project impacted on school practice of the participating teachers. The paper is organized as follows. In the first section, we describe the Italian secondary school context. In the second section we describe how the Chain Reaction project was implemented in the Italian educational context. In the third section, we briefly discuss the main barriers that we encountered in the implementation in classroom practice of the Chain Reaction project. In the fourth section, we present how the Chain Reaction activities impacted on local teaching practice. Finally, we discuss what we have learned from the Chain reaction Project and some future steps we will implement.

1. REFORMS AT SECONDARY SCHOOL LEVEL IN ITALY

The implementation of inquiry practices and their effectiveness for learning are being increasingly debated due to recent reforms that tried to innovate teaching practice in Italy. Such efforts were mainly pushed to the front on the wave of unsatisfactory results of Italian students in international surveys as, e.g., IEA's Trends in International Mathematics and Science Study (TIMSS)[1], and were nested in the organization reported in Table I. Similarly, to previous school curricula developed in Italy since the first decades of the 21st century, the proposed reforms are implemented through specific guidelines issued by the Italian Ministry of Education for each type of school stream.

The guidelines list the expected student learning outcomes for each subject. These outcomes, in the latest version[2], are organized around three transversal dimensions: *knowledge*, *skills* and *competences*. The *knowledge* dimension refers to the content students should know, for example, "Newton's second law". The *skills* dimension refers to specific abilities related to the content that the students are expected to acquire in a short-term time unit: in particular, they represent the ability to apply knowledge to solve problems, for example, predicting the motion of a body on an inclined plane. The *competence* dimension refers to a larger and more complex area of abilities that students are required to develop throughout their studies. Such abilities are often interdisciplinary to school subjects, since they do not concern a specific discipline, but rather refer to methodological and personal attitudes in different situations. Examples include the ability to support an argument or to understand a written text, as well as the ability to interpret a plot or a table. As such, students are expected to acquire such abilities over many years.

As a consequence, topics (or sequences of topics) to be taught are only broadly indicated in the official policy documents; the extent and breadth, to which a specific content area should be taught (for instance, focusing on related processes and models), is left up to teachers. As an example, we report an excerpt concerning energy:

Table 1. Sketch of the Italian educational system

<table>
<tr><th>Type</th><th>Students' age</th><th colspan="2">School Streams</th><th>Cycle</th></tr>
<tr><td rowspan="3">Compulsory</td><td>6-10</td><td colspan="2">Primary</td><td rowspan="2">First</td></tr>
<tr><td>11-13</td><td colspan="2">Middle</td></tr>
<tr><td>14-16</td><td rowspan="2">Lyceums
Art, music and dance
Classical (gymnasium)
Language and human sciences
Scientific
Science and Technology</td><td rowspan="2">Vocational and Technical
Business and Management
Craftsmanship
Economics
Technology</td><td rowspan="2">Secondary</td></tr>
<tr><td>Not Compulsory</td><td>17-18</td></tr>
</table>

By the study of mechanical phenomena, students start to familiarize with the concepts of work and energy, in order to discuss a first treatment of the principle of conservation of total mechanic energy. (p.59)

In principle, the new guidelines only suggest a framework for the design of specific teaching activities, which must be contextualized and adapted by school institutions and teachers, depending on the specific circumstances of their students. However, in most of the cases, more specific learning objectives are outlined (examples concern physics):

Through the study of geometric optics, students will be able to interpret the phenomena of reflection and refraction of light, and the functioning of the most important optical instrument [...].(p. 27)

The study of thermal phenomena will be completed with the study of the laws for a gas, familiarizing with the conceptual simplification of the perfect gases and the kinetic theory; in this way, students should understand how the Newtonian paradigm can connect the microscopic and macroscopic fields. (p. 343)

Within such a framework, it is no surprise that IBSE is not explicitly mentioned either as a suitable teaching approach or as a content itself. However, the guidelines suggest that the teaching of biology, chemistry and physics should include some aspects of Scientific Inquiry (SI) for example:

- Observing and identifying different phenomena;
- Formulating an explanatory hypothesis using models;
- Formalizing a problem by applying math to solve it;
- Identifying the significant variables to describe a phenomenon;
- Reflecting upon collected data and their analysis during a laboratory experience;
- Validating a model;
- Assessing the scientific and the technological issues concerning the society in which students live.

The introductory part of the guidelines clarifies the approach further.

[....] students must achieve a good knowledge of the fundamental subjects of physics and natural science (chemistry, biology, Earth science, astronomy), of their procedures and research methods, and get

familiar with applied science, [...] also building on a regular use of laboratory, [students must achieve] a good command of specific languages and research methods peculiar of experimental science (p .7)

Furthermore, the knowledge of research methods specific to science disciplines is also highlighted.

[...] at the end of the first two years a student should have a good understanding of basic disciplinary knowledge and, at a basic level, of the methodology characteristic of natural science, in particular of Earth science, chemistry and biology. These different disciplinary areas are characterized by their own ideas and methods, but they are all based on the same strategy of the scientific research which refers also to the dimensions of observation and experimentation (p.28)

However, despite the broad scope of the guidelines, no specific teaching methodology for addressing the above aspects of IBSE is suggested in the guidelines (Lederman et al., 2014). A collaborative learning approach based on group work and the use of laboratory are only loosely encouraged (Lunetta, Hofstein, & Clough, 2007). In particular, laboratory activities are deemed as important to let students interact with their peers and build their own knowledge, designing and validating different strategies.

Experimental activity will accompany students throughout the first two years, bringing them to an increasing and more conscious knowledge of the subject, also throughout critical reports about the performed experiments (p.342)

The lack of specific suggestions, as well as the dearth of in-depth teacher education interventions on behalf of the Italian Ministry of Education to improve usual practices, have raised over the years the need to act at local level to help teachers deal with the implementation of such guidelines. Moreover, and in some cases most importantly, a lack of suitable spaces in schools to be used as workspace for practical activities with specific instruments led to a limited application of the above guidelines in the school practice.

Building on such background, the Chain Reaction project, as carried out by the Italian team, aimed to provide teachers with intensive training to familiarize them with IBSE practices involving low-cost materials which would allow them to implement IBSE approaches in schools where laboratory facilities were poor.

2. PROJECT SUMMARY

The activities developed and carried out for the Chain Reaction project in Italy were structured focusing on six different steps for each of the three years of the project (Figure 1).

Recruitment of schools started, for each of the three years of project, in July-September, so that activities could begin early in the forthcoming school year. The schools were chosen after a public call. Thirty teachers, graduated in mathematics, physics, biology or chemistry, were involved.

Then, the teachers who were accepted for inclusion in the project participated in a Professional Development (PD) course, totaling 30 hours, from October to March. In each year of the project, the course was organized through 10 meetings of three hours each, aimed at developing teachers' knowledge and understanding of inquiry practice and at showing them the available teaching modules. These modules

Figure 1. Chain reaction project implementation in Italy

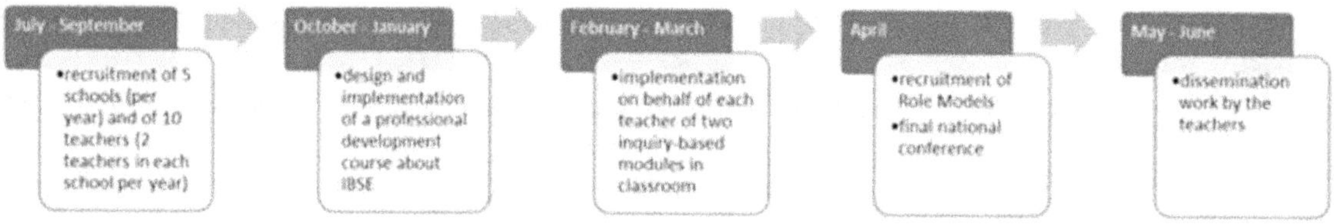

were based on the original *Earth and Universe Project Research Briefs* (EUPRBs[3]), for implementation in their schools. The modules were organized in the form of Teaching-Learning Sequences (TLSs, Meheut & Psillos, 2004). Generally, TLSs are teaching materials that feature a didactic transposition of a scientific topic with specific didactic indications for teachers on how to implement the proposed activities. TLSs can include the description of the theoretical framework and expected or exemplar empirical research results.

After the PD course, each teacher delivered two TLSs to their students. The first implementation was mainly aimed to let students (and teacher as well) familiarize themselves with inquiry practice, while the second aimed to foster the understanding of specific IBSE aspects addressed in the TLS. Activities in schools took place roughly in the period February – March, simultaneously with the last meetings of the PD course which could then offer support to the teachers during the implementation process. This arrangement also provided teachers a chance to:

- Give feedback about how their students reacted to the activities;
- Engage in a discussion about the difficulties encountered during the activities;
- Suggest possible ways to improve the activities.

To further support teachers in their practice, specific suggestions about how to implement the activities were featured in the teacher guide developed for each TLS. However, teachers were given enough freedom to manage delivery time. In general, each TLS lasted from a minimum of 5 hours to a maximum of about 15 hours.

Role Models, chosen from early-career university researchers in biology, chemistry, physics and mathematics, were also involved. They visited the participating schools once to support the students during their classroom inquiries. In particular, the Role Models gave advice on how to perform the investigation including the data collection and analysis. In addition, they discussed with students some aspects of their own scientific research.

At the end of the in-class activities, some selected groups of students presented a final report at the National Conference, which took place every year at the University of Naples, on April 15th 2014, on April 17th 2015, and on March 8th 2016. During these events, students presented their final work through a poster session and oral presentations. Every year, one group was eventually chosen each year to represent Italy at the International Conference (information about participating schools, final conferences dates, and students' final works can be retrieved from www.chreact.eu).

After the PD course and the in-class delivery of the TLS, some of the teachers involved in the project decided to share their work and ideas with their colleagues by organizing training meetings in their own schools or disseminating their experience in specific seminars or events. Some of them decided to continue developing the TLSs and began to integrate inquiry methodology more regularly in their usual teaching practice, thus recognizing its educational value.

PD Course Design

In-service PD courses in Italy are usually organized around a series of lectures, after which the teachers are required to write a short essay about the covered topics. Practical activities are rarely implemented within such courses[4]. In a marked difference to such a traditional PD model, Chain Reaction provided a consistent number of hours (about 30 per year) to teachers' training and preparation for classroom implementation of the TLSs. The aim of the PD course was to familiarize teachers with scientific inquiry practice and to improve their pedagogical content knowledge about IBSE. This strategy was informed by the awareness of the lack of an appropriate training about IBSE (Blanchard, Southerland & Granger, 2009) especially in the Italian school practice. However, while PD courses may improve teachers' attitudes towards, and beliefs in, inquiry (Carleton, Fitch & Krockover, 2007), such positive attitudes are not always transferred to teachers' practice (Marshall & Smart, 2013). We devoted the first three meetings to an introduction to the Chain Reaction IBSE approach, providing teachers with some relevant results of research in literature and involving them in the same activities they would have proposed to their students. Teachers were also required to read, in groups, some fictitious research documents. Their task was to decide if the described studies and investigations were conducted according to IBSE methodology, and to explain their point of view to the rest of the teachers. The aim of the activity was to encourage peer interaction and to reflect on some issues related to IBSE that could be a source of misconceptions for students as, for example, the existence of a single scientific method (Lederman et al., 2014).

The teachers were also asked to complete a portfolio to report their comments and reflections about the activities, as well as suggesting improvements to the course. Starting with the fourth meeting, teachers were involved in laboratory sessions trying out the same activities to be implemented in their classrooms. This gave teachers the opportunity to engage in practical work with their peers, but also to let them experience possible difficulties that students could have encountered during the implementation of such activities, for example, in the design of suitable and feasible experiment.

Choice of TLSs

We chose seven of the original EUPRBs and transformed them into TLSs by adding students' worksheets and detailing further the teachers' guide. Some of the first-year participating teachers were involved in the process in the second and the third year to provide suggestions and comments. After thorough discussions, we designed new contexts and added further scientific topics to meet the Italian guidelines requirements and better match the TLSs with teachers' expectations and practice. Finally, new support materials (videos, tables, articles retrieved from the web) were included in the revised versions of the original modules. The main changes carried out are described hereafter[5].

- **Green Heating:** We essentially kept the first text (Green Heating 1), discarded the second one (Green Heating 2) and opened Green Heating 3-5 in form of students' worksheets. More specifically, we designed the worksheets to emphasize the difference among *research question, hypothesis,* and *prediction* in the process of IBSE. The following requests were added: - answer the question: *"Which color should be used on a surface to best absorb e.m. radiation?"*. Students for instance can measure the temperature rise of a liquid (water or oil) in some containers (glass, for example) differently colored (including black) and kept at the same distance from a source of e.m. radiation, as an ordinary light bulb. They could also test the hypothesis that *"a solar panel whose surface is black can heat water better than a panel whose surface is white, because black surfaces absorb more heat"*. This can be made, for example, by wrapping black adhesive tape around a can containing water or oil, and some white tape around a similar one, putting them at the same distance from a source of e.m. radiation and monitoring temperature at regular time intervals;

- **Green light:** We largely adapted the original module, changing completely the scenario since the proposed one was rather old and too different from our national situation. Hence, we placed the students in a virtual parliament where they were asked to argument about decisions to adopt the compact fluorescent bulbs instead of incandescent ones. Moreover, students were asked to provide counter-arguments to support their decisions, through posts on a political blog. The aim of the task was to improve students' argumentation skills and to familiarize them on how to use evidence to support their claims or construct a critique of another's thesis (Jimenez-Aleixandre & Erduran, 2008). Additional resources were also changed. More specifically, after summarizing the most salient features of EU legislation about compact fluorescent lamps (CFLs), students were required to design an experiment by which they could collect data about the real benefits of CFLs. Then, students were asked to respond to criticisms based on the proposed public-opinion poll, and to summarize all the collected evidences to produce a blog post promoting the use of CFLs. The resulting TLS was proposed to teachers who were asked to design a suitable scenario starting from the EU normative about fluorescent lamps. The main reason for such changes was that we and the teachers had small concerns about the described functioning of the incandescent and compact fluorescent lamps. In particular, the teachers judged the original description of the mechanism too qualitative for the students' level. Our changes aimed at including some details about the physical mechanism underlying the functioning of CFL lamps. The teachers expressed some concerns about the formalization of the relationships between the efficiency of the two kinds of lamps, which was hard to quantitatively measure. The teachers also suggested to evaluate the costs of disposal of both type of lamps.
- **Plants in space:** According to the suggestions given in the original EUPRB, pupils were asked to measure the effects of light on plant growth, by designing an experiment to investigate the relationship between the e.m. radiation and the rate of photosynthesis. We suggested the following experiment: a pondweed is locked up in a funnel connected to a burette filled up to a predetermined level with water, then inserted in a water tank and finally illuminated by an incandescent lamp; the quantity of oxygen released is proportional to the decrease in the water level measured with the burette. Effects due to different light wavelengths may be observed by using filters made of transparent colored sheets. The main reason for such changes was that we had small concerns about the practical implementation of quantitative and qualitative experiments about photosynthesis (e.g. time needed for experiment, oxygen measurement and CO_2 measurement). Hence, to let students gain more insight about the phenomenon, we included in the module some simulations and designed worksheets using the Predict-Observe-Explain strategy (White & Gunstone, 1992). In the second year, however, we decided to propose simulations only as an option: teachers could decide to use these additional resources, if judged useful and/or necessary. Such further change was done to make the inquiry activities more open and to let teachers organize their activities autonomously.

- **Collision Course:** We slightly changed the module, introducing some qualitative experiments regarding the impact of the spheres of different mass on sand and powder to simulate impact of celestial bodies with the Earth. We introduced a new context in which the students played the role of scientists of the Star Centre, who were asked by their Director to investigate some evidences of possible impact of asteroids with Earth. We added to the activity some research questions and added some videos about asteroids' impacts on Earth to further stimulate students' interest. Students had to investigate about possible impacts of asteroids with Earth, evaluating the possibility of building an asteroid shield. Students were finally asked to write a report about their investigations, also including evidences emerging from their experiments.
- **Mars-ology:** This module was mainly chosen by the teachers to put more emphasis, during the implementations, on experimental activities. We introduced a new context in which students represented scientists of the Institute of Planetary Research, who had been contacted by the NASA in order to investigate about possible evidences about volcanic activity on Mars. The activities were focused on the study of tables in which data about Earth volcanoes and Martian volcanoes were reported. Students were also asked to design their own experiment to estimate the lava flow viscosity, and to report, in their final work, results obtained by their research.
- **Out of sight, out of mind:** We introduced a new context, in which students acted as members of a non-profit association who should write a letter to their Mayor in order to raise awareness about the waste situation in their region. We added some documents concerning the Italian legislation about waste disposal criteria, and others by national associations for the environment about the situation of waste burnings. Experimental activities consisted of the study of the permeability of different soils, using different grounds and different liquids to simulate the behavior of leachate. Students were asked, at the end of activity, to report all their investigations in the letter to be sent to politicians. When the resulting TLS was proposed to teachers, some concerns emerged about the excessive length of the resources (about 15 pages overall) given to students. However, we decided to keep the new designed materials, because the provided texts gave all information students needed to plan, organize and implement the activities
- **ET Phone Earth:** We completely changed the context and the scenario, adding some scientific documents and articles to the module. Students were asked to discuss in small groups and to argue in favor or against extraterrestrial life. A newspaper article about Drake's equation[6] was added, in order to let students reason about the probability of existence of planets similar to Earth.

Evaluation of the TLSs

To collect primary data about the effectiveness of the TLSs, we engaged the teachers who had implemented the same TLS (on average 2 teachers) in a post-class focus group so to collect comments and raise issues about the work done in meetings after the classroom implementations. During the focus group we also discussed with the teachers how they organized their work during school practice, their perception of inquiry and the difficulties they had encountered.

We also investigated students' knowledge about IBSE before and after classroom implementation using a modified version of the *Views About Scientific Inquiry* (VASI) questionnaire (Lederman et al., 2014). We first compared the IBSE aspects targeted in the original questionnaire with the aims reported in the new national guidelines (see background section). The comparison was carried out with the help of the teachers involved in the administration of the questionnaire and the in-classroom delivery of the TLSs. As a result, we selected the following aspects of IBSE, listed below in alphabetic order:

1. All scientists performing the same procedures may not get the same results;
2. Explanations are developed from a combination of collected data and what is already known;
3. Research conclusions must be consistent with the collected data;
4. Scientific data are not the same as scientific evidence;
5. Scientific investigations all begin with a question but do not necessarily test a hypothesis;
6. There is no single set and sequence of steps followed in all scientific investigations.

The adapted VASI instruments featured eleven questions. To validate the new questions, we pre-tested the whole questionnaire with a sample of about 20 students of the second year of a secondary school and then discussed the results with the teachers involved in the research study. Then, we administered the questionnaire to about 900 students during the three years.

To analyze students' answers to the adopted version of the VASI questionnaire, the following five categories were introduced:

- *Informed*, if the answer was correct and overall fully congruent with the targeted response for a given aspect of SI
- *Mixed or Partially correct*, if the answer was incomplete or not completely coherent with the adopted view of a given aspect of SI.
- *Naïve*, if the given answer was in contradiction with the adopted view of a particular aspect or corresponded to a known misconception about SI.
- *Unclear*, if the student's answer was only "yes/no", without any further justification.
- *Not given,* if the student did not answer

The categories were slightly adapted from the original study by Lederman et al. (2014). The complete data analysis procedure can be found in Galano et al. (2015).

To let students acquire the expected knowledge about IBSE, they worked in small groups (3-4 persons, see for example Figure 2) and were asked to fill in the designed worksheets to scaffold their knowledge about targeted IBSE aspects. For instance, they were asked to carry out their investigations and to discuss with the rest of the class their findings. Each group worked independently, using only the provided materials. Students were advised not to use any textbooks or other documents, to avoid possible inconsistencies between the proposed approach and that of the textbook.

Evaluation of IBSE Teachers' Practice

As part of our ongoing research activity on teacher education, we investigated the way in which the teachers implemented inquiry activities in their school practice. We focused on the *transformations* occurring during class-work and interaction with their peers within the training course context. The term *transformation* is used to describe teachers' selection and re-organization of the features of a didactic innovation (Pintò et al., 2003) such as the original designers' objectives (Van Den Akker 1998), or the

Figure 2. Students working during IBSE activities

use of specific teaching approaches, technologies, and languages (Sassi, Monroy &Testa, 2005). The concept of teachers' transformation is symmetrical to that of fidelity of implementation (McNeill et al., 2018) recently introduced in educational research. We prefer to use the term transformation since we want to focus on how teachers actively adapted or modified the teaching activities related to the above inquiry aspects in their school practice. Inquiry-based approaches are particularly likely to undergo transformations and modifications since teachers may:

- Find significant challenges in implementing them (Luft & Pizzini 1998);
- Hold contradictory beliefs with respect to inquiry pedagogy (Windschitl, 2003);
- Have never experienced themselves this approach (Kleine et al., 2002).

Consequently, IBSE approaches have often not naturally been adopted by teachers, but rather imposed from a research or policy level onto classroom level. Since IBSE is not merely a teaching method but involves also aspects of how the Nature of Science is conceptualized (Lederman, 2006; Schwartz & Crawford, 2006), discrepancies between what is expected from teachers and what is really implemented by them (Capps, Shemwell & Yound, 2016) are not surprising. These should be investigated in the wider frame of transfer of knowledge. In order to do this within the *Chain Reaction* activities, videos were collected during classroom implementations and analyzed as follows (for more details, see Zappia et al., 2015).

1. Development of a grounded coding category system, using as unit of analysis of the videos a time frame of ten minutes in order to reveal significant teachers' actions and students' reactions to teacher's triggers
2. Design of a scoring rubric to assess the emerging categories to evaluate the extent to which a specific inquiry aspect was accepted or transformed with respect to TLS designers' intentions. The finalized rubric featured three levels (low – the aspect was modified; medium – the aspect was partially adopted; high – the aspect was adopted)

Overall, the results of the analysis of the students' questionnaire show that the percentage of informed responses only slightly increased between pre- and post-test in the majority of the targeted aspects. Moreover, students' achievements depended primarily on how the teachers enacted the TLSs. In particular, we found that the majority of positive gains have been related to implementations carried out by teachers who adopted most of the IBSE aspects in their practice (Galano et al., 2015; Zappia, 2018).

3. BARRIERS/OBSTACLES

First of all, data collected during the Chain Reaction project helped us investigate teachers' attitudes towards the curriculum guidelines described in the background section, as well as towards using an IBSE approach in their teaching. Data from the interviews and post-instruction discussions suggest that teachers often struggled to integrate the suggestions of the new guidelines into their classrooms. Some of the interviewed teachers claimed that they had been given very little time to accommodate their practice to new ministry requirements, and most of them did not receive proper training about specific aspects of the guidelines. In particular, what teachers often struggled to understand was how students could reach the necessary standards in only a few hours a week. They were often anxious about the time spent on transferable skills when they had so much other curricular material to cover.

Second, concerning IBSE and the way it was implemented in our TLSs, teachers complained that the proposed activities required an excessive amount of time spent on students' observations and on letting them conduct their own investigations, which, at first, could lead to "wrong" conclusions. From the interviews, it emerged that, in the latter cases, teachers were tempted to provide the "right" answer to the students before they could finish the task. In other interviews, some teachers claimed that, due to the number of students per class (about 30) and to the lack of specific teaching assistants who could support such kind of activities, it had been difficult to implement the proposed IBSE methodology in their daily school practice. In other cases, teachers pointed out that one of the most important obstacles was the time constraint, which did not allow them to deepen the topic as required in the TLSs. For instance, some teachers claimed that the discussion of the TLSs' initial scenario was too long (about one hour), compared to the scheduled week-time for science teaching (two or three hours, depending on the school stream).

Due to the aforementioned reason, the main difficulty was related to recruiting volunteer teachers, willing to participate to the project. Despite the interest of comparing their practice with that of colleagues of other European countries, the requirement of having at least two implementations of TLSs (overall 20 hours) prevented many teachers from being involved in the project. The main justification for declining to participate was that the teaching time allowed for out-of-textbook topics is usually very limited in their school. In particular, many topics targeted in the EURPBs are not usually taught in the first two years of secondary schools. Therefore, in order to harmonize the Chain Reaction activities with those of their usual practice, the teachers had to organize, in many cases, extra school time meetings with their students for the implementations. Such meetings were not always easy to organize, since classrooms were not always available and/or students were involved in other activities. Another reason was that the PD course sessions were held in the afternoon when teachers were often involved in other school extracurricular duties.

Overall, such findings suggest that implementation of IBSE in the Italian context was affected by both teachers' attitudes and beliefs about scientific inquiry and by local constraints of the specific school.

4. IMPACT

Students' Views About IBSE

First of all, the project allowed us to investigate students' views about scientific inquiry at the beginning of the secondary school stream. As expected from the Italian results in international assessment probes, students' views did not generally correspond to accepted views in literature. This may be due to the usual textbook presentations, which often underline the existence of only one scientific method, the so-called Galilean one. In some cases, science textbooks also suggest the existence of two experimental methods, one based on induction, the other based on deduction. In both cases, textbooks underline that there is a fixed sequence of steps to follow during a scientific research. Despite the explicit recommendation not to use any textbooks during the classroom activities, students could have used the textbooks *before* them, thus influencing the questionnaire's (negative) results. After the implementation, we found some improvement, though not significant (Galano et al., 2015). Such evidence confirms the teachers' perceptions that to implement IBSE in normal teaching practice would require extensive teaching time, collaboration/support from the other teachers in the school and deeper changes in the current content-laden science curriculum. However, one aspect that teachers considered positive was that the TLS helped students discuss the results and conclusions obtained by the different working groups. In this way, students could explore the difference between results from a single experiment and results coming from more and/or different experiments. In general, while the collected results suggest that the TLS implementations were not sufficient to have a significant impact on students' overall views about IBSE, further investigations are needed to understand if the proposed TLSs could foster students' understanding of at least some aspects of IBSE. To this end, we plan to modify some of the TLSs to make them more suitable to target specific IBSE aspects, and to implement them with a new sample of teachers and students.

Teachers' Implementation of IBSE

Another interesting result is that the students, who mostly improved their views about IBSE, were taught by teachers, who did not transform the proposed approach during practice; for instance, they did not change the aims of activities nor the balance between the different aspects of inquiry targeted in the

TLSs. The fact that some of the participating teachers transformed the aims of the TLSs to adapt them to their practice and did not implement the activities according to IBSE principles was a quite unexpected result given the 30-hour training course that all the teachers attended.

It is possible that the course was not sufficient to provide teachers with a solid knowledge base about IBSE in order to effectively implement the TLS. In particular, we noticed differences between what teachers declared when discussing the IBSE during the course sessions and what they actually did when delivering the TLSs in their classes. Hence, more research is needed to investigate the factors underlying teachers' adoption or transformation of the IBSE approach. To this aim, we plan to investigate which IBSE aspects teachers mostly adopt or transform after training periods devoted to the explicit teaching of IBSE as content.

Involvement of Teachers in Training Activities

During the three years of activities of the Chain Reaction project, about 10 out of the 30 involved teachers decided to share their work and ideas with their colleagues, organizing seminars and training meetings in their schools. For instance, in the first year of the project, two of the involved teachers of a vocational school organized a short course for their colleagues of about four meetings of 2 hours each. The aim was to disseminate what had been done during the PD course about IBSE. The course involved 10 science teachers and featured three activities:

- A presentation of the main features of IBSE;
- A discussion about the school practice implementation of the proposed TLSs;
- A discussion about the revised version of the questionnaire used to investigate students' knowledge of IBSE.

Hence, the teachers became trainers themselves. They reported that the participants in the meetings specifically asked for more information about IBSE, since they had no experience of it. One experienced teacher (about 25 years of teaching experience) reflects in retrospective:

My involvement, as well as that of my students, has triggered in the colleagues who participated in the training meetings a self-evaluation process of science teaching as it happens in the classroom.... I am sure that such experience will bloom in the coming years

Such evidence suggests that to improve ownership of curriculum reforms, teachers should have opportunities to be active agents in their teaching context (Kirk & MacDonald, 2001). The Chain Reaction project allowed participating teachers to became trainers themselves, thus increasing their expertise in IBSE.

Moreover, they were interested in knowing whether students could independently organize their own work during activities. Finally, the participants were willing to know students' learning outcomes obtained after the TLSs, and to discuss about possible strategies to improve the teaching of scientific disciplines, from an integrated viewpoint. The same teacher explained to her colleagues:

The students learned about viscosity, and about models in physics and geology. But also, they learned to search effectively the Internet and became able to use the presentation software to make transitions, upload videos, extract pictures and prepare animated gifs... they were genuinely aware of the importance in participating to the project... they were enthusiastic and responded well to my stimuli.... they made suggestions about the aims and the setting of possible experiments... since the first steps some of them appeared to be more very motivated and committed....

The teacher's comments confirm previous research results (Nelson, 2001) according to which reform-based change to the usual teaching is positively affected by evidence of improved student's learning outcomes about the addressed content knowledge.

Involvement of Teachers in Dissemination Activities

In addition to the National Conference days, all the involved schools organized at least one day to disseminate results during, for instance, *School Open Days.* In such events, researchers involved in the Chain Reaction activities were invited to present to parents, teachers and principals the main features of the project. Another kind of event in which the Chain Reaction activities were disseminated was scientific fairs. In one of such sessions, teachers who participated in the project discussed their experience during the National and International Conference, as well as the value of the training course about IBSE.

5. CONCLUSION AND FUTURE DIRECTION

Our involvement in the Chain Reaction project allowed us design and refine the chosen inquiry-based TLSs. Moreover, we had the opportunity to involve approximately 30 teachers that were interested in acquiring competencies about scientific inquiry to meet the requirements of the new curriculum guidelines. One of the teachers recalls:

My colleague and I found this course extremely valuable: not only did we learned the problem solving and inquiry-based approach basics, but also we understood how to implement such an approach in classroom with the pupils in order to give them space to carry out their own investigations. An important tool we used was the "portfolio" that we started to use during the course to plan the possible strategies and that proved to be very useful when designing the students' activities

Then, she acknowledges:

I think I've gained a lot from this project... At the beginning, I thought that it was impossible to apply the IBSE methodology every day, unless the curricula were completely modified, and the classrooms equipped with ICT. Now I think that the most important thing is to improve our interaction with the class. I understood that students were able to do things that I've never believed they could do... they have been able to make predictions and hypothesize solutions, and at the same time, they learnt how to communicate the results of a research in different languages and professionally with audiovisual means... they also understood the importance of drawing on previous work... with usual teaching, it would have been impossible to achieve all such things...

Overall, we feel that we had been involved by the participating schools in their relevant process towards innovating teaching practice: such endeavor is still at the beginning and more efforts are needed to support the teachers and their students. The designed TLSs, which include detailed teachers' guides, may be a useful starting point for other teachers that will attend our training programs.

Another relevant aspect is the students' attitude towards participation in the National and International conferences. They were enthusiastic in reporting their work in a conference-like environment to peers from other schools. Our results are coherent with much of existing literature, which shows that IBSE-based approaches are effective in fostering students' positive attitudes to Science.

A final positive outcome was the involvement of teachers and students in participating in the international conferences. The students were given the chance to deepen their knowledge about IBSE and to be involved in IBSE activities with students of the same age from other countries, comparing reasoning strategies and acquired competencies. At the same time, teachers exchanged practices with foreign colleagues looking for solutions to similar problems. This made us re-think our model of PD programs so that, in the future, we will include sessions devoted to group work with teachers from different schools to allow them the opportunity to exchange their experience. We believe that such peer-work will be a fruitful way to continue the chain reaction generated during the last three years.

REFERENCES

Blanchard, M. R., Southerland, S. A., & Granger, E. M. (2009). No silver bullet for inquiry: Making sense of teacher change following an inquiry-based research experience for teachers. *Science Education, 93*(2), 322–360. doi:10.1002ce.20298

Capps, D. K., Shemwell, J. T., & Young, A. M. (2016). Over reported and misunderstood? A study of teachers' reported enactment and knowledge of inquiry-based science teaching. *International Journal of Science Education, 38*(6), 934–959. doi:10.1080/09500693.2016.1173261

Carleton, L. E., Fitch, J. C., & Krockover, G. H. (2008). An In-Service Teacher Education Program's Ef-fect on Teacher Efficacy and Attitudes. *The Educational Forum, 72*(1), 46.

Galano, S., Zappia, A., Smaldone, L., & Testa, I. (2015). Secondary students' views about scientific inquiry. *Il Nuovo Cimento, 38*, C, 3, 94.

Jimenez-Aleixandre, M. P., & Erduran, S. (2008). Argumentation in Science Education: An Overview. In Argumentation in Science Education: Perspectives from Classroom-Based Research. Springer.

Kirk, D., & MacDonald, D. (2001). Teacher voice and ownership of curriculum change. *Journal of Curriculum Studies, 33*(5), 551–567. doi:10.1080/00220270010016874

Kleine, K., Brown, B., Harte, B., Hilson, A., Malone, D., & Moller, K. (2002). Examining inquiry. *Principal Leadership, 3*(3), 36–39.

Lederman, J. S., Lederman, N. G., Bartos, S. A., Bartels, S. L., Meyer, A. A., & Schwartz, R. S. (2014). Meaningful Assessment of Learners' Understandings About Scientific Inquiry—The Views About Scientific Inquiry (VASI) Questionnaire. *Journal of Research in Science Teaching, 51*(1), 65–83. doi:10.1002/tea.21125

Lederman, N. G. (2006). Syntax Of Nature Of Science Within Inquiry And Science Instruction. In L.B. Flick & N.G. Lederman (Eds.), Scientific Inquiry and Nature of Science (pp. 301-317). Dordrecht: Springer.

Luft, J. A., & Pizzini, E. L. (1998). The demonstration classroom in-service: Changes in the classroom. *Science Education, 82*(2), 147–162. doi:10.1002/(SICI)1098-237X(199804)82:2<147::AID-SCE2>3.0.CO;2-E

Lunetta, V. N., Hofstein, A., & Clough, M. (2007). Learning and teaching in the school science laboratory: an analysis of research, theory, and practice. In N. Lederman & S. Abel (Eds.), Handbook of research on science education (pp. 393-441). Mahwah, NJ: Lawrence Erlbaum.

Marshall, J., & Smart, J. (2013). Teachers' transformation to inquiry-based instructional practice. *Creative Education, 4*(2), 132–142. doi:10.4236/ce.2013.42019

McNeill, K. L., Marco-Bujosa, L. M., González-Howard, M., & Loper, S. (2018). Teachers' enactments of curriculum: Fidelity to Procedure versus Fidelity to Goal for scientific argumentation. *International Journal of Science Education, 40*(12), 1455–1475. doi:10.1080/09500693.2018.1482508

Méheut, M., & Psillos, D. (2004). Teaching-Learning sequences: Aims and tools for science educa-tion research. *International Journal of Science Education, 26*(5), 515–535. doi:10.1080/09500690310001 614762

Nelson, T. H. (2001). *A Science teacher's wisdom of practice in teaching inquiry-based oceanography*. Paper presented at the annual meeting of the American Educational Research Association, Seattle, WA.

Pintò, R., Ametller, J., Couso, D., Sassi, E., Monroy, G., Testa, I., & Lombardi, S. (2003). Some problems encountered in the introduction of innovations in secondary school science education and suggestions for overcoming them. *Mediterranean Journal of Educational Studies, 8*(1), 113–134.

Sassi, E., Monroy, G., & Testa, I. (2005). Teacher Training about Real-Time Approaches: Research-based guidelines and materials. *Science Education, 89*(1), 28–37. doi:10.1002ce.20041

Schwartz, R. S., & Crawford, B. A. (2006). Authentic Scientific Inquiry As Context For Teaching Nature Of Science: Identifying Critical Element. In L.B. Flick & N.G. Lederman (Eds.), Scientific Inquiry and Nature of Science (pp. 331-355). Dordrecht: Springer.

Van Den Akker, J. (1998). The Science Curriculum: Between Ideals and Outcomes. In B. J. Fraser & K. G. Tobin (Eds.), *International Handbook in Science Education* (Vol. 2, pp. 421–447). Dordrecht: Kluwer. doi:10.1007/978-94-011-4940-2_25

White, R., & Gunstone, R. (1992). *Probing Understanding*. London: The Falmer Press.

Windschitl, M. (2003). Inquiry Projects in Science Teacher Education: What Can Investigative Experiences Reveal About Teacher Thinking and Eventual Classroom Practice? *Science Education, 87*(1), 112–143. doi:10.1002ce.10044

Zappia, A. (2018). *How secondary school science teachers implement scientific inquiry in classroom practice* (PhD thesis). Camerino, Italy: Università degli Studi di Camerino, School of Advanced Studies.

Zappia, A., Galano, S., Smaldone, L., & Testa, I. (2015). Science Teachers' Transformations When Implementing Inquiry-Based Teaching-Learning Sequences. In C. Fazio & R. M. Sperandeo (Eds.), *E-Book Proceedings of the GIREP-MPTL 2014 International Conference: Teaching/Learning Physics: Integrating Research into Practice* (pp. 897 - 906). Palermo, Italy: Università degli Studi di Palermo.

ENDNOTES

1 https://nces.ed.gov/timss/

2 Guidelines can be found at http://www.indire.it/lucabas/lkmw_file/licei2010/indicazioni_nuovo_impaginato/_decreto_indicazioni_nazionali.pdf

3 The *EUPRBs* can be found at www.chreact.eu

4 For further details see http://www.indire.it/en/area-di-ricerca/training/ (in English)

5 The complete collection of adapted TLSs (in Italian) can be found at www.chreact.it

6 https://www.ilsole24ore.com/art/tecnologie/2010-08-27/lequazione-drake-ieri-oggi-101908.shtml#continue?refresh_ce=1

Chapter 7
Inquiry and Chain Reaction in Germany:
Challenges, Opportunities, Activities, and Results

Manuela Welzel-Breuer
University of Education Heidelberg, Germany

Nicole Marmé
University of Education Heidelberg, Germany

Sönke Nils Graf
University of Education Heidelberg, Germany

Jonathan Glaser
University of Education Heidelberg, Germany

Ann-Katrin Krebs
University of Education Heidelberg, Germany

ABSTRACT

Chain Reaction in Germany was implemented as professional development (PD) project. Starting each year by a PD course the Chain Reaction philosophy, aims and model have been introduced, and the teachers were familiarized with six selected Earth and Universe Pupil Research Briefs (EUPRB), which were given as the practical basis of the IBSE approach. The chosen EUPRBs did match the German curricula regarding content and research focus. The teachers have been adapting these ideas to their own scenarios and their specific school conditions. They delivered student-focused lessons using the EUPRBs or complementing it. Using critical thinking and problem-solving skills, students at secondary level have been working within scientific scenarios they have been personally interested in. Their work has been published via posters and oral presentations that have been shared on the school site as well as on national "Express Yourself Conferences." These annual conferences celebrated science and the work of the students. In this chapter, challenges, opportunities, activities, and results are reported.

DOI: 10.4018/978-1-5225-5439-4.ch007

BACKGROUND: SCIENCE EDUCATION WITHIN A FEDERAL SETTING

Before starting to describe the project implementation of Chain Reaction itself, it is necessary to explain the political background in Germany, since this considerably affected the whole realisation process.

Germany consists of 16 federal states. Each federal state is autonomously responsible for its education and science policy. That makes nationwide and even sometimes regional innovation a challenge. Nevertheless, besides differences between the states, there are also commonalities. In the first part of this section, we will explain this rather complex background focussing on conditions for science teacher education at universities, teaching science at schools and for Inquiry Based Science Education (IBSE) in German classrooms.

Despite the German federalism, a Standing Conference of the Ministers of Education and Cultural Affairs (KMK) of all German federal states was installed in 1948. The KMK is instrumental in the coordination and development of the country's education. Its "…consortium of ministers is responsible for education and schooling, institutes of higher education, research, and cultural affairs, which formulates the joint interests and objectives of the 16 federal states" (Kultusminister Konferenz, 2017, para. 1). The KMK's agenda is to address "educational, higher education, research, and cultural policy issues of super-regional significance with the aim of forming a joint view and intention and of providing representation for common objectives" (Kultusminister Konferenz, 2017). Respectively, the Standing Conference of the 16 ministers regularly discusses the country's issues in the field of education, implements expert groups for the development and assurance of common standards, and publishes recommendations to support the achievement of high national standards in education. The present standards for science education in German schools were published by the KMK in 2005.

Looking at these standards, science education plays an important role in the field of secondary education in Germany. It is therefore provided nationwide with science offered disciplinarily in the form of biology, chemistry, physics, and geography. It parallels the form of more general- and inquiry-oriented interdisciplinary subjects, including science and technology at the secondary level. Preceding this practice, the children at the primary level are getting to know basic science phenomena around their everyday life and are becoming familiar with experimental activities.

It is customary in Germany that all teachers at primary and secondary levels teach at least two subjects. They have or have yet to study these different subjects at university level, as well as pedagogy and education disciplines. These studies occur in two phases. For instance, two state exams (first at the university level and then at a state seminar) existed up until 2015 in the German state Baden-Württemberg, where the project was implemented. After 2015, mainly Bachelor's and Master's degrees are earned at the university. While German teachers (or student teachers) have to choose and combine two to three subjects, examples of combinations include for instance physics and sports, chemistry, and the German language, or biology and chemistry.

Teacher training programmes are offered at Bachelor's and Master's levels during pre-service teacher education at German universities. Despite intense communication about the quality of education within professional networks of scientists, politicians, and science educators in this decentralised system, a strong heterogeneous picture is present in pre-service teacher education and inquiry. In the German state of Baden-Württemberg, for example, lower secondary teachers are educated at universities of education. These universities are more specialised in teacher education, including a large focus on pedagogy, didactics, methods of teaching, and in-school practical experiences. In contrast, common universities in

Baden-Württemberg and their "gymnasium" (high school) teacher training programmes, aiming at the education of higher secondary teachers who will teach in this specific type of school called gymnasium, are much more focused on scientific content. As a result the majority of gymnasium students are prepared to proceed to at Universities. This logically results in an inhomogeneous picture concerning the individual qualification profile of German secondary teachers with respect to their teaching approaches.

That brings us to the question of what is meant by science education and how should it be taught at German schools. For all that, according to the national science education standards (Kultusminister Konferenz, 2005a, b), teachers are often bound to a state-specific curriculum with an emphasis on competences versus content (Kultusminister Konferenz, 2005a). The science education standards focus on the main principles of the science subjects (or discipline), describe the competences and necessary knowledge gained by students through a certain stage of learning (e.g., by the end of grade 10), aim at systematic and interdisciplinary learning, and have a cumulative character (see Kultusminister Konferenz, 2005a, p. 6). Schools should deliver knowledge, skills, and capabilities (or competences) to enable critical thinking, decision making, and responsible and creative action. It also should raise freedom and democracy, tolerance, and respect (Kultusminister Konferenz, 2005a, p. 7). This generalised description can be seen as the underlying premise to all educational striving inside the 16 different federal states and their teacher education systems.

According to KMK (see Kultusminister Konferenz, 2005a, p. 16) competences for the field of science education require that students obtain disciplinary competences as they:

- Apply skills to manage a situation by using available knowledge and purchase required knowledge
- Understand the main interrelations within a certain context
- Use appropriate strategies to solve problems
- Apply skills while acting in everyday life and scientific contexts
- Include experiences in their actions

In physics education for example, KMK standards describe the basic education within four fields of competences (see Kultusminister Konferenz, 2005b).

1. **Knowledge:** Physical phenomena, concepts, principles, facts, and laws related to matter, interaction, system, and energy
2. **Obtainment of Scientific Findings:** Experimental and other scientific methods (or use of models)
3. **Communication:** Proper and professional exchange of information
4. **Assessment:** Identify and assess physical facts and phenomena within different contexts

Additional dimensions of action and contents describe the students' learning processes.

Having these descriptions, it is clear that the KMK standards do not explicitly provide teaching methods or strategies to be used in the schools, such as, for example, Inquiry-Based Science Education (IBSE). To interpret and apply these standards and to translate the given intentions, is part of the teachers' daily work. This is the same for all subjects respectively disciplines taught at school.

Even though it is not mentioned, IBSE is a part of Germany's national science curricula and, in accordance with the above described science education standards, it is a central feature of modern science teaching in Germany. The country's 16 federal states have implemented IBSE for more than a decade.

The given standards implicitly postulate an IBSE approach to science education in schools. Consequently, the 16 states follow the recommendations. However, each state realises this with its own emphasis, interpretation, and support.

According to these open and competence-oriented standards and the freedom of each German state to apply them, the German educational context is most likely to fit with the intentions and aims of the Chain Reaction project in terms of curriculum requirements or regulations. Based on this, German teachers should be able to find ways to implement IBSE in the use or adaptation of Earth and Universe Pupil Research Briefs (EUPRBs) in the classroom. Teachers should also have enough freedom to develop alternative research briefs or project ideas together with their students – and they are accustomed to doing.

Nevertheless, or due to this very open situation, and due to every federal state defining its own curriculum, studies show that the implementation and practice of IBSE across Germany is heterogeneous. This is especially true regarding the level of autonomy given to students in science classrooms and the openness of themes (Baumert et al., 2003; Höttecke, 2010). For example, the Third International Mathematics and Science Study (TIMSS) results indicated a dominance of teacher-centred practices using experiments and demonstrations (Baumert et al., 1997; Baumert, Bos, & Lehmann, 2000; Drechsel & Prenzel, 2011). Even the 2015 TIMMS results show a very slow development of inquiry-based teaching practices within German schools (see Rieser et al., 2016). Science classrooms (or school laboratories) are well equipped; teaching and learning materials are available. However, the quality, amounts, and use depend on the state or region. In some instances, it also depends on the level of urban quarters and type of school. Within German schools it is the given team of teachers who at last decides how to realise the science curriculum.

Professional development (PD) programmes for in-service teachers (regarding inquiry approaches) are regularly available to teachers. German teachers are required to meet a set number of PD hours in a period of 1 to 4 years depending on the federal state. However, this obligation is enforced in only some of the federal states (in our case e.g. in Hesse) (Lipowsky, 2013, Hessisches Kultusministerium, 2017). Thus, a "common" level or background of professional development does not exist in Germany.

Participants in the Chain Reaction project were recruited from three of Germany's federal states: (1) Baden-Württemberg; (2) Rhineland-Palatinate; and (3) Hesse. Considering the aforementioned information, the Heidelberg Chain Reaction team were faced with diverse educational contexts, settings, professional backgrounds, and experiences. Accordingly the team was prepared to individualize the PD course and to acknowledge diversity and openness as described in the following project summary.

PROJECT SUMMARY

The Chain Reaction project in Germany was implemented as a professional development project (PD). Its aim was to enrich and improve IBSE approaches in 15 partner schools of the Rhine-Neckar region (comprising the above mentioned three federal states) from 2013 to 2016. These partner schools were recruited via the existing professional networks of the German team and public advertisement placed at the website of the university and the project itself. Thanks to well established links and good relationships to schools and teachers within the region, partner schools could easily be found.

Starting the Chain Reaction project, the German team prepared an introductory PD course for the participating teachers. On one hand, this course would establish the project's philosophy and aims. On the other hand, participating teachers would become familiar with a set of selected inquiry-based

EUPRBs, which during the course were given as a concrete and practical basis of the Chain Reaction approach. To effectively select suitable EUPRBs, the ten teachers of the first-year partner schools were provided with the themes available and asked about their interests, beforehand. Using this information, the researchers selected and translated six of the given EUPRBs into German for the Heidelberg PD course. The six themes were: (1) Out of Sight, Out of Mind; (2) Collision Course; (3) Green Heating; (4) Plants in Space; (5) Cosmic Website; and (6) Portable Hydroelectric Power Plants (PHEPPs). The chosen EUPRBs represented scenarios to reflect real world academic science, support inquiry-based teaching and learning of science, and involved engaging topics appealing to female and male genders. These EUPRBs also matched the German curricula with respect to targeted the competences, the content and research foci.

The Teacher Briefings

The first step in each project year required that ten teachers of five schools attend a one-day session of the Heidelberg University of Education's practice-oriented and interactive courses. During the session, the participating teachers were introduced into organisational matters of the project, IBSE theory and practice during which they became familiar with the possibilities of adapting IBSE using the given EUPRBs. The course's duration was 7,5 hours. The structure of the Block 1 partly changed for year 2 and 3 by starting with the own experiences of the participating teachers of using IBSE. Table 1 gives an overview of the teacher briefing sessions in the years 1 to 3.

Teachers were introduced by members of the German project team to the aims of the project, the IBSE model used and familiarised with the six selected EUPRB resources. They examined the kits prepared (see Figures 1 and 2), necessary to realize the chosen EUPRBs, with material for experimentation. They also used them to try out the research tasks proposed and extended those by investigating the themes along own additional research questions. During the sessions the teachers discussed with their peers the possibilities of implementation of the EUPRBs into their classroom activities according to their individual situations (see Figure 3). Next, the participating teachers adapted the ideas to their individual scenarios and specific school conditions, and finalized that process with a presentation of

Table 1. Program teacher briefing (Years 1-3)

<table>
<tr><th></th><th colspan="2">Year 1</th><th colspan="2">Year 2 and Year 3</th></tr>
<tr><td>09:00</td><td>Come together</td><td>Informal welcome with coffee and tea</td><td>Come together</td><td>Informal welcome with coffee and tea</td></tr>
<tr><td>09:30</td><td>Block 0</td><td>Organisational matters</td><td>Block 0</td><td>Organisational matters</td></tr>
<tr><td>10:00</td><td>Block 1</td><td>Introduction: IBSE theory and practice</td><td>Block 1</td><td>Individual reports and introduction on using IBSE in one's own lessons</td></tr>
<tr><td>12:00</td><td colspan="2">Lunchbreak</td><td colspan="2">Lunchbreak</td></tr>
<tr><td>13:00</td><td rowspan="2">Block 2</td><td rowspan="2">Understand and familiarize with EUPRBs</td><td>Block 2</td><td>Understand and familiarize with EUPRBs</td></tr>
<tr><td>15:00</td><td>Block 3</td><td rowspan="3">Feedback and reflection. Optimizing for one's own group of students</td></tr>
<tr><td>15:30</td><td colspan="2">Teabreak</td><td></td></tr>
<tr><td>16:00</td><td>Block 3</td><td>Prospects and open questions</td><td></td></tr>
<tr><td>16:30</td><td colspan="2">End</td><td colspan="2">End</td></tr>
</table>

their implementation ideas to the other course participants. Thus, the PD day ended up with concrete ideas respectively scenarios for implementation activities.

It was important to the German team to present the teachers with an experience that they later had to convey to their students – a classroom setting. In this way teachers themselves have the chance to become immersed into the potential of IBSE and EUPRBs not only theoretically but also affectively. The idea that teachers that are being trained in what they themselves have to teach need to experience it inside their training is called a "Pädagogischer Doppeldecker" (pedagogical double decker) and first was mentioned by Geißler in 1985 and has since been an important aspect of German teacher training

Figure 1. Experimental kits are prepared for introducing and examining EUPRBs at teacher briefings

Figure 2. Teachers trying, testing and discussing the EUPRBs

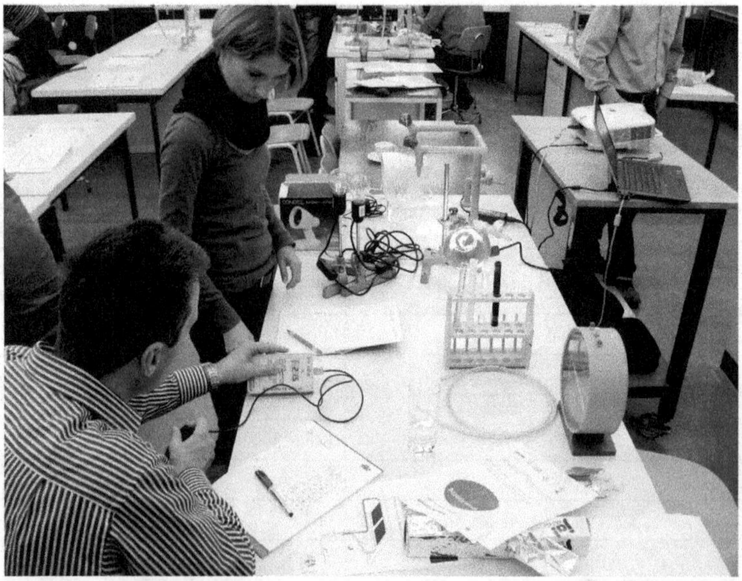

(Geißler, 1985; Wahl, 2013). Hence, the teachers' ideas for own research projects developed on the basis of the given EUPRBs and displayed by themselves in form of posters during the PD session were useful. In addition, they matched the teachers individual schooling contexts (see for instance Figure 3).

Indeed, feedback indicated that teachers found the session inspiring to their upcoming tasks and that they found it helpful to get a feeling of how the project might feel to their students. However, the teachers in this situation felt somehow uncertain whether or not students would be "hooked" in the same way.

Application of the EUPRBs

After the introductory PD course, each teacher delivered a series of student-focussed lessons (over a period of up to three months) implementing projects suggested by the EUPRBs or their own research ideas which are initiated by the EUPRB approach. Using critical thinking, reasoning, and problem-solving skills, students in the 14 to 16 age groups worked together to research scientific scenarios of personal interest. How the students set their ideas in scene is presented in paragraph 3.2.

In the final stages of their research activities the participating students summarised their work through posters and oral presentations. These products were showcased on the school site, as well as the university campus' National Express Yourself Conference to celebrate science and student work. There, the students received qualified feedback on their results and their presentations through peers, teachers, guests, and science experts. Finally, one outstanding project (the winner of each national conference) for each partner country was then presented and discussed at an annual International Express Yourself Conference.

The one-day national events were planned as an annual highlight for the participants as they collaborated on presentations, discussions, and experiments. The national events in Germany involved practising scientists from the University of Heidelberg, the German Cancer Research Centre, and the Heidelberg

Figure 3. Teachers' first project ideas after familiarising with the EUPRB "Plants in Space"

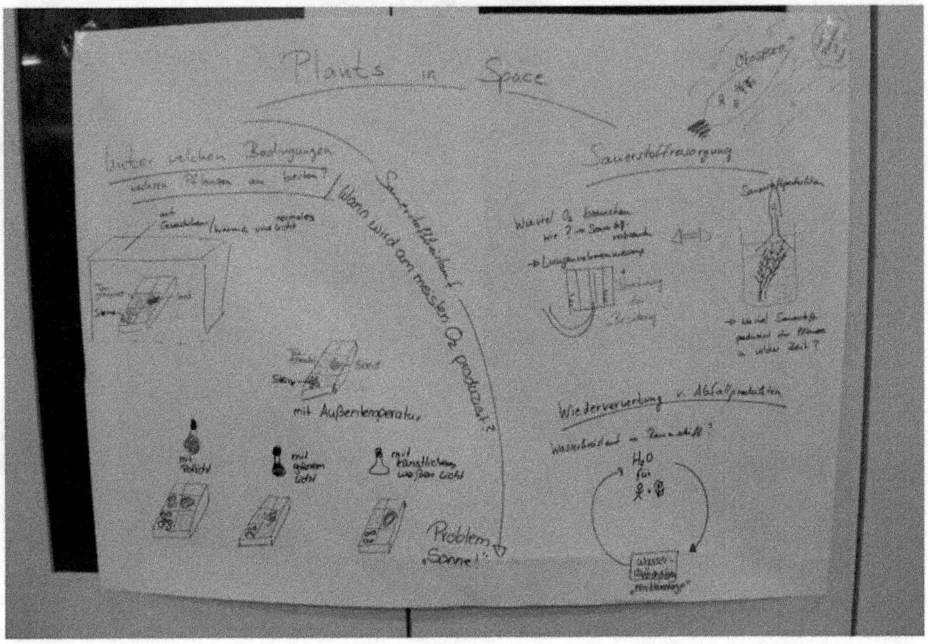

University of Education who acted as role models and experts to the students. These science experts shared their work to inform and encourage young people to consider a career in science-related fields. Acting in assessment teams with participating teachers, the role model scientists identified projects to participate in the two-and-a-half-day International Express Yourself Conferences in Sheffield (2014), Heidelberg (2015) and Plovdiv (2016).

For the second International Express Yourself Conference in Heidelberg the German team also introduced an additional inquiry day for the participating students. The idea behind that was to also use the international conference for inquiry activities. Thus, the inquiry day ended up in designing physical chain reactions within international student teams and using a variety of materials such as balls, dominoes, cardboard tubes, plastic cups, rain gutters, cords, paper, rods and stands, adhesive tapes, and others (see Figure 4). This additional day was also part of the third international Express Yourself Conference in Plovdiv and moderated by the German team, since it was much appreciated and allowed for concrete and motivating interactions of the international students and for intensive collaboration while solving problems together. It appeared, that this day was a welcomed highlight supporting initial team building.

In order to realise a formative assessment producing hints for improvement during the project period, the German team organised a final feedback meeting shortly after each international conference. The feedback meetings brought together the teachers of the participating schools and allowed them to discuss experiences and challenges as well as exchange ideas for improvement and sustainability.

The following sections will detail the key features of the Chain Reaction project in Germany in more detail.

Figure 4. Building a "Chain Reaction" using a variety of construction materials

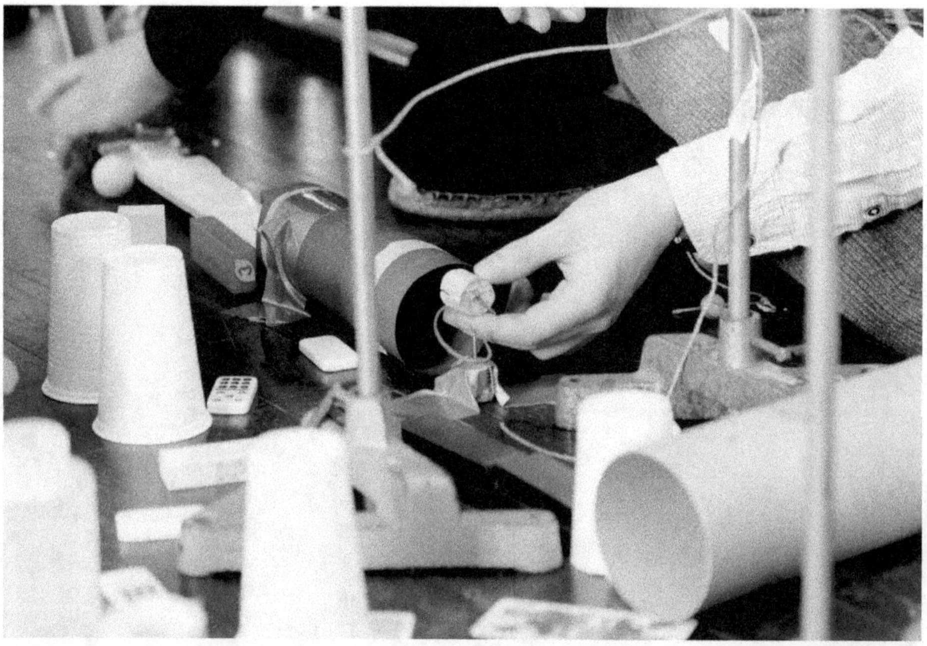

PROJECT IMPLEMENTATION

The Introductory Professional Development Course

The Heidelberg strategy implemented and supported an effective theory-practice transfer through a period (2.5 hours) of inquiry-based activities for participating teachers during the introductory PD course. This very practical and active approach was used as literature suggests that teachers often know *why* inquiry implementation is important. However, they do not know *how* to optimise its use in the classroom (Lipowsky, 2009, 2010, 2013). One argument supporting this finding seems to be a systemic problem already reported in the early 2000s by Merzyn (2004): German Universities used to offer a more theory-oriented programme for teacher education, but lacked a solid theory-practice transfer and subsequent reflection. On the international level, Lortie (2002) reports the same deficit. Also, the TIMMS data for Germany show that inquiry science education which encourages the students to understand, analyse, investigate or solve science problems autonomously, is not the norm (Baumert et al., 1997, 2000; Rieser et al., 2016).

Accordingly, fifteen of the 30 participating German teachers stated that they had limited knowledge of how to apply inquiry in practice. In order to reach an effective theory-practice relationship, the participating teachers needed to gain a clear understanding of how to implement the ideas discussed during the PD session. Therefore, it was necessary to familiarise the teachers with the nuances of inquiry, including observing phenomena, thinking about and asking challenging questions, developing an appropriate methodological design and proper data collection, reflecting on and discussing findings, and formulating implications and dissemination of research results.

After a theoretical introduction into the principles of IBSE and the concept of scientific literacy according to Rocard et al. (2007), Gräber (1999) and Höttecke (2010) and their discussion, the EUPRBs of the Chain Reaction project were presented and tested. In order to realize an effective theory-practice-transfer the learning context for the teachers was organised in a way that a typical school context was modelled in which the teachers had to take the roles of students while inter-acting with the EUPRBs. In small groups, the teachers went into the – already equipped – classroom, worked directly on specific EUPRBs, investigated the material, studied the problems formulated, found their own research questions, searched for experimental approaches, did experiments, and discussed initial results. Thus, they acted authentically in inquiry-based activities, practised teamwork and scientific communication as described in Roth (1995, xiii). Through this phase the teachers were impressed by the richness of the ideas represented through the EUPRBs and encouraged to look for further possible investigations.

Adhering to Lipowsky's (2009, 2010) findings, the PD course also focused on ways to support teachers who experienced strict professional limitations within the school and science curricula requirements. Hence, the project's researchers encouraged participating teachers to work in tandem or teams to implement inquiry approaches using various scenarios (for example, a science club[1] extracurricular activity, or in a re-organised series of lessons). An advantage of this approach was that science clubs circumvent strict curricular requirements; students could choose topics and approaches to investigating issues. This provided students with greater autonomy compared to the normal science curriculum.

About a third of the participating teachers wished to move to an extracurricular field as they did not have to account for curricular connections and additional work needed to implement the projects. One possible reason for that could be that teachers' narrow interpretation of curricular freedom might limit creativity in their lessons, which resulted in interdisciplinary education. Incidentally, this issue was a notable reason that teachers did not participate in otherwise attractive projects and activities offered or initiated by science educators.

As already described the Chain Reaction project offered 11 EUPRB resources to match the interests and options of the curricula, schools, teachers, and students from partner countries. At the beginning of the project, the researchers asked the first-year project German teachers for preferred themes out of those. The six themes named were: (1) Out of Sight, Out of Mind; (2) Collision Course; (3) Green Heating; (4) Plants in Space; (5) Cosmic Website; and (6) Portable Hydroelectric Power Plants (PHEPPs). The researchers translated and adapted the complete briefs based on these six themes. Resource kits for these EUPRBs were purchased and tested for function and applicability by the German project team prior to the PD courses. Two in-house developed themes - Solar Updraft Power (to be built and tested empirically by doing measurements) and Organic Cosmetics (to be produced and investigated according to their biochemical features) - were added to the set (resources see Seeberg et al., 2011, Jannack et al., 2011, Marmé & Knemeyer, 2015).

To summarize, during the first phase of the project, the participating teachers became familiar with the project approach as well as with concrete and tested materials for use in their schools by implementing phases of investigation and discussion (hands-on and minds-on). The idea was that the participants could understand the philosophy of the IBSE approach in a practical manner in order to be able to plan their own classroom activities accordingly.

Post PD-Course Implementation

What happened in the schools and classrooms following the introductory PD course? About half of the teachers (and students) participating in the project used the original templates and realised some of the six EUPRBs given within their schools. However, inspired by the provided EUPRBs, other teachers in collaboration with their students decided to go their own way and designed and published individual research projects according to the IBSE philosophy and the circumstances within their schools. It is assumed that a certain number of German teachers participating in the project are used to adapting ideas, and not using those exactly the same way they were designed, in order to find an individual approach within their own learning contexts. Thus, grasping the fundamental ideas behind the IBSE approach, they searched for their own special ideas and used sources beyond the scope of the offered EUPRBs. Additional resources extended beyond the earth and space science context. The self-created or individually interpreted themes focused on energy issues, adapting sustainable technologies in schools, and features of the human body. Those projects have been for instance:

- Solar Works: Generation of Energy from Dye-Sensitized Solar Cells

The aim of this project was to gain and measure energy from a self-constructed dye-sensitized solar cell. The students studied the scientific literature, bought Rhodamine B and different materials to be used as conductors. They documented their experiments and compared systematically their results.

- Generation of Energy from Sound

In this project experiments were conducted in order to generate electrical Energy from sound. The background here was the noise cars on highways produce.

- Green Light in Our School

The students of this group measured and calculated the energy demand and costs of the common lamps used in their school building and tested the efficiency of alternative light sources which produce a comparable brightness. The project resulted in a change of the light equipment within their school.

- Construction of a MultiChAcc (Energy Charger for a Mobile Phone)

The students followed the question, how different types of sustainable energy sources like small solar collectors, dynamos for bicycles and a paddle wheel for wind and/or water can be used to recharge a mobile phone. Knowing this through own experimental activities and measurements they constructed and tested an adapter in order to transfer the energy between the components (see Figure 5).

- Green Gym: Sustainable Energy Production in a Gym

In a Gym a lot of energy should be available based on the activities happening there. The students did observations and measurements and tried to find ways to make use of it.

- Fuel Made of Organic Waste

The students aimed on developing ideas for alternative and sustainable fuel production and realized a number of experiments using banana and apple peels (organic waste) and yeast in order to gain ethanol which then was be tested for the use as fluid fuel.

- The Performance of the Human Senses

The focus of this project has been the human senses and their interaction. The students of this project documented own psychological like experiments the link between sense of taste and the colour of food.

The students' themes were accompanied by their empirical scientific studies made up of theoretical and experimental products and insights. It could be observed that the students intensely searched for theoretical information and practical solutions. The students' motivation throughout the project work was on a high level. The spectrum of students' themes throughout the Chain Reaction project demonstrated the explicit ownership of the teachers and students based on the PD courses offered at the beginning of each project year.

Using this strategy of seizing opportunity, the Chain Reaction project teachers offered students greater autonomy in theme selection, efficiently managed their time both inside and outside of the school setting, and controlled the purchasing and testing of relevant materials. The researchers also observed that some students forced their teachers to dedicate more time to the project than was initially planned.

Support of Communication Skills

The German project team also supported the development of the students' communication skills focussing on scientific presentations for student conferences. A written guide on how to make good presentations was provided by the coordinator in the English language.

Figure 5. Students working on their themes – here it is the use of solar energy for a MultiChAcc

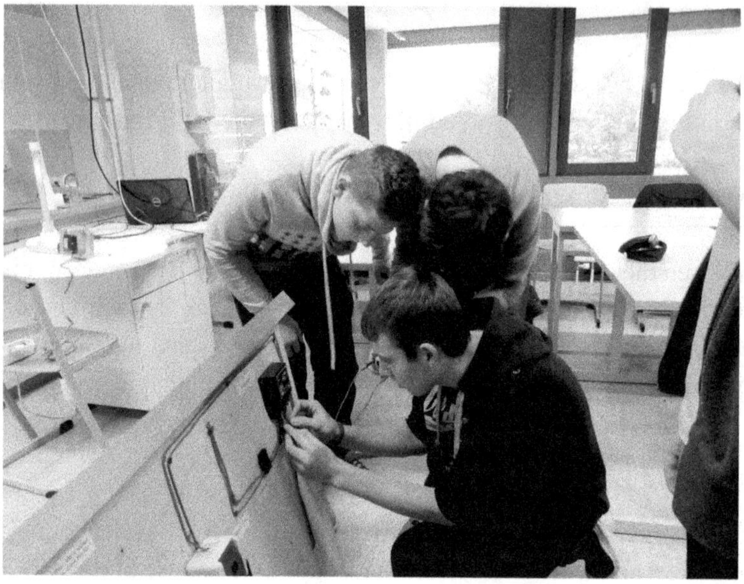

The team translated it into German and distributed it using the project website.

By offering access to an extra on-line course for "scientific writing" (see Flechsig et al., 2017) as well as "scientific presentations" the German team aimed at emphasising teachers' and students' presentation and writing skills in order to give them tools.

1. Increase their and their students' awareness of language as an "integral part of science and science literacy" and "as a means to doing science and to constructing science understandings [...] also an end in that it is used to communicate about inquiries, procedures, and science understandings to other people so that they can make informed decisions and take informed actions" (Yore et al., 2003, p. 691).
2. Increase their and their student's level of reflection, the level of depth, and the level of abstraction. Learning science and acting in scientific contexts therefore aims at communicating science in oral and written form. Leisen (2008, p. 6) reports several effects connected to writing. It supports
    a. The development of ideas
    b. In-depth involvement
    c. Increased realization
    d. Increased precision
    e. Focus on the essential
    f. Development of one's own style
    g. Support for self-reliant learning.

Students' oral and written project presentations at the national and international conferences were consequently supported by structured and focussed visualizations, real and virtual experiments and precisely formulated working steps and results.

Support On-Site

Throughout the project, the individual schools were visited once to twice by members of the German team to support the student teams as well as to answer questions on the side of the students as well as the teachers. Together, the emerging projects have been discussed and intermediate results appreciated. These visits were organised on a regular basis alongside the digital support from our role models and our team. As well as the support of the project teams in the schools, the German project team also aimed to show interest and presence towards the students – so it was clear to them that the research that they were doing took place in a context of curious non-school related people, also representing a scientific community. In addition, the project team expected to support self-confidence of the students and appreciation of their intensive work as well as support constructively the preparation of the public presentations on school site and at the national conferences. In some cases, the project members could give feedback directly on the presentation skills of the students.

The German team aimed at Chain Reaction to also take a broader scope in terms of democratic education. To agree to Cohen who states in 1996:

In an idealised deliberative setting, it will not do simply to advance reasons that one takes to be true or compelling: such considerations may be rejected by others who are themselves reasonable. One must find instead reasons that are compelling to others, acknowledging those others as equals, aware that they have alternative reasonable commitments, and knowing something about the kinds of commitments that they are likely to have…if a consideration does not meet these tests, that will suffice for rejecting it as reason. If it does, then it counts as an acceptable political reason. (Cohen, 1996, p. 100)

Presenting an idea in public and winning support, while not necessarily a part of the scientific process itself – is part of science within a democratic society, that also decides how it wants to distribute resources. That is why presenting one's own research in front of society's representatives and fellow researchers as well as interested citizens was one of the priorities set by the German team.

Presentations at National Conferences

The national express yourself conferences were planned and advertised as an annual highlight for the participants as they collaborated on presentations, discussions, and experiments (see Figure 6).

The national events in Germany, which took part in the main hall of the University of Education Heidelberg, were opened by the vice chancellor and pro-vice chancellor for research and international affairs of the University, involved practising scientists from the University of Heidelberg, the German Cancer Research Centre, and the Heidelberg University of Education who acted as role models and experts for the students. These science experts (PhD students and advanced researchers) shared their work in the form of expert presentations and discussions to inform and encourage young people to consider a career in science-related fields. Thus the students could follow approaches to and results of contemporary and authentic research conducted for example in the field of political geography (in order to understand and prevent human disasters), in the field of research on technological challenges like energy storing and energy revolution, of medicine on the impact of proteins and antibodies or viruses, and on actual research in the field of astronomy titled "In search of terra 2: Is there a second Earth out there?".

Figure 6. Proceedings of the national Express Yourself Conference 2015

Acting in assessment teams with participating teachers, the role model scientists also identified projects to participate in the International Express Yourself Conferences. Science teachers and secondary students of the region have been invited in order to inspire them for the realisation of IBSE within their classrooms and to participate in the project for the following years. Stakeholders representing the city of Heidelberg and educational institutions could get an impression of the valuable and engaging work of the students.

The German team put a lot of emphasis into organising the national conference as a model of a real scientific conference in a way that all active individuals got visible: conference material has been distributed at the registration desk, everybody wore name badges, the conference halls have been equipped professionally (see Figure 7). Officials have been invited and a high-level representative of the University – the vice chancellor and the pro-vice chancellor opened the conference. Local print media and a TV station were present and reported on the conferences. Within this setting, ten outstanding groups of each school presented their Chain Reaction projects within two conference strands, supported by moderation through a researcher of the German team. The scientific jury has been following the presentations attentively, posed questions, discussed the presentations and results and gave individual feedback. One group of each strand was selected by the jury for the final, which was then celebrated as panel presentations in the grand hall.

Student groups not selected for presentation at the stages could display and discuss their projects in the form of a scientific poster session (a poster award was announced). This poster session was arranged to take place during the lunch breaks – both offered in a well organised area of the grand hall.

For the second International Express Yourself Conference the German team also introduced an additional inquiry day for the participating students. The idea behind that was to also use the international conference for inquiry activities. Thus, the inquiry day ended up in designing physical chain reactions within international student teams and using a variety of materials. This additional day allowed for concrete interactions of the participating international students and for collaboration while problem solving. It appeared, that this day was a welcomed highlight supporting initial team building and languages were not problematic.

Figure 7. Students following the presentation of the finalists at the national Express Yourself Conference

Figure 8. Poster presenting the research of students in 2016

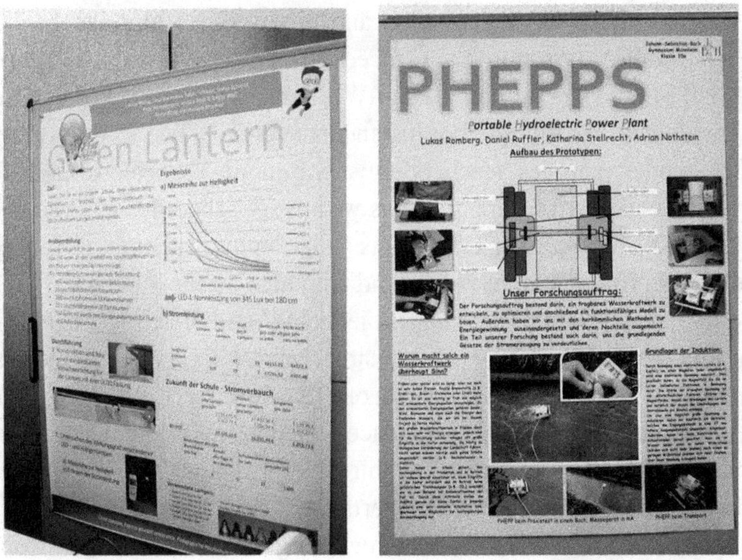

(a) (b)

BARRIERS AND OBSTACLES

In summary, the chain reaction project PD model was effectively used and adapted into practice by participating teachers and their students in Germany.

Evaluation and anecdotal evidence show that engaging teachers in the Chain Reaction project was a generally positive experience. One special example for the enthusiasm ignited in the participating teachers is the following: one of our teachers, who was partner of the first year was so excited about the project, that she decided, as she was about to change her workplace to a new school, to simply import the project with its ideas to her new school. So, immediately after arriving at the new school she talked to her new colleagues and recruited two new teachers, who then applied with us and were accepted for the second year. For us this was a special privilege as we were able to win over three colleagues inside the new school, which of course has a bigger chance to lead to sustainable changes in the school culture. In addition, this teacher asked us, whether she too could join her new colleagues in some of the project sessions, as she planned to proceed with the IBSE projects on her own. Of course, we agreed and thus, she was our first and experienced "veteran".

However, there are specific problems to address.

Firstly, despite of a successful advertisement for the project, it was increasing difficult to recruit teachers in the second and third years of the project through public calls. According to the responses of school headmasters and teachers to the call of participation, this was due to a heavy workload and involvement experienced by teachers of the region. In particular the Metropolitan Region Rhine-Neckar is well-supplied with already existing opportunities to join science projects. For example, the schools are engaged in regional, national and international science activities like Hector Seminars (workshop offers for gifted students), Heidelberger Life Science Lab of the German Cancer Research Center, "Jugend forscht" (national competition mainly in science fields), COMENIUS projects, International Young Physicist Tournament (IYPT), science days (like "explore science") and science nights organized by cities and foundations, and others. The schools of the region are invited to participate in all these activities and programmes. Consequently, they are busy and occupied at a high level.

Not to forget that in parallel, the schools were faced with the development and application of new curricula introduced in 2015 in the federal state Baden-Württemberg. This fact of a professional overload is well known and communicated in Europe. For instance, the OECD TALIS report mentions: "Across the participating countries and economies, teachers' most commonly reported reasons for not participating in professional development activities are conflicts with work schedules and the absence of incentives for participation." (OECD, 2014, p. 86). Taking this situation into account, it, indeed, seems to be necessary to activate existing professional networks and convince teachers personally to take the chance to participate in innovative and relevant educational projects and/or PD offers (personal dimension). Another option could be to offer the participants "win-win situations" with respect to motivationally rewarding situations according to the Self-Determination Theory (Ryan & Deci, 2017) (e.g. new competences to be gained, possibility of interdisciplinary experiences, meeting with other practitioners, possibility of intercultural exchange, European perspective). A third vector of approach could be the enhancement or support with material and personal resources. Nevertheless, to be able to understand the main obstacles and issues better, greater insight would be needed regarding possible incentives from the perspective of the teachers. Further research is strongly recommended. Knowing more about that, could help to solve individual, structural and political problems.

We had to draw from our existing networks of teachers extensively to overcome this challenge, also we noticed that it was necessary to actively reduce the worries and concerns of teachers to be able to cope and master the project. Like other chain reaction teams (see Ireland) also our teachers reported a fear of being unable to do experiments with their students due to curriculum restraints and time pressure.

As shown by Klusmann et al. (2006) especially dimensions of classes that describe a teacher's ability to adapt to the pupil's needs are heavily influenced by a teacher's workload. Overloaded teachers "focus on the ensuring of an undisturbed class, avoiding the risks of an adaptive, individualised, cognitively activating class; in the long run at the expense of quality of instruction." (Klusmann, Kunter, Trautwein, & Baumert, 2006, p. 171). We believe that this research to be highly relevant for the inquiry based science education – growing responsibilities and increasing workloads (see Heise, 2009, p. 10; OECD, 2014b, p. 86), are likely to prevent many teachers from adopting more student centred forms of teaching science and instead cling to more teacher centred forms of science teaching. In this manner an increasingly high workload might act as a "killer of innovation" – possibly leading to less engagement on the side of the students and frustration on the side of the teachers.

As of yet the German team was uncertain of how to interpret the impressions it received from directly communicating with the teachers: based on the research mentioned above, as well the team's own experience, the description of increasingly high workloads is an issue of real concern, however conversations with project teachers indicated that it could be possible that teachers are unaware of the freedom they are given by the curriculum. Also, it could be possible, that the increased workload leads to less time to reflect and gain an objective view of curricular demands – resulting in a subjective handling of the curriculum. Further research in this direction is highly advisable, since a debonding of real curricular demands and teachers' actions will lead to unknown results.

Secondly, the researchers experienced a similar issue in recruiting role models. PhD students and young researchers are occupied with their research projects and qualification processes. As they are forced to regularly present results at conferences, to publish and network with their scientific communities, they are always busy. Once again, personal professional networks helped to solve this problem in Germany. One argument used by the German team was the responsibility researchers have for the next generation to make science available for all – underlining the arguments for its importance for and acceptance by society, its specific approaches, methodologies and results.

Thirdly, there were issues which remained unresolved because of their specificity concerning the project agenda. The project asked for a certain rhythm of work: the PD courses are to be delivered in autumn, project phases are to be carried out by February, National Express Yourself Conferences are to be offered in March, and the International Express Yourself Conferences had to take place in May of each project year. This tight schedule was not easy to follow and to synchronize with the participating schools and teachers. The period between the national and the international conferences was extremely short – especially for the winning students of the national conferences in order to prepare their presentations to be presented in English.

A fourth issue was the lack of storage space for greater amounts of project equipment and bigger and more student-produced artefacts, than usual. Therefore, the researchers had to look for creative solutions introducing strategies matching the structure of common school buildings in Germany. To realise inquiry-based science education at an acceptable level, a certain space and flexibility is needed. Perhaps, an improved co-operation between schools and local informal educational institutions or opportunities could allow this to happen.

Finally, like teachers, students lacked time to engage in inquiry-based projects. Work-related pressures, including examinations and subject assessments, caused students to list extracurricular inquiry work as a low priority.

IMPACT

At the 2014's National Express Yourself conference, students participating in the project answered a questionnaire to evaluate the working phase of the project. It referred to the themes Out of Sight, Out of Mind (used by 35%), Green Heating (27%), PHEPPS (17%), Collision Course (10%), Cosmic Website

(6%), and other topics (5%). The students had to rate the quality of the project work in terms of autonomy experienced, team work realized, support received through the teachers and project members, the enjoyment and the results gained. Fifty-three students took the survey, including 28 female and 24 male students. Most of the students (80%) were between 14 and 16 years of age and grades 9 or 10. This evaluative data in the project's first year provided evidence that the majority of students highly appreciated working autonomously on their inquiry projects as well as valued rich and authentic contexts through inquiry-based teaching and learning materials like the EUPRBs (see Menrath, 2014; Welzel-Breuer et al., 2017). That underpins an impact on a motivational level.

Evaluation results also showed that the chosen EUPRBs were appropriate for IBSE. Most students (96%) answered that they were strongly self-directed during the project. The option "the project allowed me to work and to do research independently" had 67% "totally agree" and 28% "largely agree" responses. Furthermore, 95% of the students confirmed that they independently planned and carried out the project. A few students stated that the teacher supported their work significantly (13%) or influenced students' decisions (20%).

The evaluation showed that the topics and provided materials were appropriate for German schools. According to students' opinions, more than 80% felt that "the material was helpful," "enabled them planning experiments," and "did not lead to excessive demands".

More than 90% of the students were satisfied with their results and had fun during the project. The researchers were pleased by "fun" as a strong motivational factor to do science. Most of the students (85%) agreed that the Express Yourself Conference played an important role in motivation.

The quality of the German student teams' presentations at the national and international Express Yourself Conferences convinced the Chain Reaction team members, role models, and participating teachers of the quality of the students learning experiences. It demonstrated that students can work on their own hypotheses and use the results to positively reflect on their research questions. This result could be seen as a specific trademark of the Chain Reaction project. The products, feedback data and researchers' observations during the period of project implementation suggest that this type of work allowed participating students to understand the basic criteria of science.

The students emphasised the enjoyment and motivation gained through their inquiry projects. In all three final feedback sessions, participating teachers stated that almost all the participating schools' students were encouraged through the project to increase participation in regional and/or national science competitions and fairs. Participating teachers also reported that the students felt appreciated and acknowledged based on their work and engagement at the conferences. The quality of the university's conference setting strongly supported this positive statement. Students especially valued the conference publicity, authenticity of the academic atmosphere, experience from peers' presentations, and networking and nationwide connections. The poster sessions in each student conference attracted high praise from participants who valued the opportunity to share their work. Data also suggests that participating students gained elaborate presentation skills from their involvement.

In the follow-up meetings, the German team organised at the end of each project year, the teachers were asked for their feedback and about the sustainability of the participation. Here, the German team could come to know that the teachers appreciated the PD opportunities, particularly the initial briefings with a strong theory-practice link. The quality of the EUPRBs and related material kits were described as attractive, convincing, and successfully implemented. Participating teachers were impressed by the professional atmosphere at the Express Yourself conferences. Most participating teachers used the conference and final meetings to discuss ideas and proceed with inquiry practice in their school.

Teachers reported that the project results could change their perspective on the aims and practice of science education and the possibility to put scientific knowledge to good (and sustainable) use. In one project school, electric lighting was changed to a more sustainable form – thus also showing the project's potential to increase true participation and allowing science to give valuable impulses for the democratization of schools. In another example, the school found ways to control waste disposal in the schoolyard – also as a sustainable result of the Chain Reaction project. According to the team's assessment and feedback by students as well as teachers one of the projects' main impacts lay with a change in perspective on the side of the students: With a project like this it seems the students felt they had changed positions away from the receiving end of science towards becoming scientists themselves with their own projects. The teachers also noticed this change of attitude on the side of the students and were happy to have been given a choice of conceptually predesigned projects either to pursue or to model their own projects after. Participating teachers found themselves either reading on or communicating concrete knowledge on real research. After the project, most teachers supported profile formation at their school.

Participating teachers also expressed the need for regional networking opportunities. By connecting with colleagues, participating teachers and students could generate more project ideas for national science competitions.

As a result of this positive experience and feedback, the EUPRB-kits are still offered and regularly used at the Heidelberg University of Education within project oriented pre- and in-service teacher training courses. The acquired knowledge that sprung from the project is actively and consciously implemented in regular seminars within the Heidelberg University of Education. Knowledge gained through the Chain Reaction project is continuously used in the university's primary and secondary teacher education to this day. Some of material inside the project inspired scientists in the field of primary education think about more engaging ways of teacher training.

CONCLUSION AND FUTURE DIRECTION

The Chain Reaction project impacted the scientific reasoning of all participants. It was clear that learning occurred through individual practice and in-service teacher education. The researchers summarise that it is possible to implement real changes to the practice of science education.

At the teacher education level, project ideas have been implemented sustainably through regular pre-service science teacher training at the Heidelberg University of Education's Centre of Education for Sustainable Development. The Chain Reaction team is part of this centre. Team members include discussions regarding the philosophy of chain reaction and its approach to IBSE in their teaching offers (e.g., seminars on sustainable development, methods of teaching, and school practical). Doing so, the experimental kits and the complete EUPRBs are to be used regularly for pre-service and in-service teacher education.

Teamwork from the project created close contacts and working groups across disciplines inside and outside of the university. Currently, seminars on themes like water or renewable energy are being designed across disciplines to intensify IBSE implementation in science at the primary level.

Participants gained important content on the modes and ideas of IBSE application in practice. Participating teachers continue to implement the offered EUPRBs and implement ideas in their professional practice.

Throughout the project, several scientific papers were produced by the team and their Master's students. These papers were shared with international and German scientific communities.

Overall, a real chain reaction took place. This was evident through the international communication on IBSE within the consortium and the establishment of a continued process. Project implementation experienced a constant reflective process. The cultures of practice in partner schools and teacher education received inputs and impulses. Therefore, new networks could flourish.

REFERENCES

Baumert, J., Artelt, C., Klieme, E., Neubrand, M., Prenzel, M., Schiefeie, U., . . . Weiß, M. (Eds.). (2003). *PISA 2000 – Ein differenzierter Blick auf die Länder der Bundesrepublik Deutschland* [PISA 2000 – A critical look at the federal states in Germany]. Opladen, Germany: Leske + Budrich. Retrieved from https://link.springer.com/content/pdf/bfm%3A978-3-322-97590-4%2F1.pdf

Baumert, J., Bos, W., & Lehmann, R. (Eds.). (2000). TIMSS/III. Dritte Internationale Mathematik – und Naturwissenschaftsstudie. Mathematische und naturwissenschaftliche Bildung am Ende der Schullaufbahn, Band 1: Mathematische und naturwissenschaftliche Grundbildung am Ende der Pflichtschulzeit; Band 2: Mathematische und physikalische Kompetenzen am Ende der gymnasialen Oberstufe. Leske + Budrich.

Baumert, J., Lehmann, R., Lehrke, M., Schmitz, B., Clausen, M., Hosenfeld, I., . . . Neubrand, J. (1997). TIMSS – Mathematisch-naturwissenschaftlicher Unterricht im internationalen Vergleich. Deskriptive Befunde. Opladen, Germany: Leske + Budrich.

Cohen, J. (1996). Procedure and Substance in Deliberative Democracy. In S. Benhabib (Ed.), *Democracy and Difference: Contesting the Boundaries of the Political* (pp. 95–119). Princeton, NJ: Princeton University Press.

Drechsel, B., & Prenzel, M. (2011). Internationale Vergleichsstudien [International comparative studies]. In E. Kiel & K. Zierer (Eds.), *Basiswissen Unterrichtsgestaltung – Unterrichtsgestaltung als Gegenstand der Wissenschaft* (Vol. 2, pp. 31–49). Hohengehren, Germany: Schneider Verlag.

Flechsig, A., Knemeyer, J. P. & Marmé, N. (2017). *Schreibtraining - Wissenschaftliches Schreiben im naturwissenschaftlichen Unterricht* [A training on writing – scientific writing within science education contexts]. Doi:10.13140/RG.2.2.22283.69923

Geißler, K. A. (Ed.). (1985). Pädagogisch-psychologische Grundlagen für das Lernen in Gruppen: Lernen in Seminargruppen [Pedagogical-psychological basics for learning in groups: Learning in seminar groups]. Tübingen: DIFF.

Gräber, W. (1999). "Scientific Literacy" - Naturwissenschaftliche Bildung in der Diskussion ["Scientific literacy" – Science education in discussion]. In Qualitätsentwicklung im naturwissenschaftlichen Unterricht. Fachtagung am 15. Frankfurt am Main: GFPF.

Heise, M. (2009). *Informelles Lernen von Lehrkräften: ein Angebots-Nutzungs-Ansatz* [Informal learning of teachers: An offer-value-approach]. Münster: Waxmann.

Hessisches Kultusministerium. (2017). Retrieved from https://la.hessen.de/irj/HKM_Internet?rid=HKM_15/HKM_Internet/nav/475/47570bc2-4148-5431-79cd-aae2389e4818%26_ic_uCon=f2640339-835f-5431-79cd-aae2389e4818.htm&uid=47570bc2-4148-5431-79cd-aae2389e4818

Höttecke, D. (2010). Forschend-entdeckender Physikunterricht. Ein Überblick zu Hintergründen, Chancen und Umsetzungsmöglichkeiten entsprechender Unterrichtskonzeptionen [Inquiry based physics education. An overview on background, chances and possibilities for implementation of corresponding teaching concepts]. *Naturwissenschaften im Unterricht Physik*, *21*(119), 4–12.

Jannack, V., Seeberg, S., Knemeyer, J.-P., & Marmé, N. (2011). Windhösel Kraftwerke - ein Unterrichtsprojekt zum Thema Regenerative Energien [Windhösel powerplants – a science project for schools on sustainable energies]. *Naturwissenschaftliche Bildung als Beitrag zur Gestaltung partizipativer Demokratie. Jahrestagung der GDCP in Potsdam*, *2010*, 312–314.

Klusmann, U., Kunter, M., Trautwein, U., & Baumert, J. (2006). Lehrerbelastung und Unterrichtsqualität aus der Perspektive von Lehrenden und Lernenden [Teacher's burden and quality of teaching from a perspective of teachers and learners]. *Zeitschrift fur Padagogische Psychologie*, *20*(3), 161–173. doi:10.1024/1010-0652.20.3.161

Kultusminister Konferenz. (2005a). Aktivitäten der Länder zur Weiterentwicklung des mathematisch-naturwissenschaftlichen Unterrichts [Activities of the German states concerning a further development of math and science education in schools]. *Beschluss der Kultusministerkonferenz vom 18.03.2005*. Retrieved from http://www.kmk.org/fileadmin/Dateien/pdf/PresseUndAktuelles/2000/nawi.pd

Kultusminister Konferenz. (2005b). *Beschlüsse der Kultusministerkonferenz. Bildungsstandards im Fach Physik für den Mittleren Schulabschluss (Jahrgangsstufe 10)* [Decisions of the Standing Conference of the Ministers of Education and Cultural Affairs of the Länder in the Federal Republic of Germany. Educational standards for physics education at the intermediate secondary level (grade 10)]. Retrieved from https://www.kmk.org/fileadmin/Dateien/veroeffentlichungen_beschluesse/2004/2004_12_16-Bildungsstandards-Physik-Mittleren-SA.pdf

Kultusminister Konferenz. (2017). *The Standing Conference of the Ministers of Education and Cultural Affairs of the Länder in the Federal Republic of Germany*. Retrieved from https://www.kmk.org/kmk/information-in-english.html

Leisen, J. (2008). Lesen ist schon schwer genug, dann auch noch Schreiben? [To read is difficult enough yet, then is writing to be added?]. *Naturwissenschaften im Unterricht. Physik*, *19*(104), 4–10.

Lipowsky, F. (2009). Unterrichtsentwicklung durch Fort- und Weiterbildungsmaßnahmen für Lehrpersonen [Development of instruction through professional development measures for teachers]. *Beiträge zur Lehrerbildung*, (27), 346-360.

Lipowsky, F. (2010). Lernen im Beruf. Empirische Befunde zur Wirksamkeit von Lehrerfortbildung [In-service teacher trainings. Empirical results on the efficacy of teacher education programmes]. In F. H. Müller (Ed.), *Lehrerinnen und Lehrer lernen. Konzepte und Befunde zur Lehrerfortbildung* (pp. 51–70). Münster, Germany: Waxmann.

Lipowsky, F. (2013). Lehrerfortbildung neu und weiter denken. *AFB-Expertentagung zur Reform der Lehrerbildung in Hessen*. Retrieved from https://assets05.hessenspd.net/docs/doc_45511_2013731128.pdf

Lortie, D. C. (2002). *Schoolteacher: a sociological study*. Chicago: University of Chicago Press.

Marmé, N., Glaser, J., Graf, S., & Welzel-Breuer, M. (2017). Schüler-Forschungsthemen im EU-Projekt Chain Reaction [The EUPRBs of the European project Chain Reaction]. In C. Maurer (Ed.), *Implementation fachdidaktischer Innovation im Spiegel von Forschung und Praxis. Gesellschaft für Didaktik der Chemie und Physik, Jahrestagung in Zürich 2016* (p. 540). Universität Regensburg.

Marmé, N., & Knemeyer, J. P. (2015). Kosmetikprojekt Marmétics verbindet außerschulisches Lernen mit dem regulären Schulunterricht [Marmétics combines informal learning and formal learning in the subject Sachunterricht]. *Naturwissenschaften im Unterricht. Chemie*, *26*(147), 42–43.

Menrath, M. (2014). *Schülerbefragung zu Inquiry Based Science Education im Rahmen eines EU-Projektes* [Student survey on inquiry based science education within the Chain Reaction project] (Unpublished thesis). Ruprecht-Karls-Universität Heidelberg, Germany.

OECD. (2014). *TALIS - The OECD Teaching and Learning International Survey*. Retrieved August 12, 2014, from http://www.oecd.org/edu/school/talis.htm

Rieser, S., Stahns, R., Walzebug, A., & Wendt, H. (2016). Einblicke in die Gestaltung des Mathematik- und Sachunterrichts. In TIMSS 2015. Mathematische und naturwissenschaftliche Kompetenzen von Grundschulkindern in Deutschland im internationalen Vergleich. Münster: Waxmann.

Rocard, M. (2007). *Science Education NOW: A renewed Pedagogy for the Future of Europe*. Brussels: European Commission. Retrieved from https://ec.europa.eu/research/science-society/document_library/pdf_06/report-rocard-on-science-education_en.pdf

Ryan, R. M., & Deci, E. L. (2017). *Self-determination theory: basic psychological needs in motivation, development, and wellness*. New York: Guilford Press.

Seeberg, S., Jannack, V., Knemeyer, J. P., & Marmé, N. (2011). Marmétics - ein interdisziplinäres Kosmetikprojekt in der virtuellen Lernstadt Lucycity [Marmétics – an interdisciplinary project oft he virtual learningspace Lucycity]. *Naturwissenschaftliche Bildung als Beitrag zur Gestaltung partizipativer Demokratie. Jahrestagung der GDCP in Potsdam*, *2010*, 306–308.

Wahl, D. (2013). *Lernumgebungen erfolgreich gestalten: vom trägen Wissen zum kompetenten Handeln* [Successfol realization of learning contexts – from inert knowledge to competent action]. Bad Heilbrunn: Verlag Julius Klinkhardt.

Welzel-Breuer, M., Marmé, N., Graf, S., Glaser, J., & Krebs, A.-K. (2017). Chain Reaction: ein FP 7-Projekt und seine Ergebnisse - Die Herausforderung: forschendes Lernen in den Unterricht integrieren [Chain Reaction: an FP7 project and its results – the challenge of integrating inquiry-based learning in science lessons]. In C. Maurer (Ed.), *Implementation fachdidaktischer Innovation im Spiegel von Forschung und Praxis. Gesellschaft für Didaktik der Chemie und Physik, Jahrestagung in Zürich 2016* (p. 536). Universität Regensburg.

ENDNOTE

1 Clubs or AGs (Arbeitsgemeinschaft) are voluntary in nature and are usually founded by a teacher with interested students engaging in certain, often extracurricular activities.

Chapter 8
IBSE Training Feedback and Its Impact on the Design of the Next Training Program:
Insights for Trainers

Kallia Katsampoxaki-Hodgetts
University of Crete, Greece

Stylianos Terzakis
University of Crete, Greece

Nikolaos Chaniotakis
University of Crete, Greece

ABSTRACT

An inquiry science-based education is commonly followed in a variety of educational contexts around the world and is a key parameter in various national curriculum guidelines. The impetus of this chapter is to record the initial and final reactions of science teachers participating in a series of one-year action research and training program that took place in the University of Crete (UoC) in 2013-2016, identify their perception of the first training course, and explore the impact this data had on the program's redesign for the following training session by the technical board. Teacher reactions and responses regarding what they thought had, and had not, worked well in their classes were taken into account prior to re-designing the training program that the new teachers were going to join the following year. Looking into the general benefits as well as challenges, the authors also examined the overall effect of the UoC IBSE training program to participants as reported by both students and teachers.

DOI: 10.4018/978-1-5225-5439-4.ch008

INTRODUCTION

Inquiry Science Based Education (IBSE) is commonly followed in a variety of educational contexts around the world and is a key parameter in various National Curriculum guidelines (Katsampoxaki-Hodgetts et al, 2014, National Research Council, 2012). Despite its conceptual underpinnings being embraced by many teachers and trainers as a means of fostering critical thinking, empowering the students with science related skills and increasing students' preference for science-related careers (Saltiel, 2012) and despite evidence that inquiry can address the needs of diverse students (Atwater, 1994), there are still limited numbers of testimonials that the Chain Reaction IBSE case-based scenarios could provide a ready-made "recipe for success" applicable to almost everyone.

On the one hand, traditional science teaching concepts often make use of compartmentalised input such as data, formulas, equations or theories, that are difficult for students to understand and use appropriately in the right context (Trna, J. et al., 2012). As such misconceptions of scientific concepts and memorisation of theories and practices that are not linked to everyday life take their toll on student motivation and their interest in the studying of science (Osborne and Dillon, 2008). On the other hand, the effect of collaborative inquiry in science education has been well documented in the literature. When students use inquiry-based student-centred approaches that emphasise students' own ideas and questions (Colburn, 2000), they seem to develop team-working skills and skills for critical thinking, problem solving, evaluation and communication which are integrated in their inquiry. Students conduct research similar to scientists and their scientific literacy skills appear to be improving (Gormally et al., 2009) despite pedagogical and conceptual challenges (Bell et al, 2010).

Although there is extensive literature taking into account student feedback in re-formed curricula (Kazempour et al., 2012) and considerable literature indicating the difficulty of assigning new roles to science teachers (Abd-El-Khalick, F., 2004, Crawford, 2000; Crawford, 2007; Capps and Crawford, 2012), there seems to be limited literature regarding the impact of teacher feedback and its role to the implementation of new IBSE training programmes. For example, Trna J. et al .(,2012) provide a comprehensive description of their training programme but they fail to go beyond that; there is little reflection on their training programme and their professional development input and approaches. In the same vein, Anderson (2007) and Abd-El-Khalick et al. 2004 focused on inquiry as content and subject-matter to be learned. A similar approach is followed by Juntunen and Aksela (2013) who describe their framework of in-service training course in terms of how the input was delivered to the students with little reference to the effectiveness of the training impact on teachers' feedback and how these feedback was used for training purposes.

The impetus for our study is to record initial and final reactions of the science teachers participating in a series of one-year action research and in-service training programmes that took place in the University of Crete (UoC) in 2013-2016, identify their perception of the first training course and explore the impact this data had on the teacher training programme re-design by the Technical Board. Bianchini and Colburn's (2000) conceptual framework in which the teacher acts as a researcher, Jeanpierre et al (2005) professional development characteristics with a focus on both content and process, and Vygotsky's (1978) social constructivism in education were instrumental in designing our in-service teacher training programme in which teachers had to convey the nature of science effectively to students.

Teacher reactions and responses regarding what they thought had, and had not, worked well in their classes were taken into account prior to re-designing the training programme for the new teachers that were to join the following year. Looking into the general benefits as well as the challenges, we also examined the overall effect of the UoC IBSE training programme to participants as reported by both students and teachers. Then, in an attempt to be in sync with teacher's beliefs and needs, we re-constructed our training programme taking into account constructivist, cultural and cognitive perspectives (Keys and Bryan, 2001).

CONTEXT

The Greek educational system is divided into three broad levels, primary, secondary and tertiary. Secondary education comprises two stages: Gymnasio (variously translated as Middle or Junior High School), a compulsory three-year school, after which students can attend Lykeion, an academically-oriented senior high school (Moschochoritou et al, 2014).

As such, the target age group (14-16 year old) for the *Chain Reaction Project* (2013-2015) would involve students in the last year of Gymnasio (Table 1) and the first year of Lykeion (Table 2).

Teaching Science in the Greek Junior High School (Gymnasio) and Senior High School (Lykeion)

The first three years of Gymansio are compulsory and attended by 12-15 year old students. Science classes are taught by teachers who are deemed to be part of a generic science field including Physics, Chemistry, Biology and Geography (Table 1).

Physical science subjects are to be accompanied by a minimum number of laboratory activities per course. They are generally considered to be discrete subjects and are taught accordingly. Yet, in 2003,

Table 1. STEM in the Greek educational system; Gymnasio

Gymnasio	Physics	Chemistry	Biology	Geography
Year 1	1 hour/week	-	2 hours/week	2 hours/week
Year 2	2 hours/week	2 hours/week	-	2 hours/week
Year 3	2 hours/week	1 hour/week	2 hours/week	-

Table 2. Science courses available in Lykeion

	Physics	Chemistry	Biology	Project
Year 1	3 hours/week	2 hours/week	1 hour/week	2 hours/week
Year 2	2 hours/week	2 hours/week	1 hour/week	2 hours/week
Year 3	3 hours/week	2 hours/week	2 hours/week	-

a cross thematic curriculum framework for compulsory education supplemented the General Education programme (1999). In this, "a holistic approach to content learning should be followed, whereby cross-disciplinary connections and relationships rather than delineations between academic disciplines are promoted." (Greek Pedagogical Institute 2003).

In this vein, emphasis is not to be placed on the content knowledge or products but on the processes or skills and competencies through which academic knowledge can be acquired. As such, there is a plethora of resources promoting cross-disciplinary lesson plans and materials provided online by the "Digital School" resource bank (Digital School, Greek Ministry of Education). Yet, it is still up to the teachers' discretion whether these practices are implemented or not due to lack of formalised observations during lessons and formalised teacher evaluation.

As far as the teaching and learning model is concerned, the majority of lesson plans and practices remain teacher-centred and content based. Essential competencies such as critical thinking and inquiry learning are often viewed with suspicion as there is little training or experience to rely on. As such, students and teachers alike tend to reproduce the content knowledge as presented in the school textbooks. Regionally, school advisors often organise workshops and seminars in order to suggest and promote student-centred models including group work and inductive learning. Yet again, their application depends on the discretion and the disposal of the teachers themselves (Moschochoritou et al, 2014).

Regarding the implementation or incorporation of new technologies and student-centred approaches, no quantitative or qualitative data is available. Sadly, evaluation is still considered to be "a weak point" in the Greek education system (Papagueli-Vouliouris, 1999:1). Teacher promotion is based on seniority (Papagueli-Vouliouris, 1999) and training opportunities as opposed to class observations and teacher external assessment. This could explain why there is a widely held impression that the traditional school still prevails allowing little room for the implementation of anything different.

Another key feature in the teaching of physical sciences is experimentation. In Gymnasio there is only one laboratory available to all students regardless of student numbers, year or grade. Equipment ranges from basic to satisfactory depending on the financial budget of each school, the initiatives and responsibility of which are taken by the relevant Board of each school. Fortunately, this is not the case in Lykeion, where laboratories are funded by the EU and there are set standards for an up-to-date physical science laboratory.

Laboratory student and teacher training is also provided by EKFE centres (Physical Science Laboratory Centres). In these centres, students and teachers can be trained and supported in order to design, set up and conduct experiments in the laboratory. Operating on a regional basis, EKFE centres provide a variety of training to permanent and newly appointed teachers and organise an exceptional range of experiments and seminars aiming at a continuous renewal and updating of professional training. Video demonstrations of experiments can be found on various EKFE sites online.

Generally speaking, however, laboratory work in Gymnasio is rarely satisfactory as teachers often resort to demonstrations of laboratory work due to inadequate time for experimentation, an overburdened timetable with limited science teaching hours and poor infrastructure. Teacher reluctance and the Ministry's perfunctory interest regarding student experimental time and work does not encourage the incorporation of new laboratory practices. As such experimentation is yet again up to the personal beliefs and tenacity of the teacher.

During the three years of Lykeion, given that the average number of students in a class is 27, the curriculum is academically oriented and seeks to prepare students for Higher Education entry examinations. The prevailing teaching model is also teacher-centred and didactic with experiment demonstrations by the teacher being very common. Yet, despite the discouraging conditions and limited training opportunities, and depending on the ability, tenacity and pedagogical content knowledge of the teacher, Lykeion teachers often create a learning environment that allows critical thinking and student cognitive development to take place. Students do not only observe experiments. They can identify, classify, question, make projections and explain the reasons for their choices carry out experiments when appropriate and record their observations for later review.

In the cross-thematic curriculum in Lykeion, "project work" was introduced and established in 2011 two years prior to the implementation of the Chain Reaction Programme which may explain some teachers lack of experience with IBSE. During project work, students in groups of five implement collective projects on topics of their interest. It is expected that learning is student-centred and that students discover connections between content knowledge and their everyday experiences. Suggested topics are related to the content covered and they last for about four months while activities aim to deepen student understanding of content knowledge and develop their cognitive and social skills and attitudes, necessary for their active involvement in future life.

It is worth noting that in the last two years of Lykeion, students choose an arts or science orientation having little or no interest in General Education science courses. Consequently, low student motivation and interest renders the teaching of physical sciences even more difficult.

RATIONALE OF CHAIN REACTION INITIAL IMPLEMENTATION

Participants

In each year of the three year programme ten science teachers from a variety of science disciplines voluntarily joined the Chain Reaction programme. Participating teachers had mainly majored in Chemistry (40%), Biology (30%), Physics (20%) and other science related subjects. At the programme onset, only four out of ten participating teachers had experienced similar inquiry focused professional development (PD) courses, four out of ten had participated in more than two inquiry based courses and three out of ten had no prior experience of inquiry focused professional development.

There were also 250 students participating in Chain Reaction, 60% of which were female.

Reaching Out

Teacher recruitment was addressed from a variety of different routes. Initially we informed the central offices of the Secondary education, we announced our project via social media, general e-mails, or directed e-mails to schools, school principals, individual teachers, and existing contacts. In fact, the encouragement of school principals/advisors made a noticeable difference in attracting the attention of teachers and convincing them to participate to the programme and their contribution was essential in making sure that enough teachers and schools participate in Chain Reaction each year of the project.

Professional Development Rationale, Teacher Feedback and Follow-Up Modifications

Initial Professional Development: Rationale and Outline

Dissemination/Training events were organised in UoC premises and local schools and included talks and workshops on the five out of the six EUPRBs we had translated (http://www.chreact.gr/?q=node/14). During the professional development (PD) sessions, teachers were acquainted with the project's objectives and teacher expectations as well as general guidelines and practical advice regarding IBSE. In this, the goals, objectives, requirements and expectations of the Chain Reaction project were presented and discussed. Teachers were given a chance to conduct some relevant experiments and briefs regarding soft research skills that foster critical thinking and IBSE. A pertinent example of this can be found in the activities involving the "Six Thinking Hats" designed by Edward de Bono as a tool that promotes discussion, facilitates critical thinking and maximises collaboration. Other activities entailed the importance of making mistakes in science (card game), the importance of peer-feedback and relevant norms, a sample lesson on scientific method with independent/dependent variables and control groups (using sticky notes).

Given that the introduction of "science -related projects" were very recent and an unexplored area in the terrain of Physical Sciences and following a draft preliminary questionnaire to identify teachers' needs, the technical board (TB) assumed that all teachers required initial support, especially for delivering the EUPRBS in the classroom. In fact, this process was mandatory for our teachers in Greece, since most of them did not have any formal training focused on IBSE during their initial teacher education. Each year of the programme, teachers attended two face-to-face training sessions. However, upon completion of the first year programme, evaluation data indicated that teachers required even more scaffolding to support their implementation of the Chain Reaction programme (Katsampoxaki-Hodgetts et al, 2014).

Initial Teacher Feedback on Professional Development

The participating teachers were concerned that our PD sessions did not include a scheme of work with itemised tasks and objectives before deadlines upon which they should plan their action. They needed something more concrete than to simply be provided with scenarios and resources or suggested options of various plans of action. They also asked for an extended list of suggested experiments for each theme (EUPRB) and specific instructions as to how to facilitate experimentation in compliance with IBSE. Another recurring issue was that of the students' knowledge gap and students' hesitation or even refusal to undertake a classroom inquiry on their own (with minimal teacher intervention) and what follow-up actions should be taken.

When teachers described the insights they gained during their participation, we found a range of responses. Most of them highlighted a degree of difficulty regarding IBSE implementation, with some of them feeling confident that they could support an inquiry approach again and some of them suggesting they would not. Students' difficulty in conducting their own classroom inquiries and lack of coordination and planning of their actions reported by teachers were often juxtaposed by the degree of responsibility and meticulousness prevalent in students' work. Also, some teachers found their own need to prepare well before the plenary presentations and discussions as a shortcoming or a weakness, and confided knowledge gaps such as how to design reliable questionnaires and/or how to interpret observations of a sample through a biology microscope.

Initial Professional Development Sessions Modifications Following Initial Teacher Feedback

Taking into account teacher requests and reported challenges, the Technical Board (TB) re-designed the PD sessions to reduce teacher preparation time and coordinate their work. Teachers were provided with a generic yearly scheme of work with learning goals and objectives expected to be achieved by students and teachers alike. Based on the visits to schools, and teacher briefing sessions, we suggested a comprehensive briefing procedure, as described above, in four sessions. First, they had to understand the aims of the general IBSE open inquiry methodology. Following this, they should have a good understanding of the current state of knowledge about the implementation and usefulness of IBSE. Then they had to know the issues that arise during the implementation of an inquiry-based approach to science education.

Additionally, in years 2 and 3, there was an online forum and online course running alongside the face-to-face PD sessions respectively. In each case, teachers were provided with additional substantial support materials supplementing the EUPRBs. (link for the forum http://www.chreact.eu/forum/ link for the online course https://opencourses.uoc.gr/courses/course/view.php?id=348). As part of our scaffolding procedure, the TB, scienctists as role models and peers visited each school twice during the implementation of the EUPRBs in school classrooms. The first time, teachers presented the Chain Reaction programme to the students, they explained what EUPRBs are and how students should undertake the associated tasks. The technical board answered any question that they might have, gave them a presentation describing scientific methods and organised discussions with role models regarding science applications and implications.

The second TB visit aimed at receiving feedback from the students, analysing their progress and in many cases watching presentations of their work up until that point. They also provided feedback while the teachers were present and help with students' presentations, including the style and content in order to prepare them for the national conference. The role model activity included both teacher and student focused sessions. During the student focused sessions, students would ask the role models questions they had prepared at home which often examined the scientist's motives to take up such a professional path, their science ethics, personal setbacks, research challenges and how they overcame them. The teacher focused section involved a presentation of a research project purpose, methods and method selection, results and impacts as well as the challenges encountered by the researcher before meeting their research objectives. The section was of particular importance as it was designed to raise awareness of issues pertinent to critical thinking and inquiry such as the distinction between fact and opinion, source evaluation and reliability and plagiarism, and how teachers can use feedback in order to enhance teaching presence.

Teachers were also encouraged to share their experiences of their IBSE projects both by giving presentations at the National Student Conference (in parallel sessions) and by submitting articles in open-access or peer-reviewed journals disseminating research conducted by their students following the EUPRB guidelines. In fact, concurrent teacher sessions about IBSE practices were organised during the second and third Greek National Express Yourself Conference. In these dissemination events and national conferences, among other teachers and researchers, many Greek teachers involved in the second and third year of the project gave oral presentations focused on their experiences, talking about different issues they encountered during the coordination and implementation of Chain Reaction activities in their classes or EUPRB adaptations they made during lessons.

Eventually, all teachers participating in the Chain Reaction programme presented papers and workshops at the aforementioned conferences. Topics included: IBSE challenges and student misconceptions, IBSE approaches and methods, EUPRB modifications, Inquiry and technology. Some of the speakers drew upon the challenges students and teachers alike faced, and some on the experimental procedures followed in each EUPRB. Prior to their presentations, they had been asked to send an abstract on their talk so that it was available online. Teacher presentations were recorded and subsequently used for the online IBSE course provided by the University of Crete in year 2 of the programme. (See link here: (https://opencourses.uoc.gr/courses/course/view.php?id=348)

The duration of the national conference was a full day and the participating teachers attended lectures regarding the programmes requirements and IBSE practices around the world. They were given the opportunity to explore the materials required for the experiments suggested in the student worksheets and devise an initial plan of action regarding one case-scenario after discussion and elaboration with peers. During the project, they were given guidelines and supporting material regarding IBSE practices. For instance, they were provided with guidelines as to how team members can give constructive feedback and how teachers can pose open questions that facilitate further inquiry. Teachers were also free to collaborate or ask the support of an external science teacher whose main role was to provide peer mentoring. Peer mentors acted as facilitators offering ideas for investigations, experiments and school field trips.

DATA COLLECTION

Before implementing the second year PD, we asked year one teachers some questions in an attempt to understand their views of potentially beneficial changes to the programme; i.e. what worked and what additional support they might need. Twelve (12) open ended questions and three closed-questions were developed to gather the teachers' views through a survey:

1. What did this method have in common with traditional classroom practices?
2. What new skills or knowledge did you acquire by the end of the project?
3. What changes would you suggest for your students to benefit more from a new IBSE project?
4. What did you like most in this project?
5. What did you like the least in this project?
6. What difficulties did your students come across during the project implementation?
7. What difficulties did you (as teachers) come across during the project implementation?
8. What were the project's weaknesses?
9. How did the training board cooperate with you?
10. The instructors gave you advice / ideas / support material useful for the implementation of research work?
11. What kind of support would you prefer (other than the one provided)?
12. If you had the opportunity, would you participate again in Chain Reaction project?

Qualitative data was also gathered in the form of a reflective diary on behalf of the teachers who were asked to report possible barriers and suggestions for improvement.

Alongside the teacher's survey another questionnaire was designed and distributed to students asking for their views of science, the scientific method, teacher contribution to their science education and future aspirations (Katsampoxaki-Hodgetts et al, 2014).

FIRST YEAR CHAIN REACTION FEEDBACK: IMPACT ON STUDENTS, TEACHERS AND SUBSEQUENT TRAINING PROGRAMME

As we came into the project at an early stage of IBSE development in our country, first year involvement provided a real insight into possible difficulties in implementing inquiry approaches in schools. A pleasing response, identified through the survey of teachers, was that they valued their involvement in the programme highly and their observations reflected the objectives of the programme. Despite the challenges, when asked whether they would participate in this project again, the vast majority of teachers (78%) and all (teacher) peer mentors gave a positive response. The teachers also reported that students benefited from conducting an inquiry as it helped them to create a firm understanding of how science is conducted by professional scientists engaged in research and development activities. Additionally, the teachers indicated that group work and peer-feedback seemed to be leading to increased student motivation, and development of inquiry skills such as date collection and analysis skills while improving group solidarity and collaborative working.

Students' latent skills, novel ideas or implementation skills during experimentation was another feature visible in our data. To start with, a comparison between teacher and students' responses gave rise to a range of relevant pedagogical issues:

Just like their teachers during training, students initially did not know exactly what to do unless being explicitly told. They often blamed their teacher for being indifferent or unable to support them as they had not realised how complex research could be. According to their initial views, experiments entailed implementation of instructions provided by their teachers. This perception was changed at the end of the Chain Reaction programme (Katsampoxaki and Chaniotakis 2015). Students and teachers appeared to be more confident to take initiatives and ownership of their "research" projects. In fact, some teachers produced their new EUPRB's (briefings) thus adding to the repository of IBSE briefings.

Research skills and practices during experiments and knowledge of theory were hand in hand and in some cases the former preceded the latter. Students had difficulty finding and evaluating reliable open-access sources. They could not always discern facts from views and interpretations, and they could not evaluate sources as reliable or unreliable. They could not understand University level papers and their lack or prior knowledge often impeded progress as they had to do additional research in order to understand basic but key scientific notions. As they could not make working associations between theory and experiments, and because reliability or replication of experiments were hardly conducted in projects prior to Chain Reaction, teacher feedback was an essential component during inquiry so that students could apply the experimental method in a reliable way. The aforementioned challenges seemed to be overcome after the implementation of the Chain Reaction Programme (Katsampoxaki and Chaniotakis 2015) as students seemed more aware of fine nuances between reliability and bias, facts and opinions and independent and dependent variables.

PROJECT IMPLEMENTATION: HOW THAT FEEDBACK WAS USED TO UPDATE THE TRAINING PROGRAM IN THE SUBSEQUENT YEARS

Taking into account the aforementioned issues we devised a training programme for the second year of the project (Table 3). A major change in our training programme was the inclusion of peers as mentors in the training sessions in attempt to foster collaborative environments and diverse training experiences for the teachers. Along with induction, dissemination events, field trips and in-school events, we also made use of online forums and an online course with guided or unguided discussions so as to maximise the impact of the project and to foster dialogue and debate.

All in all, a series of induction and dissemination events, six workshops and three National Conferences followed by training sessions in order to inform the schools about chain reaction and to recruit 10 teachers from five new schools took place three times a year. Over three years, approximately 450 students and 400 science teachers were involved in the project.

Concurrent sessions on IBSE by science teachers who had already implemented the programme were organised during the National Conferences. In these sessions, 25 of the previous Chain Reaction teachers had the opportunity to share their experience and practices and promote further discussion regarding IBSE teaching approaches and challenges. The conference was open to university students and graduates, pre-service and in-service teachers, parents and the wider community and leading policy makers and science education professionals. Approximately 150 participants attended each National Express Yourself Conference in Greece. All in all, 20 Role Models facilitated the conference organisation and evaluation of student teams.

In-school events were organised and monitored before each National Express Yourself conference in which students who had completed the Chain Reaction briefs presented their work to other students and teachers in their school and at UoC.

Fields trips were also found to be successful. All students involved (n: 450) were given the chance to go on a field trip at least once to interview university researchers, specialise scientists or technicians on site, witness how each site operates and retrieve relevant information regarding their research brief.

In October 2014, teachers were invited to participate in an open access online forum that addressed issues relevant to IBSE. The impact of the online discussion was substantial, especially for the first four weeks of the training programme, as the conversations were read by other secondary teachers that were interested in running IBSE projects in the near future. Topics included: general principles of IBSE, active reading and writing techniques, raising student awareness of experimental method procedures (e.g. identifying independent and dependent variables, using legitimate and reliable resources) and student monitoring and assessment. In fact, up to date, the average number of views in each section of the Forum ranged from 2000 to 3000.

A new open access ten week online course designed by the Technical Board of UOC was for the training of the new science teachers as action researchers. Its syllabus (Figure 1) aimed at fostering critical thinking, empowering participants with real-life science related skills expected to increase students' preference to science-related careers. The rationale behind this new course was to allow teachers to take ownership of their own IBSE content knowledge (CK) and pedagogical knowledge (PK), practices and materials. Teachers were also asked to take a learner role similar to that of their students; they would act as critical thinkers by evaluating existing practices, evaluate current resources and modify recent practices so as to make them a better fit for their own circumstances.

The online course objectives required that teachers attending this course should learn about the general framework and approaches of inquiry learning while examining the experiences of other teachers who had implemented research work previously in their own classrooms. Taking into account a scaffolded knowledge integration framework (Linn et al., 2014) teachers should be able to implement existing teaching scenarios or design and synthesise their own to teach science and offer them for discussion, peer-review and assessment before implementing them in their own classrooms. The course, based on the experience and the material created under the Chain Reaction project, was expected to encourage sustainability of IBSE practices. This course has currently been recognised as one of the best online and distance learning courses available in Greece in 2015/2016 (https://opencourses.uoc.gr/courses/course/view.php?id=348)

In order for the resources to be available to everyone, online teaching and learning materials were made freely available via the project website (www.chreact.gr). The website was updated in 2015/2016 and can be reached by any science teacher who is interested in IBSE briefs and ways of implementation.

STUDENT CONTRIBUTION

Student contribution, ownership and involvement was found to have generally improved.

Examining representative samples of work (Andreadakis 2016), it was noticed that participating students successfully posed scientific questions, recorded observed data, provided written quantitative evidence graphically in support of their findings and drew rational yet tentative conclusions (see Appendix).

Teacher Insights Following Student Response to the Programme

Barriers and Challenges Observed by Teachers

What was of particular importance was a list of important issues reported in teacher diaries (Appendix). Generally, teachers' concerns raised issues pertinent to student lack of prior knowledge, source validity, setting up experiments, avoiding spoon-feeding the students and understanding and using dependent and independent variables during experimentation.

Requisite Prior Knowledge

Another issue that teachers addressed was that student prior knowledge was not always requisite for the tasks at hand. Students came across a number of new scientific and technical terms they were not familiar with. For example, terms that might need to be addressed in just one research brief were: photosynthesis, tropism in plants, light, colour, energy, light intensity, oxygen consumption and dietary needs (Appendix). As such, as well as active reading activities suggested in the general notes, teachers were advised to encourage students to use simple experiments, such as how feeling the warming effect of different light colours with their hands can help relate colour to energy. Likewise, observing direct sunlight during class time could be a starting point using window transparent glass, sunglasses or coloured plastic films. In order to understand photosynthesis, teachers were advised to start by asking students about the parameters that they think could affect plant growth. Teachers noticed many issues that need further clarification before student research.

Source Validity (in Search of Secondary Data)

Source validity appeared to be another barrier during student search of secondary data. When looking for information, students often stumbled upon resources that were not reliable or provided inadequate data. Teachers were advised to design web quests in order to encourage students to cross-check data and raise their awareness of internet safety and criteria they can use in order to evaluate the reliability of sources available online. These were authenticity, objectivity, timeliness, relevance and efficiency following the Kleinnpell et al model (2011) which encourages students to examine whose interests are promoted by a publication, whether information is biased, if it entails an opinion or a fact supported by evidence, whether it is up-to-date or obsolete, and whether it is relevant and useful for students.

Setting Up the Experiments

Teachers also noticed considerable student difficulty in setting up the experiments. In more than one case, students were confused by the vast array of plants to choose from. As research time was limited, bibliographical data on plant nutritional value were found to be a good indication for selection. Alternatively, students could choose plants that were easily accessible to them. Additionally, students were encouraged to use everyday instruments and devices in order to make measurements of the plants' growth and height. They could also take into consideration plant growth under natural and artificial light and note that choice of light source (LED's or Fluorescent light-bulbs) might take up more research time. Ambient temperature was another factor that could be tested but resources available to students did not always allow for successful monitoring or resilient control.

Designing the Method

Another challenge for students was with regards to designing the method. The introduction of the idea of standardization-bench mark was deemed very important. When a suggestion of making a measurement or a simple experiment was made, for instance, teachers should encourage different students to conduct it in several groups so that they can compare their results in the following plenary session. What students needed to notice and discuss was why their results might be similar or different, and what could a possible discrepancy reveal about the actual experiment. Students needed to pay attention to hidden or inferred information. For example, if there are different lamps of the same colour, will they provide the plants with the same energy?

Understanding Dependent and Independent Variables

Moreover, lack of student understanding of dependent and independent variables invariably caused experiment reliability issues. Before students setting up an experiment, teachers should make sure that students were aware of the scientific method and that they can distinguish qualitative and quantitative data. They could use classroom tasks that allow students to practice writing suitable hypotheses and identifying dependent and independent variables and control groups. The use of sticky notes proved to be very effective to break down the different variables. Also, making students aware that scientists should only tweak or change one of the variables at a time is of great importance. At the same time, students should decide on the factors that need to be kept constant during the experiment. Light, water, soil, type of plant and temperature are key parameters determining plant growth. So, for example, to measure the effect of light on plant growth, students should cultivate the same plant in the same substrate with the same amounts of water added and at the same temperature. They should also be made aware that they should be using the same or equivalent instruments to conduct their measurements.

Given the data, students often had to organise them in a table or graph. Students needed help to choose and construct suitable types of graphs. Common mistakes included the lack of legends and captions under Figures, mislabeling the axes of a graph or missing units of measurements.

Keeping Track of Research Progress

Keeping track of research progress also seemed to be a daunting task for students. As there would be multiple groups of students working on different research questions at the same time, teachers reported that they often felt disoriented and lacked control over what was going on. One way to keep track of progress was asking every team to present their experiments, data and conclusions in weekly plenary sessions. Working in small groups, teachers could invite students to share the information they have collected so far and to identify initial patterns. Also, making the most of blogs, students could post summaries of steps taken, experiments, data, setbacks, further questions and possible future actions. No matter what the inquiry path followed, it is important to bring the class together to "check in" and share and discuss the main ideas of the question at hand.

Overcoming an Urge to Spoon-Fed Students

Teacher's also reported their difficulty in overcoming an urge to "spoon-feed" their students. One of the challenges for teachers was deciding how, when, and when not, to intervene. Teachers could use guiding questions to help students draw up their plan of action, monitor their progress and consolidate their knowledge with giving them the answer. A useful resource can be found in the Appendix: [http://www.edu.gov.on.ca/eng/literacynumeracy/inspire/research/CBS_InquiryBased.pdf.]

Asking Questions and Critical Reading

Other major challenges teachers reported included asking open-ended question and provided practice opportunities for students' critical reading. The most significant difficulty entailed effectively implementing open investigations. Students were not accustomed to this teaching method. It would be much easier if they had learned open inquiry and critical reading strategies from other classes to be able to ask more questions and deal with new knowledge critically.

With teachers often prompting students with open-ended questions, students brainstormed and raised numerous research questions which could not be pursued due to time limitations. Students often had difficulty focusing only on some variables at a time. Some groups had so many independent variables that it was difficult to determine causal relationships. Then, there was the question of how many measurements should be taken. They also had to talk about experimental uncertainty; thus, a simplified reading on hard science was used to explain statistical analysis. Any difficulties in analytical chemical methods were overcome with the help of their teacher and the Chain reaction technical board.

Degree of Guidance vs. Taking Own Initiatives

The degree of teacher guidance vs. student taking own initiatives was also debated by teachers. Initially all students decided early on to get involved in the program showing great enthusiasm and full commitment to the subject of research (this is important in relation to global hunger which is a sensitive issue). However, numerous students needed more encouragement to take initiatives. Terzakis (2015) reported that their motto in all the meetings was *''If there is something you know, communicate it. If there is something you do not know, search it.''*

In essence, there should be gradual release from a highly structured investigation to a less guided one and finally towards a completely open type of investigation. For example, students used web quests for the first experimental determination of nitrogen content was considered to be a structured inquiry activity. Following this, additional experiments in their EUPRB's were conducted by students as open inquiry investigations. It was reported that the student brief initial questions and the parallel readings through which students had to take up the role of the researcher really helped towards a smoother student initiation and implementation of the research project. Also, the initial Web-quests gave students good impetus and momentum to greater interest in many aspects of the subject with a holistic and systemic approach.

Making Most of Mistakes, Errors and Setbacks

In an attempt to help students realise that science has progressed based on trial and error and that mistakes should not be penalised but used as pedagogical insights, teachers also used observation, peer discussion and student scaffolding. Terzakis (2015) reported that he tried to find a suitable pedagogical way to deal with student mistakes. For example, during the construction of a hydroponic facility (from the brief "Feed the World by Terzakis"), students initially placed it at a dark shady point and later on they moved closer to the window. Before moving it, students were asked to decide if the distance from the window and the angle of falling light plays a role in the plant growth.

IMPACT OF IBSE TRAINING SESSION AND THE CHAIN REACTION PROGRAMME AS A WHOLE

Despite teacher initial apprehension and hesitation to participate in a forum sharing their teaching practices together with the reported lack of facilities, previous experience and the pressure of a very tight time frame, a number of benefits were identified:

- Multifunctional role of training board and teachers during the training; by using peer-feedback our resources multiplied and teachers were less resistant to feedback from peers
- Effective and transparent communication via forums and online course facilitated teacher work and reduced work load. Teachers could reach the information without having to ask for it.
- Feedback provided by a variety of participants (not just the TB): Teachers and students reported feeling empowered by the whole process.
- Addressing issues before they came up and on the job. Inviting teachers to do action research and present their findings during local conferences improved teacher training as potential problematic issues were reported, addressed and resolved faster.

- Colleagues seemed knowledgeable regarding new technologies and seemed eager to participate in such educational practices.

As a result of these changes, a final Chain Reaction Public Seminar was organised as a two-session event in the PanHellenic Conference on "Teaching approaches and experimental teaching in science education", which took place in Pyrgos, Aristotle University of Thessaloniki, in Greece on 16-17 April. About 200 science teachers attended the event most of them working in secondary schools. Issues highlighted during this session were confirmed by participating teachers' action research during the Chain reaction programme. Participating teachers recognised the beneficial value of this approach, stressing that student experience was the key for success. Accordingly, teachers and students acquired a better sense of accomplishment regarding their ability to coordinate and pursue a science related course of action, respectively, and students' perception of physical sciences improved notably. Other teachers presenting attested that increased cooperation among students and between students and teacher was also observed (Korakaki 2016).

Another science teacher supported that a "bottom-up" planning supported by academic research through action research could lead to a possible involvement of teachers in the development of an educational approach that they themselves would implement (Kelefiotis 2016). Although teacher practices regarding IBSE varied considerably regarding open inquiry, all teachers supported that the briefings provided (EUPRB's) allowed them to practice IBSE successfully but they argued that student improvement can only be noticeable given ample time.

As it was suggested by Stavgianoudakis and Nikolaou's presentations (2014, 2015, 2016), in their second year of Chain Reaction IBSE implementation, students seemed to have more consolidated views regarding their choice of future career. In fact, whilst in year one students were not sure of whether they should follow a science career (44%), the following year there was a considerable increase in favour of it, (35% over 19%), whilst the negative answers remained the same. When students were asked what they liked the most during the project, after a year, they seemed to prefer working in teams (60%) whereas in the second year, they seemed to be divided between working in teams (35%) and presenting their findings in the plenary sessions (30%). More specifically, part of the typical procedure in class during the project was that students would discuss in teams and plenary sessions coordinated by the teachers so as to clarify confusing issues, share research-related updates and plan their next course of action. Interestingly, the percentage of students participating in discussions almost doubled the second year (Question 3: 80%). In addition, regarding student attitude towards mistakes we noticed a clear shift between the two years, 72% and 90% of students attesting that it was not difficult to admit it when they found out there was an error with 72% in 2014 and a surprising 90% in 2015 (Question 4). Another notable difference between the two years was that the second year science teachers' emphasis on improving student scientific literacy, knowledge and practical use of dependent and independent variables, graph and table writing and presentation of their findings. In fact, 74% of students in 2015 over only 16% in 2014 attested the use of graphs and tables during findings. A similar improvement in students' confidence to use science literacy skills after participation in the inquiry labs has also been documented in the literature (National Research Council, 2012).

All the above can be indicative not just of an attitude shift on behalf of the students and teachers but also evidence of student ownership and participation increase due to familiarity with this approach.

THE CHAIN REACTION LEGACY: IBSE TOOLKIT

The Chain reaction consortium compiled a concise IBSE Toolkit containing a brochure and poster. This brochure and poster was used extensively in all subsequent conferences UOC attended. Overall, 1000 brochures were handed out to science teachers in various national conferences in Greece. The kit contained an extensive description of various aspects relevant to IBSE drawing upon the following issues: **"What is IBSE?"** which provides a definitive general framework of the approach. **"How it can be used in class?"**, which encapsulates key steps of implementing and teaching approaches and tasks, a lesson plan and challenges that teachers implementing the EUPRBS are expected to encounter. Additionally, the section **"Communicating science"** provided a succinct guide (template) as to how student can design and prepare oral presentations, slides and scientific posters. The kit also contained online elements. The toolkit was designed to make implementation of IBSE for science teachers manageable, sustainable, easily accessible and cost-effective:

- **Manageable:** It provides a short account of how to deliver a specific EUPRB used as an example, i.e. Green light, breaking up its delivery into smaller steps with think-clouds attached.
- **Sustainable:** Following a similar pattern to that of the delivery, it also provides a short step-by-step account of how a teacher can "create" their own EUPRB.
- **Accessible and Cost-Effective:** This toolkit can be added as a resource to the existing website(s) so that there are no printing and publishing expenses. It can reach everybody at minimal cost.

Such legacy should be promoted by policy makers throughout Europe so that inquiry based learning is accessible, cost-effective and implemented even in the most remote or rural areas.

CONCLUSION

IBSE requires interdisciplinary content, process and pedagogical knowledge for maximum impact. For this reason, open access courses based on the experience and the material created under the Chain Reaction project can now be used to increase the sustainability of IBSE practices in Greece.

As we have already stated in our EU policy report (Katsampoxaki-Hodgetts et al, 2016), based on the qualitative and quantitative data gathered during the Chain Reaction project, a uniform IBSE inquiry policy should be created, disseminated, and implemented within every national curriculum so that standardisation is achieved. This way, student, teacher, and trainer empowerment is facilitated, allowing for high quality and consistent approaches, practices and materials to be disseminated top-down and bottom up.

Training practices should also foster collaborative environments for training teachers and encourage peer-feedback as a key feature. Teachers repeatedly reported that they would feel more confident to discuss setbacks or challenges they faced with a colleague rather than a school advisor. Training approaches should also encourage more teachers to try the method as a lack of familiarity made early rejection more likely (Korakaki, 2015). Teachers should be equipped with all the support necessary to engage in action research as it provides real and relevant insights into what works best in their discipline with students (Kelefiotis, 2015).

Policy makers might be able to overcome concerns regarding IBSE implementation, lack of extensive training, and minimal top-down support, if more teachers are empowered and trained in order to create their own inquiry materials and a more cohesive approach is implemented consistently throughout Europe. Implementing inquiry-based practices only for one year is not adequate (Katsampoxaki-Hodgetts and Chaniotakis, 2015) as students who took part in it for two years were found to have much more consolidated views regarding the nature of science, the role of scientists and the scientific method. All in all, teachers involved in this venture reacted favourably to the involvement of real researchers in common school practices (Katsampoxaki-Hodgetts, Booth and Chaniotakis 2016) and suggested that Teacher Training Professional Development similar to that provided by Chain reaction stakeholders and top-down support should be provided in schools on a regular basis.

REFERENCES

Anderson, R. (2007). Inquiry as an organising theme for science curricula. In S. Abell & N. Lederman (Eds.), *Handbook of research on science education* (pp. 807–830). Mahwah, NJ: Lawrence Erlbaum Associates.

Atwater, M. M. (1994). Research on curricular diversity in the Classroom. In D. L. Gabel (Ed.), *Handbook of Research on Science Teaching and Learning*. New York: Macmillan.

Bell, T., Urhahne, D., Schanze, S., & Ploetzner, R. (2010). Collaborative Inquiry Learning: Models, tools, and challenges. *International Journal of Science Education*, *32*(3), 349–377. doi:10.1080/09500690802582241

Bianchini, J., & Colburn, A. (2000). Teaching the Nature of Science through Inquiry to Prospective Elementary Teachers: A Tale of Two Researchers. *Journal of Research in Science Teaching*, *37*(2), 177–209. doi:10.1002/(SICI)1098-2736(200002)37:2<177::AID-TEA6>3.0.CO;2-Y

Capps, K., & Crawford, B. A. (2012, August 26). Inquiry-based instruction and teaching about nature of science: Are they happening? *Journal of Science Teacher Education*.

Colburn, A. (2000). An Inquiry Primer. *Science Scope*, *23*(6), 42–44.

Crawford, B. A. (2000). Embracing the essence of inquiry: New roles for science teachers. *Journal of Research in Science Teaching*, *37*(9), 916–937. doi:10.1002/1098-2736(200011)37:9<916::AID-TEA4>3.0.CO;2-2

Crawford, B. A. (2007). Learning to Teach Science as Inquiry in the Rough and Tumble of Practice. *Journal of Research in Science Teaching*, *44*(4), 613–642.

Gormally, C., Brickman, P., Hallar, B., & Armstrong, N. (2009). Effects of Inquiry-based Learning on Students' Science Literacy Skills and Confidence. *International Journal for the Scholarship of Teaching and Learning*, *3*(2), 16. doi:10.20429/ijsotl.2009.030216

Hofstein, A., & Tuan, H. (2004). Inquiry in Science Education: International Perspectives. *Science Education*, *88*(3), 397–419. doi:10.1002ce.10118

JeanPierre, B., Oberhauser, K., & Freeman, C. (2005). Characteristics of Professional Development That Effect Change in Secondary Science Teachers' Classroom Practices. *Journal of Research in Science Teaching*, (42), 668–690.

Juntunen, M., & Aksela, M. (2013). Life-cycle analysis and inquiry based leaning in chemistry teaching. *Science Education International Journal.*, *24*(2), 150–166.

Katsampoxaki-Hodgetts, K., Booth, J., & Chaniotakis, N. (2016). *Chain reaction outcomes on EU IBSE policy*. Retrieved from https://drive.google.com/file/d/0BwmqSk-8ziILdTA1TmJtRzlFU2M/view

Katsampoxaki-Hodgetts, K., & Chaniotakis, N. (2015). IBSE Case Study: Student perceptions of science and inquiry after a two year IBSE project In a Greek secondary school. *Scientix National Conference, ScienceView Proceedings.*

Katsampoxaki-Hodgetts, K., Fouskaki, M., Siakavara, K., Moschochoritou, M., & Chaniotakis, N. (2015). Student and Teacher Perceptions of Inquiry Based Science Education in Secondary Education in Greece. *American Journal of Educational Research*, *3*(8), 968–976.

Kazempour, M., Amirshokoohi, A., & Harbour, W. (2012). Exploring Students' Perceptions of Science and Inquiry in a Reform-Based Undergraduate Biology Course. *Journal of College Science Teaching*, *42*(2), 38–42.

Keys, C. W., & Bryan, L. A. (2001). Co-Constructing Inquiry-Based Science with Teachers: Essential Research for Lasting Reform. *Journal of Research in Science Teaching*, *38*(6), 631–645. doi:10.1002/tea.1023

Kleinpell, Ely, E. W., Williams, G., Liolios, A., Ward, N., & Tisherman, S. A. (2011). Web-based resources for critical care education. *Critical Care Medicine*, *39*(3), 541–553. doi:10.1097/CCM.0b013e318206b5b5 PMID:21169819

Linn, M. C., Bell, P., & Davis, E. A. (2004). *Internet environments for science education.* Lawrence Erlbaum Associates Publishers.

Moschochoritou, R., Siakavara, K., & Katsampoxaki-Hodgetts, K. (2014). *How sciences are taught in Greece*. University of Crete. Retrieved from http://www.chreact.gr

National Research Council. (2012). *A Framework for K-12 Science Education: Practices, Crosscutting Concepts, and Core Ideas. Committee on a Conceptual Framework for New K-12 Science Education Standards. Board on Science Education, Division of Behavioural and Social Sciences and Education.* Washington, DC: The National Academies Press.

Osborne, J., & Dillon, J. (2008). *Science education in Europe: Critical reflections.* London: Nuffield Foundation.

Saltiel, E. (2005). *Inquiry-Based Science Education: Applying it in the Classroom.* Methodological Guide.

Trna, J. (2012). Implentation of inquiry based science education in science teaching training. *Journal of Educational and Instructional Studies in the World* , 2 (4), 2146–7463.

Vygotsky, L. S. (1978). *Mind in society.* Cambridge, MA: Harvard University Press.

APPENDIX

Example of Issues Reported in Teachers' Diaries:

The following sections are more examples of issues that participating teachers felt they should clarify before student research.

1. Photosynthesis: Photosynthesis is the process used by plants, algae and certain bacteria to harness energy from sunlight and together with some basic chemical elements produce more complex compounds. Oxygenic photosynthesis is the most common and is seen in plants, algae and cyanobacteria. In this, light energy transfers electrons from water (H_2O) to carbon dioxide (CO_2), which produces carbohydrates. In this transfer, the CO_2 is "reduced," or receives electrons, and the water becomes "oxidised," or loses electrons. Ultimately, oxygen is produced along with carbohydrates.
2. Days and nights in space: Light is needed to power the plants for production of organic energy molecules. Light dependent reactions (light reactions) are initiated by coloured pigments, mainly green coloured chlorophylls. Light-independent reactions (dark reactions) make use of these organic energy molecules providing the energy to fix the CO_2 (carbon dioxide) into carbohydrates. These facts must be taken into consideration when planning the experiment.
3. Tropism: Tropism is a biological phenomenon, indicating growth or turning movement of a biological organism, usually a plant, in response to an environmental stimulus. Phototropism, movement or growth in response to lights or colours of light. Geotropism (or gravitropism), movement or growth in response to gravity.
4. Light spectrum and energy: Visualisation of light spectrum can be useful in understanding the differences between different colours. Light can be separated using a prism.
5. UV vs Visible vs IR wavelengths: The UV part of the spectrum carries more energy than that of the visible spectrum. If used though, it can damage the plants. It can also damage human skin. Plants must be chosen so that they can utilise safe areas of the light spectrum. The atmosphere of the earth absorbs most of the UV radiation. Outside the spaceship there is abundant UV radiation. This can be used for some processes that require a lot of energy.
6. Dietary needs of plants: Plants, like humans, require food. Plant food is different than that of humans. Plants use simple elements or compounds to build the complex food required by humans. As such, human waste can often be used to feed plants. Urine contains a lot of ammonia, while human excrements contain organic nutrients and other micronutrients the plants can reuse.
7. Dietary needs of humans: What are the minimum food requirements for survival of people is another possible research question. Each living species requires different kinds and different amounts of food sources and facilitate or hinder energy conservation in space. In fact, the type of activity, the environment, size, and even gender play a big role. For instance, smaller humans usually spend less energy. Additionally, unwanted activities can be minimised to save energy.
8. Selection of plants as potential food source: Humans require a variety of foods and nutrients. Different plants require different food sources, and produce different foods for humans. There is a great need for the best possible selection of plants to be used. Measurement of biomass might be an indication of that. Plants should provide as much food as possible for humans with minimum energy or light required.

9. Recycling in space: There are elements that can be recycled and other that are not. What is the difference? How to recycle? What are possible dangers or **setbacks?** Students need to understand the difference between an element and a compound. Compounds can be built from elements. They require energy to be built; by plants for example. They release energy when used as food. They break down into simpler compounds or sometimes elements. Bond energy is turned into fuel for survival. Water can be recycled via the atmosphere of the spacecraft, human urine or wastewater. Water does not break down. Not all types of water can be recycled. Some are used for the synthesis of other compounds. Oxygen is used to oxidise food and release energy. Some of it is turned into carbon dioxide. Carbon dioxide can be turned into oxygen again by photosynthesis, and thus it can be recycled. Organic compounds, such as sugars made by plants, cannot be recycled. They break down into simpler compounds (usually carbon dioxide and ammonia) and they release energy. Plants recycle CO_2 and NO_3^- to build new compounds with the help of the light.

Representative sample of Student work by Andreadakis 2016.

Astral-Producers Team: Studied Protein Containing Plants in Natural and Artificial Light and in the Dark for Human Survival in Space

Student Scientific questions:

- What is the nutritional value of the proposed plant ?
- How does the wavelength of light affect photosynthesis ?
- Is artificial (LED) or natural light more effective for photosynthesis?

Summary of Work

This space investigation studied what plants can provide a space crew of eight oxygen atoms, rich in vitamins and energy and how these affected plants from the natural light, darkness and various artificial light shades. Upon studying the literature, students selected two plants of high nutritional value; lentils and beetroot. Then, they measured the growth of plants in natural light, in the dark and artificial red and green light. The analysis of the measurements and our observations showed that plant growth varies depending on the light obtained as the optimum growth had plants in natural light and artificial light while in darkness growth was retarded and plants wilted and died. For this space mission, lentils are suggested as a better option. It is a plant rich in protein and vitamins and a plant which grew in both natural and artificial light.

Online Course Syllabus

See Figure 2.

Modifications of Course Design After Teacher Feedback

See Table 3.

Figure 1. Sample of student abstract for National Conference

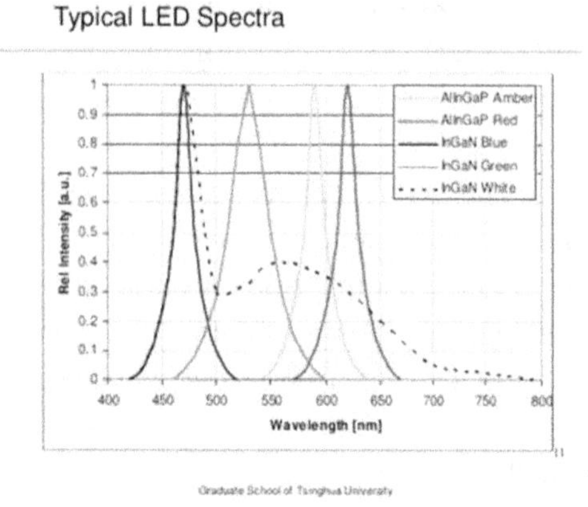

Figure 2. Syllabus of online IBSE teacher training course

Syllabus

Week 1	Induction week: Survey enquiring on teacher perception on IBSE, nature of science and scientific method.
Week 2	IBSE framework and rationale.
Week 3	The importance of posing scientific questions. Micro-teaching: how to train your students so that they grasp the notion of (in)dependent variables & control group.
Week 4	Scaffolding using the EUPRB model so as to help teachers design their own EU PRB scenario and brief.
Week 5	Evaluation and analysis of recent IBSE implementations reported by teachers.
Week 6	Formative assessment: how to assess all learning experiences/ stages.
Week 7	Evaluation of sources / internet information.
Week 8	Design challenge: design a one to two hour IBSE lesson.
Week 9	Management of technical problems impacting IBSE implementation and suggested solutions.
Week 10	Technology and IBSE: Evaluation of suggested resources.

Table 3. Modifications to the initial programme

Elements of the programme that changed	Training Programme 2013/2014	Training Programme 2014/2015
	Dissemination Event	Dissemination Event
	Induction seminars by Peer Mentors	Induction seminars by TB and Peer mentors
Training **Source of training input**	Input given to teachers by two assigned by the TB Peer Mentors and Technical Board Leaders over the phone, monthly visits and by email.	Input given to teachers through an asynchronous accessible to all forum in which peer mentors included colleagues that implemented the project the year before and the Technical board.
Training **Transparency of feedback and communication**	-	Congruent perceptions of the Nature of Science among teachers of different disciplines, knowledge gaps and misconceptions are already apparent using the online Forum.
Timetable: **guidelines and deadlines**	Role Model Visit and National Conference	Teachers had to discuss every week the reported difficult issues they could have well before these actually happened: 1. organising group work, group dynamics and students norms for civil discussions 2. Critical thinking skills and six thinking hats 3. Identifying students' knowledge gap and what to do next 4. How to design lessons so as to enable students to identify reliable resources 5. Experimental method: In what ways can teachers help students understand the method, variables and control and held them apply it in their context? 6. Field trips: supporting teachers plan and organise them.
Support **More experiments to supplement the existing ones given in EUPRBS**	None or one in each EUPRB	A list of related experiments and instructions was added to each EUPRB link on our Website.
Teacher's knowledge, skills gap	-	Through the Forum, previous participating teachers could ask or give substantive feedback to their colleagues.
Role Models Visit	Later in the project	Earlier in the project
Action Research	Optional and teacher initiated	Teachers are invited and provided with guidelines as to how they can research issues relevant to IBSE.

Chapter 9
Chain Reaction:
The Jordanian context

Anwar Batikhi
Jordan Society for Scientific Research, Jordan

Zeid Al-Bashaireh
Mutah University, Jordan

Oraib Nawash
Jordan Society for Scientific Research, Jordan

ABSTRACT

Due to the huge development in the communication sector, science education has become very challenging. Inquiry-based science education (IBSE) has been proposed by academic leaders and policymakers as having the potential to increase student involvement and provide opportunities for the development of the key skills desired by employers in the future. Thirty teachers from 15 schools participated in the project and disseminated the knowledge obtained to their colleagues through peer-to-peer knowledge sharing. Most teachers expressed an increase in their understanding of IBSE, and 66% agreed that they received high-quality IBSE/EUPRB briefings. Students showed motivation to learn science and obtained scientific approaches and steps of answering their inquiries in order understand their surroundings, improve their achievement of scientific knowledge, build positive science attitudes, improve their understanding of the nature of science, and develop IBSE skills such as scientific data collection analyses, interpretation, and how to provide evidence to test ideas and develop theories.

INTRODUCTION

Recently, due to the huge development in the communication sector and the emergence of smart technologies, science education and its approaches have become very challenging, very demanding and boring to some extent to students. This could be seen in students' lack motivation and low self confidence in learning, persistent achievement gaps in science and mathematics among many student subgroups demographic changes, large class size, broken links with other lessons, poor physical conditions of schools

DOI: 10.4018/978-1-5225-5439-4.ch009

(fewer laboratory opportunities), and the intensive curriculum allowing insufficient time allocation for science education. Thus, Inquiry-Based Science Education (IBSE) has been proposed by academic leaders and policy makers as having the potential to increase student involvement and provide opportunities for the development of the key skills desired by employers in the future. Thirty teachers from 15 schools participated in the project and disseminated the knowledge obtained to their colleagues (approximately 1440) through peer-to-peer knowledge sharing. Most teachers expressed an increase in their understanding of IBSE and 66% agreed that they received very high quality IBSE/EUPRB briefings. Students showed motivation to learn science as they obtained scientific approaches and steps of answering their inquiries in order understand their surroundings, improve their achievement of scientific knowledge, build positive science attitudes, improve their understanding of the nature of science and developed IBSE skills such as scientific data collection and analyses, interpretation, using creative thought, provides evidence to test ideas and develop theories.

In conclusion, The Jordanian participation in the Chain Reaction Project (CRP) was a priceless opportunity to spread the IBSE approach amongst schools in Jordan.

This chapter describes the Jordanian preparation and the procedures of applying Chain Reaction Project teaching and learning materials. It also presents the barriers that face the implementation or the projects' materials and how they are overcome. Moreover, the chapter sheds light on the advantages and the experiences that Jordanian science teachers, science educators, school head teachers and students gained by taking part at this project.

BACKGROUND

This section will highlight the importance of inquiry-based science education (IBSE) in the expansion of student' knowledge and understanding of scientific concepts or phenomena, and their ability to provide learners with scientific justifications and explanations. Moreover, this section discusses the situation of IBSE in Jordanian education system and Jordan's engagement and participation in the project from its early stages.

It can be claimed that teaching science in an interesting, practical and convincing approach for students is an indispensable tool to cope with the rapid development in science and information technology. Supporting that, several research studies, such as: OECD (2006) and Rocard et al.(2007), reported that as students get older, there is a decreasing interest in studying science and that the unsuitable outdated methods of teaching and learning science in schools are in part to be blamed for this Additionally, another study showed that approximately 60% of European students of secondary level stated that science learning is not interesting enough (Ministry of Education, Youth and Sports CR, 2010). Thus, the Inquiry-Based Science Education (IBSE) approach has been proposed by academic leaders and policy makers as having a great potential to increase student involvement and provide opportunities for the development of the key skills desired by employers in the future (Fensham, 1986; Linn, et al., 2006; European Commission (EC) and High-Level Group on Science Education, 2007). The term 'inquiry' indicates generally the process of acquiring or obtaining information by investigation, often personally and voluntarily carried out by the person who is very keen to find the answer to the phenomenon in question. This approach includes investigation of a problem; finding truth or knowledge that requires critical thinking, making observations, asking questions, doing experiments and stating conclusions; and thinking creatively and using insight (Hiang's, 2005). Thus, it can be claimed that students who learn science through inquiry will be able to think beyond concepts and apply them as they conduct science investigations (Figueroa, 2011).

In Jordan, the concept of inquiry-based learning was recently introduced as one of the science teaching tools in schools, yet it is still practiced on a small scale and mostly as an individual choice by interested teachers. The Ministry of Education (MoE) started to realize the importance of this approach and facilitated the implementation of some relevant projects and initiatives to help deal with this issue. Nevertheless, traditional teaching/learning approaches, where students are the receivers while the teacher is the dispenser are still the prevailing experience. Within the context of science teaching, science teachers in Jordan find it difficult in managing IBSE classes, in particular in term of providing the necessary facilities and providing the appropriate educational environment, due to a lack of training in how to implement such teaching and learning materials and the teaching and because learning materials are not not based on IBSE style. However, Qablan and DeBaz (2014) reported that the implementation IBL teaching strategies in Jordan showed that using these strategies broadened teachers' understandings and reinforced their skills in teaching science through inquiry by gaining sophisticated understanding about the role of these teaching strategies on teaching and learning. Therefore, not only does the student benefit from an IBSE approach, the teacher also does. The need is there for the development of training projects to educate teachers first about IBSE in order to keep our students motivated about science learning and to make it smoother for teachers to manage this kind of teaching. Therefore, Jordan Society for Scientific Research (JSSR) took advantage of their participation in the Chain Reaction project to spread the IBSE approach among the Jordanian teachers and students.

JSSR is a non-profit cultural association that was established in 1999 and undertakes various and varied activities to achieve the objectives of supporting scientific research projects, introducing the concept of scientific research in the curricula and the various media through visiting schools and universities, lecturing, holding workshops and holding seminars through which the importance of scientific research in development and development is presented.

JSSR has good connections with the Ministry of Education through cooperation agreements between the two parties and the scientific presentations that were voluntarily presented by the researchers, who are members within the JSSR from different disciplines, to students in schools of different ages. This positive collaborative history facilitated the implementation of the project and paved the way for the participating schools to benefit from the successive training carried out by the project's technical team. Improving the education procedures in Jordan is very challenging particularly after the influx of hundreds of thousands of Syrian refugees, more than half of whom are children at school age and 3% have disabilities. This resulted in overcrowded classes; double-shift schools; increased education costs for the Government of Jordan; hard economic conditions which invariably lead to increase school dropouts and a brain drain as people sought better pay abroad (Geopolitica, 2016). Projects on IBSE, like the Chain Reaction project, are very important to support the whole education process to mitigate the educational disadvantage that the current political situation known as "Arab Spring" brought about. These projects could improve the quality of education and makes it more interesting to learn science within crowded classrooms.

PROJECT PREPARATIONS

This section presents the schools, teachers and scientist role model selection process by the project team (also known as the Technical Board or TB), the introduction of IBSE and the EUPRBs to the teachers and school managements and the workshop to train them to practice IBSE activities. It explains the translation of EUPRBs to the Arabic Language and highlights the Express Yourself National Conference preparation and activities.

The Chain Reaction Project (CRP) focuses on a sustainable approach to IBSE. The Jordan Society for scientific Research (JSSR) took the responsibility to implement, manage, and monitor the project in cooperation with the educational authorities in Jordan.

Chain Reaction is rooted in problem solving and inquiry based learning principles, and aimed to actively involve students in their science studies, and enable them to experience the excitement and challenges of experimental and investigative science, and develop as science active citizens. Over the life cycle of the project (30) teachers, (15) schools were involved directly with the proportion of 10 teachers from every 5 schools. These teachers were then required to disseminate their developing knowledge to colleagues through peer-to-peer knowledge sharing. It was anticipated that this dissemination would in turn reach 1,440 teachers over the three-year life of the project (based on an average of eight teachers per school receiving this briefing).

Firstly, the EUPRBs were translated and modified to the Arabic language by the TB, as Arabic is the official language of teaching in Jordan. The translation process went so smoothly and accurately. Then, the TB contacted high-level management at the Ministry of Education (MoE) to introduce the project to them and guarantee their cooperation and assistance. The MoE agreed to facilitate the implementation of the project in the chosen schools, distributed all over the country. A project management team member signed the agreement between the MoE and the JSSR. After that, the TB, in cooperation with the directorates of education chose five male and female schools, taking into consideration including schools from different geographic spots in Jordan and based mainly on the desire of school to participate. Jordan is divided into 12 governates; 4 in the north (Irbid, Ajloun, Jerash and Mafraq);) in the middle (Amman, Balqa'a, Zerqa and Madaba) and 4 in the south (Karak, Tafeelah, Ma'an and Aqaba). Thus, for the first year of the project, schools from Irbid, Balqa'a, and Amman governorates were chosen. For the second year, five schools were chosen from Amman, Zerka, Ramtha, Karak and Madaba governates. For the third year five schools were chosen from private schools located mainly at Amman, Karak and Madaba governorates.

The TB contacted the head teachers of the targeted schools to guaranteed their commitment to facilitate the project implementation. Each head teacher nominated two science teachers to participate in the development and implementation of the project teaching materials within the classroom setting. Moreover, the TB visited the five schools on site and held a meeting with the head teachers and science teachers in the presence of a science supervisor (who is tasked with supervising science teachers in schools within each directorate of education). The science supervisor was asked to ensure that the science teachers had enough teaching experience to implement the project materials, and the school have enough teaching facilities, science laboratories, and information technology facilities such as PCs, internet etc.

The TB organized four to five meetings with science teachers from the five schools at the JSSR building. The teachers received initial training in how to carry out science lessons based on IBSE, problem-solving and inquiry-based learning methods in Science Education. They were briefed on how to implement the EUPRBs by carrying out several workshops and meetings that took place at the JSSR office in Amman. The EUPRBs were discussed in detail with the science teachers in order to guarantee a proper adoption of the activities in the classroom setting. Moreover, science teachers were asked to cooperate with the school's head teachers to provide time to facilitate the students work and activities as well as with the lab technicians to provide the necessary materials and learning resources.

Additionally, they were required to disseminate their training and knowledge to colleagues at their schools through peer-to-peer knowledge sharing and using existing networks of teachers to establish groups of teachers to share ideas and exchange experiences related to the project. In turn, teachers were able to build their knowledge and skills, learn independently as well as being part of a wider teacher network.

As students' perceptions are crucial, eight graders were chosen to take part in this project. Eighth graders in Jordan are 14 years old. They study science as an integrated programme while ninth grade students, in the Jordanian education system, start to study science as separate subjects (physics, earth science, chemistry and biology). It is expected that the students of this age would be able to understand the aim of the project and the process of adapting the briefs with their teacher's help. It is well known that students at this age have the ability to think critically and collect data, analyze data, and explain findings. They reach the formal operational stage of cognitive development according to Piaget's levels of cognitive development (Rocard et al., 2007). The EUPRBs adapted to match the proposed outcomes of science teaching within the context of the Jordanian curriculum needs; among which are; cultural and curricular needs of pupils, taking into consideration the facilities available in the schools' environment, especially within the rural and remote areas. In Jordan, male and female schools generally are separated, both are administered by the MOE. Also, science subject should not contradict with Islamic religious beliefs, regulations, the society traditions and style of life.

The EUPRBs were also adapted to be in line with the students' previous knowledge and scientific experiences. They were simplified so that the students could understand the concepts and facts and in turn master scientific skills and procedures. However, the PRBs were still based around the proven principles of problem-solving and inquiry-based learning in science and based on materials that have already been tried and tested as part of the original project, described in chapter 2.

The project also engaged practicing scientists (early year science professionals or Ph.D. students) who acted as Role Models for the students and shared their work to inform and encourage them to consider a career in science related fields. The inclusion of Role Models in the Chain Reaction project provided students with an opportunity to meet with practicing scientists and to learn from them in an interactive and informal context. These sessions were intended to give students, not just a first-hand opportunity to learn from scientists, but also an insight into the varied career paths available in science. The Role Models selected in Jordan were scientists, Ph.D. science education students or pure science students. One of them who worked as a science teacher did her research under the supervision of the TB. The other one is a noted researcher in the environmental sciences center at the Royal Scientific Society.

Other Role Models were recruited from local industry, academic and research institutions and Universities, to share their scientific experience with the students. They were asked to visit students at their schools and watch their work and projects and give them support in their work. Thereafter, they briefed students on how they succeeded in their academic and research work, how they encountered and tackled problems they faced, and how they worked under difficult circumstances, such as the lack of resources and materials, financial obstacles, difficulties in transportation and little social support. They were also asked to help the students to prepare their presentations and materials that were presented at the National Express Yourself conference.

The Role Models also acted as impartial experts during the National Express Yourself events. On a national level they decided which group of students would go to the international celebration event, where they issued special commendations to specific presentations. Teachers attended the events with their students and were also encouraged to network with the Role Models. This aimed to actively promote future interactions between schools and the Role Model scientists, with a view to future collaborations and school visits taking place. Role Models were also encouraged to provide feedback to the teachers on the students' work and achievements.

The National Express Yourself conferences were an exciting celebration of science, allowing students to showcase and communicate both their progress and work and to see the work of other schools involved in the project. The student presentations encouraged them to present details about the inquiry processes they engaged in and share findings whilst being creative and imaginative through their presentations.

In Jordan, the first National Express Yourself conference was held on the 5th April 2014, in Jordan University – Amman. The opening ceremony was carried out under the patronage of the Minister of Education. The audience exceeded 300 persons (students, teachers, parents, science professionals and other stakeholders). The preparation for these events started from the beginning of the year by asking each school to present two presentations that students did well during their work. Then the TB contacted the hosting university to facilitate enough places for the conference venue. In parallel, effort was made to provide logistical preparations such as transportation for students and teachers, food, conference materials and conference tables. Five Role Model judges were chosen to evaluate students' presentations and projects. The conference concluded with one school winner who would then participate in the International Express Yourself Conference. These conferences allow a selected group of students and teachers from each of the participating partners to come together on an annual basis to present their projects at a European/International level. The projects selected at the national events were chosen based on both academic and non-academic achievement and in recognition of effort and personal development.

PROJECT IMPLEMENTATION

The TB visited the cooperating schools and met the chosen science teachers and the participating students and discussed with them the aims of the project and the process of adapting the briefs with their teachers help under the TB supervision.

Most EUPRBs are simulation activities, which involve students working as realistically as possible within a real-life scenario. Students were supplied with adequate information that allow them to operate within the simulation. Simulations are 'job-centered' as the emphasis is on carrying out a certain task(s) and students behave according to the task at hand. Simulations provide a framework and philosophy through which students are not only considered as active participants but are given considerable responsibility for their actions and decisions. Within the context of the EUPRBs, simulation enables the students to gain insight into, and awareness of, the roles and activities carried out by science and engineering researchers. They behave as classroom scientists by formulating researchable questions, drawing up hypothesis, collecting and analyzing data, formalizing results discussing results, interpreting results and drawing conclusions. Teachers just followed the translated PRBs and did not create their own.

Implementation in the Classroom

Participating science teachers were asked to introduce the briefs, setting the context for the activity and stimulating student interest and 'readiness' to get involved. They explained the structure of the activity and gave the Student Activity Sheets. They were directed to take on the role of a research director or supervisor - simply giving an overview of what was expected thereby creating a sense of realism and purpose. If there was additional information about the topic it was reported in the background information for teachers, at the end of the Teachers' Notes for each EUPRBs material. The students generally took responsibility for working through the activity. Science teachers were required to coordinate changes during an activity such as arranging groups for investigation tasks, reporting back, etc. They had to facilitate students work by sitting in a discussion, helping move things forward if necessary and giving help and advice when appropriate.

Science teachers encouraged a cooperative approach, to provide a stimulating and creative atmosphere which in turn helped to develop problem-solving and communication skills, as well, as knowledge and understanding of scientific concepts. They facilitated students participation and motivation by using modern science teaching strategies based on constructivism theory such as I think we Think, think per share, round table, project work, concept mapping etc. (Tint & Nyunt, 2015). Such strategies give students an opportunity to be engaged in the learning and the teaching process and implement the PRBs according to problem-solving using an inquiry approach. These strategies were introduced to the teachers during the training session at the beginning of each year of the project.

The TB visited the teachers in their schools to follow up the process of EUPRBs implementation and activities to ensure that they delivered the use of critical thinking, reasoning and problem-solving skills to students. Also, it was crucial to find out if they implemented the EUPRBs activities in a real classroom situation, to ensure that the science practical work was done as was intended and to understand the extent of students' engagement in the IBSE activities. The TB provided support for the teachers and the students in their work and gave them guidance on how they could apply the EUPRBs in the classroom settings, how scientific data could be collected and analyzed and how to interpret data, using creative thought by providing evidence to test ideas and develop theories. The TB member helped students and teachers in their formulation of explanations of the many phenomena that can be developed using scientific theories, models and ideas. Precisely, students were asked to:

- Plan to test scientific ideas, answer scientific questions, or solve scientific problems.
- Collect data from primary or secondary sources, including ICT sources and tools.
- Work accurately and safely, individually and with others, when collecting first-hand data.
- Evaluate methods of collection of data and consider their validity and reliability as evidence.

Inquiry Skills (IS) are important for the continuity of the learning process, where learners can build a deep understanding of the scientific concept or phenomenon under consideration, to expand their knowledge of them and to provide scientific justifications and explanations (Dokme& Aydinli, 2009). Students were asked to practice IS by encouraging them to ask questions to explore the scientific phenomena and expand and deepen their understanding of it and practice scientific research by collecting and analysing data in order to come to answers supported by evidence (Omar, 2015). In summary, students learned how to:

- Recall, analyze, interpret, apply and question scientific information or ideas.
- Use both qualitative and quantitative approaches.
- Present information, develop and argument and draw a conclusion, using scientific, technical and mathematical language, conventions and symbols and ICT tools.

Barriers/Obstacles

This section describes many barriers and obstacles that faced the implementation of the project in the Jordanian schools. These obstacles are related to difficulties on how to facilitate the implementation of the project at the top management level of the Ministry of Education (MoE). For example, the bureaucratic procedures where we had to get the official agreement to carry out the project with the schools belonged to the MoE. We had to contact the director of every Directorate of Education to make them aware of the project, where in turn, the director provided us with a written official letter to the school head teacher, whom we brief on the project aims and procedures. Then we contacted the science regional supervisor to also brief them on the project aims and procedures. Finally, we visited the schools on site to meet with the science teacher to explain the aim of the project and ensure their willingness to work on the project. These bureaucratic procedures were time consuming and thus affected the project implementation by creating delays in presenting the project materials as agreement had to be made from all of these stakeholders. Yet, after several meetings and discussions, stakeholders absorbed the idea and the importance of the project. In addition, some schools hesitated to cooperate when we informed them that this project is supported by the EU. They thought that the implementation of the project materials may have negative effect on their students' school science as they thought that the differences in the education standards and the high education quality in EU would make it difficult for their students to participate and understand the project. This proved to be totally untrue when the project was implemented. Moreover, they thought that the implementation of this project may affect their school timetables. But, the project team work hard to change some of these perceptions about the project and these efforts resulted in their approval to join the project.

As Arabic is the official language of Jordan and the language of teaching and learning, it was hard for students to understand the project EUPRBs, although they were translated and modified to the Arabic language by the Technical Board member (TB). Sometimes they needed to refer to extra references on the subject matter or the procedures in order to carry out their activities. This put extra pressure on the project team to overcome these problems. The language problem also faced many students when their schools won the Express Yourself National Conference as some of them had to work hard to improve their English Language skills.

In term of required facilities, lack of materials, tools and instruments in the school laboratories especially in the remote regional schools was another limitation that faced the implementation of the project. Sometimes, it was so hard to find the necessary materials, tools and instruments. Also, at times, while the laboratory was well-equipped to properly carry out scientific activities and experiments, safety issues were prevalent in many laboratories.

Furthermore, there were obstacles related to the teacher knowledge and experience with IBSE and problem-solving methods of teaching. Teachers were not trained properly on such methods during their pre-and in-service teacher training programs. Some of science teachers were keen to use traditional methods of teaching and held negative attitudes towards the modern teaching methods. Teachers had to receive initial training in problem solving and IBSE, as well as being briefed on how to implement the

EUPRBs through several workshops and meetings that took place at the JSSR office. The project team offered teachers some incentives to take part in these workshops, such as paying for their travels and food to encourage them to participate.

Finally, there were obstacles related to cultural issues and gender. For example, some parents did not allow their daughters to be among the working group at schools as this project was an extra science curricular activity and not a part of the school science textbook. Some parents did not agree for their daughter to travel abroad to attend the international conferences, because they thought that they were too young to travel abroad. Also, some parents held the social traditional belief that girls have no right to travel alone without authorized persons (such as very close relatives).

The Impact of the Project

This section will highlight the impact of the project on the teachers' teaching styles, students' understanding of scientific concepts and the other school teachers who did not participate formally in the project. This section is supported by the results of research carried out as part of the project on the Jordanian student science inquiry skills, as perceived by the participating students.

The results showed that students did show positive cooperation and enjoyed working on the project since they are rarely engaged in IBSE activities outside of the project as they were used to learning science following the conventional way.

The Chain Reaction Project is cyclical in nature. So, for each year the project new group of teachers from different schools were recruited, ensuring the participation of a large number of teachers and students. The project provided a strong and sustainable IBSE framework both for teacher educators and for teachers, along with resources tailored to each partner's individual cultural and curricular needs. This sustainable framework emerged from the fact that the involved teachers empathized that they will continue adopting the IBSE beyond the project period. Also, the results of the project suggest students may adopt this approach during their future careers if the become teachers, scientists or as future parents. The teaching process itself is an influential process, it is hoped that the participating teacher may influence the non-participating teacher to adopt IBSE because of the positive response from students to the IBSE approach.

Implementing IBSE through the Chain Reaction Project showed to have a great impact that astonished all stakeholders of the project. For instance, the outstanding students' performance in planning, innovative and critical thinking, testing and presenting their research project showed that when the proper student-focused education method is used, best outputs are realized. Additionally; the sustainable nature of this kind of project is ensured as education as a process is sustainable by itself that runs and therefore influences the generations. The impact of the project on teachers, students and community was seen through different means including:

Case Studies

The interesting "try it yourself approach" of inquiry-based learning strengthened both understanding of scientific concepts as well as understanding of how scientists work when solving problems in order to serve their surrounding community. One student who participated in a research project about the effect of coffee dregs on seed germination and plant growth shared the result with her grandfather, a farmer in one of the rural areas in Jordan. The grandfather, in turn, repeated the experiments carried out by his granddaughter by adding the coffee dregs to his garden in the manner she described. Weeks later, the grandfather was amazed by the results of plant growth obtained by adding the coffee dregs and thanked his young granddaughter for this precious recommendation. The lesson learned from this case study is that when students learn a scientific concept by inquiry learning methods and test a hypothesis derived by him/herself, this can lead to the creation a well research practiced science student and an influencer that could convey and teach this knowledge to the others.

Related Research Study

Battikhi et al., (2016) explores the insights of the Chain Reaction project's participating science teachers in Jordanian schools regarding the impact of project on their students' inquiry skills gained by their engagement in the project compared with those who did not take part in it. The perception of participating science teachers was investigated using a five scale (options) Likert type of questionnaire including 30 teachers where 15 participants were males and 15 were females.

The results showed that the science teachers rate the degree of inquiry skills for the participating students to be very high with grand total mean (4.55) and standard deviation of (0.47). Whereas, the teachers rated the degree of non-participating student of having inquiry skills to on moderate level, with grand total mean score (3.03) and standard deviation of (0.63). This result highlights the importance of the briefs and activities of Chain Reaction project and their impact on students' affectation of inquiry skills. Inquiry based learning gives students a chance to experience science in real settings within a context that they can understand and are able to relate to, which is more effective in establishing and maintaining interest (Dokme & Aydinli, 2009).

Gender wise, there was no statistically significant differences that can be attributed to the gender of teachers on their perceptions regarding the participating students' level of inquiry skills. This may be attributed to the fact that all students had the chance to practice inquiry skills by adopting the same EURPBs and have received the same support from the Jordanian project team (TB).

- **Questionnaires tackling impact**: An evaluation questionnaire designed by the coordinator and translated into Arabic was completed by 18 teachers who participated directly in the Chain Reaction project and by the project team. The majority of teachers expressed an increase in understanding of inquiry-based learning as a result of their participation in the project and 66% agreed that they received very high quality IBSE/EUPRB briefings. Most of the teachers answered positively for the hopes listed and expected from their engagement in Chain Reaction. However, some respondents reflected negatively in terms of how the experience impacted upon their practice career and the chance at promotion within their career. This was attributed to the fact that the philosophy of IBSE is not yet adopted by the Education system and is not yet considered an extra skill preferred for teachers. As a result they considered that it will not add any value for their career. Fortunately, responses such as this were rare.

From the project technical team perspective, 100% of respondents strongly agreed that they had a high motivation for participating in the Chain Reaction project. For the personal outcomes, while all respondents hoped personally (to a great extent) to get the maximum outcome from the engagement in the Chain reaction project, 50% agreed to some extent that they received specific approaches to aspects of IBSE pedagogy. This indicates the high expectation as a partner for implementing the Chain Reaction

Project in Jordan. In relation to the wider outcomes, about 50% of the respondents hoped to see some general collaborative approaches across the participating schools and improved organizational approach to mathematics and science teaching.

- **Personal Notification:** For the duration of the project period, it was noticed that teachers, students, parents, Ministry of Education officers, non-participating schools and the project team showed a high level of enthusiasm and interaction with the project activities. This was noticed through:
    - Several non-participating schools, particularly science teachers contacting the project team in Jordan to express their interest in participating in the project.
    - The high quality of presentations in terms of scientific concepts, organization, presenting and English language speaking ability presented by Jordanian students to compete in the national and international was very remarkable and this in turn drew the attention of the project coordinator and parents.
    - Teachers expressing their gratitude for providing them with the opportunity to participate in the project and how this changed their prospective in educating science.
    - Teachers requesting another phase of the project with a special emphasis on mathematics teaching using the inquiry-based learning method.

CONCLUSION AND FUTURE DIRECTION

In this section of the chapter it is important to draw some conclusions and map the road for future directions. This section therefore presents several tracks on how to transform and enhance the experience with other teachers and students who did not participate in the project.

The project gave the participating students an opportunity to show positive cooperation and enjoy working on a project since they are rarely engaged in IBSE activities outside of this project. It is hoped that the experiences that students gained from their participation in this project will stay with them for the future learning of science, since, they are used to learning science through conventional ways of teaching. This project encouraged a cooperative approach, provided a stimulating and creative atmosphere and developed problem-solving and communication skills, as well, as knowledge and understanding of scientific concepts. Most EUPRBs are simulation activities, which involve students working as realistically as possible within a real-life scenario. students were provided with all necessary information which allowed them to operate within the simulation. Simulations are 'job-centered' as the emphasis is on carrying out a certain task(s) where students behave according to the task at hand. It is interesting to mention that within the context of the EUPRBs, simulation enabled the students to gain insight into, and awareness of, the roles and activities carried out by science and engineering researchers.

Moreover, students were able to understand the aim of the project and the process of adapting the briefs with their teachers help and the TB supervision. Simulations provide a framework and philosophy in which students are not only considered as active participants but are given considerable responsibility for their actions and decisions. Moreover, it was a great opportunity for them to prepare their own presentations and present them at the Express Yourself National Conference. It is hoped that students may also be able to transfer the knowledge they experienced from this project to their peers in other science classes or schools. This project enhanced the students' motivation to learn science, improve their achievement of scientific knowledge, build positive science attitudes and improve their understanding of the nature of science. Additionally, students learned how to develop skills with IBSE such as scientific data collection and analysis, interpretation of data, using creative thought, provide evidence to test ideas and develop theories. They learned how the explanations of many phenomena can be developed using scientific theories, models and ideas.

As noted above, students can also transfer many cooperative teaching strategies that they used in the learning and teaching process during the project implementation to their peers in their schools, such as active reading techniques, small group discussion (SGD), active writing, step by step discussion, Jigsaw method and problem solving/investigative work in teams. It is interesting that many of participating students informed the coordinator that they talked to their parents, relatives and friends about their experience in the project, and they were very proud to report that they are now able to produce their own learning materials from the surrounding environment.

Science teachers who participated in the project were asked to share the experience they gained from the project with their counterpart teachers in their schools and nearby schools. They were advised to establish a network for science teachers who participated in the project to exchange data, knowledge, experience and teaching materials and methodology. This may be of great benefit in respect of improving the science teaching learning situation in Jordan, because, as mentioned previously, science is mostly taught in Jordanian schools by using conventional method(s) of teaching. Therefore, teachers are expected to enrich their teaching materials and procedures by using the EUPRBs. The briefs knowledge, skills and procedures can support Jordanian official science curriculum with extra learning and teaching experiences and illustrations. For example, the scientific concepts in the EUPRB themes like "feed the world" and "plants in space" are new concepts for the Jordanian curriculum, these could be approached and used to enrich the curriculum using important innovative scientific procedures and problem-solving methods.

The project offered an opportunity for cooperation between the schools head teachers and the lab technicians in order to provide the necessary materials and learning resources and time to facilitate the students work and activities. It was hoped that this would improve science practical work at Jordanian basic schools.

The project team members were also influenced professionally by their involvement in project. They now have had a very interesting experience in being involved in such a large scale IBSE project. They have had the opportunity to cooperate with the project management teams and gained more experience from the coordinator (Sheffield Hallam University) on managing international EU funded project. They also had an opportunity to look at the EUPRBs and to make use of them in the training programs of science teachers.

REFERENCES

Battikhi, A., Albashaireh, Z., & Nawash, O. (2016). Chain Reaction Develops Jordanian Student Inquiry Skills as Perceived by Participating. *International Journal of Education and Human Developments*, *2*(3), 21–29.

Dokme, I., & Aydinli, E. (2009). Turkish Primary School Students Performance on basic science process skills. *Procedia: Social and Behavioral Sciences*, *1*(1), 544–548. doi:10.1016/j.sbspro.2009.01.098

Omar, N. A. (2015). The Use Of Collaborative Learning to Improve Critical Thinking of General Certificate Of Education Advanced Level Psychology Students: An Experiment, Research. *Journal of Applied Sciences (Faisalabad)*, *10*(10), 594–597.

Qablan, A. M and DeBaz, T. (2014). Facilitating elementary science teachers' implementation of inquiry-based science teaching. *Teacher Development: An International Journal of Teachers' Professional Development*, 3-21. . doi:10.1080/13664530.2014.959552

Rocard, M., Cesrmley, P., Jorde, D., Lenzen, D., Walberg-Herniksson, H., & Hemmo, V. (2007). *Science education NOW: A Renewed Pedagogy for the Future of Europe*. Brussels, Belgium: Office for Official Publications of the European Communities. Retrieved from http://ec.europa.eu/research/science-society/document_library/pdf_06/report-rocard-onscience-education_en.pdf

Tint, S., & Nyunt, E. (2015). Collaborative Learning With Think -Pair - Share Technique. *Computer Applications: An International Journal*, *2*(1), 4.

Chapter 10
Inquiry and Chain Reaction in Slovakia

Janka Raganova
https://orcid.org/0000-0002-3310-3319
Matej Bel University Banska Bystrica, Slovakia

Stanislav Holec
Matej Bel University Banska Bystrica, Slovakia

Martin Hruska
Matej Bel University Banska Bystrica, Slovakia

Miriam Spodniakova Pfefferova
Matej Bel University Banska Bystrica, Slovakia

ABSTRACT

The chapter presents how the Chain Reaction project was implemented in the context of science education in Slovakia. The educational reform highlighted inquiry as a leading approach in science education. But teachers were not prepared for implementation of such approaches. There was a lack of resources and a strong time pressure caused by a reduced number of science subjects lessons. Students' motivation to do and to study science was weak. Thus, the project team from Matej Bel University Banska Bystrica concentrated on two main issues: gaining teachers' engagement in non-traditional ways of science subjects teaching and motivating students to put their effort into inquiry-based science projects. The main problems connected with delivery of the Chain Reaction approaches in Slovak schools are discussed and the impact on secondary school students, science teachers, and teacher trainees is demonstrated. The developed pre- and in-service teacher training courses will serve as a tool of sustainability of the inquiry-based approaches in science education in Slovakia.

DOI: 10.4018/978-1-5225-5439-4.ch010

BACKGROUND

Inquiry-Based Learning in the Educational Policy of Slovakia

Schools, teachers and students in Slovakia have experienced considerable school reform since 2008. This reform maintains the traditional form of teaching science as three separate subjects (biology, chemistry and physics) at both the lower and upper secondary school levels (student age ranges 10 – 14 and 15 – 18). However, common goals between the three subjects have been formulated and the main objective is that schools students should develop an understanding of natural phenomena through targeted activities in which inquiry-based approaches dominate. The curriculum states that the development of inquiry skills including competence in observation, perception of relations between objects and phenomena, classifying, and measuring are the common goals and tasks of all scientific school subjects (Statny pedagogicky ustav, 2015b, p. 7). Approaches that promote discovery, exploration and investigation should enable students to develop basic inquiry skills and support the development of positive attitudes towards science as a way of understanding and exploring the world (Statny pedagogicky ustav, 2015a, p. 1). This demonstrates that the curricular reform since 2008 has been underpinned by a belief in inquiry as an effective approach to teaching and learning about science.

Implementation of Inquiry-Based Science Education (IBSE) in Secondary School Classrooms

Although the reform strongly emphasized students' active learning with the use of inquiry based strategies, its practical implementation in schools has been problematic.

Firstly, the numbers of science lessons at both the lower and higher secondary levels decreased dramatically after the reform, but the content was not proportionally cut to reflect this reduction in time. For example, at the lower secondary education level hours of science subjects formed 21% of the total programme before the reform and only 10% of all hours in the reformed programme (Lapitkova, 2008). Also, the number of compulsory physics lessons at secondary grammar schools decreased by 50%. Physics teacher educators felt, that one of the aims of the reform, to dedicate more hours for foreign languages, was gained at the expense of science subjects. No specific arguments were given for this decision by the key author of the reform strategy, except that "numbers of teaching hours are not important" (Zelina, 2007).

In addition, before the reform, students used to be split into two groups during laboratory work or other practical activities at physics and chemistry lessons. Two "theoretical" lessons per week when the physics (or chemistry) teacher worked with the whole class were followed with a two hour practical-oriented lesson when only one half of the class was doing experimental activities. Physics and chemistry teachers were supposed to manage practical activities of the class as a whole after the reform.

Moreover, the teachers were not prepared for the reform and were not trained in the use of an inquiry approach. They felt unable to implement the wide educational goals including the development of inquiry skills within the limited time in classrooms with too many students. A lack of available teaching materials and as well as traditionally used inappropriate methods of assessment were further negative factors obstructing the implementation of inquiry activities in science lessons in Slovakia in the last decade (Jeskova et al., 2016).

Certain problems with an implementation of inquiry-based strategies at Slovak schools were connected also with students that had not been accustomed to this kind of learning. They expected more guidance compared to the expectations of more independent inquiry-based learning so they could be resistive to IBSE at least at the beginning (Kires, et al., 2016).

Thus several national and international initiatives were designed to promote inquiry-based approaches in Slovak schools. After the EU-funded Primas (2010), Fibonacci (2010), Establish (2010) and SAILS (2015) projects, targeting western and eastern regions of Slovakia, the Chain Reaction project brought IBSE mainly to schools in the central part of the country. The project's aims and model dealt directly with the most significant issues of the inquiry-based science education in Slovakia – a lack of in-service professional development for teachers, a lack of appropriate teaching/learning resources and a lack of motivation in both students and teachers to engage in inquiry approaches.

PROJECT SUMMARY

The Chain Reaction project brought together teacher trainers and researchers from the Faculty of Natural Sciences, Matej Bel University (UMB) in Banska Bystrica with science teachers of fifteen secondary grammar schools. Most of the schools were located in disadvantaged areas surrounding Banska Bystrica; four of the schools were situated directly in the town and two were located in the eastern region of Slovakia. The sample included all types of schools: state, church and private, with the majority being state schools. Students between the ages of 13 – 17 years participated in the project.

The project team translated and adopted eight Earth and Universe Pupil Research Briefs (EUPRBs) for use in Slovakia: Collision Course, Green Light, Feed the World, PHEPPS, Cosmic Web Site, Plants in Space, Ozone Conference and Green Heating. They were presented to the participating teachers at an introductory briefing session.

Teacher Briefing

The teacher briefing programme consisted of three one-day sessions held in the laboratories of the Faculty of Natural Sciences Matej Bel University Banská Bystrica: an initial teacher briefing, a networking meeting of teachers and a final evaluation meeting (Table 1).

Table 1. The model of the Chain Reaction teacher training programme in Slovakia

Session	When
Initial teacher briefing	November 2013 October 2014 October 2015
Networking meeting	February 2014 February 2015 February 2016
Evaluation meeting	June 2014 June 2015 May 2016

The initial briefing session took place at the beginning of the school year. It was focused first of all on an introduction of ideas of the inquiry-based science education and the model of inquiry used in Chain Reaction. Then the teachers had a chance to "taste" the prepared inquiry activities and become familiar with the teaching materials that had been developed, student worksheets and needed laboratory equipment.

Besides the practical training of the use of EUPRBs in the teaching process, the briefing was geared to motivating teachers towards the crucial elements in the project necessary for success and sustainability. The expected benefits and anticipated problems that could occur during project implementation in schools were discussed. Participants in the first and second year contributed greatly with their experience to the training of new cohorts of teachers in the second and the third project year.

In-School Delivery

The teachers incorporated at least one of the EUPRBs into their science lessons; Feed the World and Green Light were the most frequently used. The in-school delivery period included various types of ongoing teacher support, such as consultations by email, by phone, visits of UMB staff and role models to the schools, visits of individual school classes and teachers to the Faculty etc.

An appointment of five young researchers – the role models – proved to be very helpful and fruitful. Their background covered all science branches – biology, chemistry, physics and geography, so teachers could contact a person with the most appropriate knowledge and skills. The role models usually visited schools, gave talks and showed relevant demonstrations in front of the classes. The following discussions with members of student research teams formed the most important parts of the role models' mission. In some cases students visited a role model in his lab at the Faculty and made investigations together.

In February, in the middle of the school year, the teachers met face-to-face second time at the networking session. Teachers reported on their progress in delivering the briefs, exchanged experience and as a group discussed the problems that they had to solve while they had been managing student teams. They were keen to learn from their colleagues and confront their experience and ideas not only with teachers of the same subject but also with colleagues of different expertise. For most of them it was the first opportunity of this kind.

Student Conferences

Each project year culminated with the student Express Yourself! research conference. Teachers had a challenging task before the conference to select one team to best represent the work they did with the students. They usually organized class or school "conferences" or student presentations to younger schoolmates. The team presentations were evaluated by a jury and/or by students, so students themselves chose their deputies for the national conference.

All the three national conferences were organized as two-day celebration events in the Congress Centre Academia in the High Tatras region. The conference venue enabled participants to visit several research institutes investigating areas with direct connections to the project EUPBRs and see how science work in real conditions. In the first day it was a visit to the Aerospace and Radiation Station Poprad – Gánovce, where students could see how meteorological measurements including measurements of ozone are done. It was followed with a visit to the Research Centre of the Tatra National Park where students could see and discuss various environmental measurements and investigations of fauna and flora of High Tatras. The excursion to the Astronomical Institute of the Slovak Academy of Sciences that was organized in the second conference day brought a closer look at observations of asteroids and other astronomical bodies. Students had a lot of questions at all of these places and expressed their real interest in the presented research areas and topics. Both students and teachers benefited of a better understanding of the breadth and depth of Slovak scientific research.

The main part of the conference programme, student presentations, provided great encouragement for students as well as showcasing the excellent work of their teachers. Students not only presented valuable scientific results but also exhibited great creativity and a strong engagement in the investigated topics.

Evaluation meetings formed the final national activities at the end of each project year. Participants of the International Conference, together with their colleagues, reflected on their experiences of project implementation. The findings gained were used to improve the project in the following years.

PROJECT IMPLEMENTATION

The Teacher Professional Development Process Creates a Chain Reaction

The implementation of IBSE in Slovakia within the Chain Reaction project faced several issues. As confirmed by previous initiatives in this field (Jeskova et al., 2016, Trna & Trnova, 2016), teachers' openness to, and belief in, IBSE can be strongly influenced by their existing beliefs and the way they were educated. The majority of science teachers in Slovakia underwent traditional teaching based on the transmission of knowledge rather than inquiry-based methods, so they were without experience of inquiry-based learning and teaching but with a strong understanding of more traditional ways of teaching science. Some teachers may have been also frustrated in the way that educational innovations have been implemented. They have felt to be forced to deploy new state educational programs including inquiry-based strategies – without professional development, relevant teaching and learning resources and, very often, without necessary lab equipment and materials.

Therefore, one of the main project challenges was to gain teachers' interest and support, to overcome their initial scepticism prior to introducing them to the Chain Reaction model and inquiry approaches. This was especially difficult in the first project year, so we built upon our previous links with creative, enthusiastic teachers who were open to learning and implementing new educational strategies.

Teacher recruitment was much easier over the next two years when we could present our first-year project results, including first-year student presentations. We were able to demonstrate that all fifteen participating teachers from the first year together with their students implemented at least one of the prepared inquiry-based projects within science subjects successfully. First-year teachers provided very positive reports of their participation in the project, appreciating support they received from the project team.

The project's popularity grew among schools across the region and was enhanced through a promotional presentation from student teams at the European Researchers' Night event organized in Banska Bystrica on Friday, 26th of September 2014, by the Slovak Organization for Research and Development Activities. Two school teams from Jan Chalupka secondary grammar school in Brezno introduced their research projects conducted within Chain Reaction to the hundreds of other students, teachers and a general public as well as shared with an audience their experience gained at the national and international conferences. Thus achievements among this first group of schools caused a "chain reaction" effect and helped to easily recruit the next set of teachers.

The effect of this 'chain reaction' was also applied to the in-service professional development programme by incorporating the transfer of knowledge and experience between the previous and new project partner schools. The teachers from the first two years contributed to the professional development of new teachers by sharing their experience with new participants, explaining issues that occurred while working with students as well as demonstrating the use of selected briefs in the classroom.

Three professional development and networking sessions were supplemented by various types of ongoing support: consultations, visits to schools, and student visits at the faculty. Even the experienced teachers were keen to invite role models from the faculty to work with students – as they found this element particularly motivating for participating students. Such visits were appreciated especially in cases when students solved a complex project, for example Ozone conference that integrated knowledge from several disciplines – physics, chemistry, biology and geography. As the participating teachers did not have cross disciplinary expertise they welcomed the opportunity for students to ask questions and discuss their projects with young researchers with expertise in different disciplinary fields. The EUPRB, Cosmic web site, provided another example of the complex, and rather difficult theme, where teachers lack relevant content knowledge. Lectures given by a young female nuclear physicist or by a researcher directly participating in a CERN project served not only to explain difficult topics but also appeared to inspire the participating girls and boys.

In some cases members of the project team had to use their expertise and help teachers solve concrete technical problems, such as finding out why a model of the portable water power station created by students did not work. Checking the model by a telephone call with the unhoping teacher showed that students had used an isolated copper wire and the isolation had not allowed current to flow.

Strong links between the project team and participating teachers were developed through this type of collaboration. Most of the teachers felt fully engaged in the Chain Reaction project and their commitment to delivering the project effectively within their schools was clearly demonstrated.

Express Yourself! Conferences as a Motivating Factor to Students

Another important project outcome grew from an initial lack of students' interest in science subjects, their weak motivation to engage with science and to study science in the future (Centrum vedecko-technickych informacii Slovenskej republiky, 2012 – 2016). Our experience indicates that as students increase in age there is a decreased interest in studying science (Trna, Trnova & Sidor, 2012). This situation reflects a general trend in many other countries globally (OECD, 2006).

However, the implementation of the EUPRBs within science subjects encourages, and in many cases requires, significant student involvement and engagement beyond standard science lessons as a number of the EUPRBs demand longer time allocations than schools plan for science. For example, within the Feed the World activity, students have to investigate the germination, growth and development of beans over several weeks while Cosmic Web Site requires significant hours of programing. The Chain Reaction model recognizes that this can be problematic for teachers. Therefor many of the EUPRBs can be used flexibly with teachers often choosing elements of an EUPRB to use for an inquiry rather than opting to complete the full brief.

The participating teachers worked with all students in their class, regardless of previous interest and performance. Each student became a member of a research team and acquired responsibility for team results. Besides the attractive topics of the briefs, unusual methods and cooperation with role models, the teachers also suggested that student conferences are a strong motivational tool. Having the opportunity to present a research project and represent the school encouraged students to maximize their effort through engagement in the inquiry-based activities and to do extra work in science subjects beyond classroom time. Therefore, we decided to organize the national Express Yourself! conferences as two-day events in a place that would be attractive to young people and enabled us to prepare additional interesting activities for the participants. The congress centre of the Slovak Academy of Sciences located in High Tatras and neighboring several research institutes was chosen as the best venue. The centre provided all needed facilities at the professional level and we expected it would help a smooth running of the conference. Our young researchers were looked after as real scientists.

This strategy proved to be successful. The conferences invited teams consisting of girls and boys that had never before participated at any science focused event or competition. For many students it was also the first opportunity to stay in a hotel and be treated as important scientists.

Student presentations provided a celebration of all the work that students, together with their teachers, had engaged in. Great attention was given to poster sessions where all participants could informally discuss their projects and demonstrate their achievements in detail. Conference guests, including students from partner schools from the previous project year, contributed to the ‘chain reaction’ effect of the conference. They were able to ask questions and discuss the briefs with their colleagues in a professional way as well as share their experiences of working on the project.

Both the teachers and students voiced their appreciation of the conference environment which not only aimed to reflect a professional research conference but also aimed to establish a sharing, friendly, and supportive atmosphere. The teachers recommended keeping the conference format and to continue organizing student conferences to support inquiry-based science education in the future.

BARRIERS/OBSTACLES

Overall, we experienced positive results from the Chain Reaction project in the Slovak context despite some difficulties that teachers had to negotiate during delivery of inquiry-based activities in their classrooms.

All of the teachers identified time as the main problem. The teachers found it difficult to incorporate a number of the EUPRBs into school curricula and impossible to complete them during one 45-minute lesson. This was particularly problematic when the class had only one lesson of a science subject per week. This meant that students had to engage in some experimental work after classes in their own time. Thus in some cases, only groups of the most motivated students were able to complete their research.

Another problem which teachers identified was in managing the experimental component of students’ projects in classes with large numbers of students. Health and safety concerns together with classroom management issues proved difficult for teachers to negotiate. Additionally, some schools had limited materials and equipment for experiments which often limited the scope of inquiry activities.

Some complications were caused by the Chain Reaction model itself. The project target group, 14 – 16-year students, was rather problematic in our context as most of the students had just begun their study at upper secondary school and had gaps in their scientific knowledge and limited experience with the scientific method. However, when teachers worked with older students they tended to teach the more theoretical knowledge that is examined at school-leaving exams and entrance exams at universities rather than supporting inquiry.

The majority of EUPRBs require application of cross-curricular relationships between science subjects. Consequently, rigidly-defined physics, chemistry and biology curricula have not supported easy incorporation of these interdisciplinary problems into lesson programs. On the other hand, addressing this barrier somewhat led to the closer collaboration of teachers of different subjects which led to a positive cross-departmental impact.

IMPACT

Impact on Students and Teachers

Despite the obstacles stated during the implementation of Chain Reaction in Slovakia the project had a very positive impact on teachers, students and schools as a whole. Positive responses to the project were expressed by teachers on various occasions—evaluation sessions, formal and informal interviews as well as in the final project questionnaires. Even within one school year, teachers could see an increased student motivation to learn science subjects. The teachers were especially positive about the involvement of students who they identified as having weak performance in science subjects, children with learning difficulties and poor socio-economic conditions. The self-confidence of students increased and, in some cases, teachers were surprised how group projects impacted positively on peer to peer interactions and group collaboration.

In particular, positive impact of the project on students' engagement in physics was clearly expressed as shown through this response:

The interest of students to learn physics has increased significantly – the students chose optional physics lessons and physics themes for their projects, are interested in participation at physics competitions and other physics activities, and several students decided to study physics at university (Polakova, 2016).

The teachers stated that they joined the Chain Reaction project with great expectations. They expected to increase their enthusiasm for science teaching and learning, to improve their lesson practice and gain a more positive attitude to their professional development. The data gained from evaluation questionnaires show that teachers' expectations were met across all of these and even exceeded in the case of their attitude to the professional development and improvements of the classroom practice. This positive experience of teachers with their participation in Chain Reaction resulted in a strong desire to participate in the project beyond the project duration.

Impact on the UMB Initial Teacher Training Program

The project also had a tangible impact on the Initial Teacher Training programme at the Faculty of Natural Sciences, UMB. The science teacher trainees could experience the project approaches directly through their involvement in selected project activities as well as in the lessons where inquiry-based methods were incorporated.

The project was introduced first to the students at the Masters study level who got an opportunity to participate in a preparation of the first national conference and thus to gain practical experience and skills in organizing such events. It was most important for them to attend the conference itself when they could see the results of the project in action and become familiar with a content of all EUPRBs used in Slovakia through student presentations. Teacher trainees actively participated in the discussions after the individual team presentations and in the poster section, as well as in the evaluation of the presentations and the selection of the team for the international conference. Evening workshops of young scientists and observations of teacher – students interactions, as well as their direct interactions with members of student teams were also a great experience for teacher trainees. Their very positive reactions encouraged us to include also students of Bachelor study level into the conference organization team of the second project year. Their experience were later discussed and analysed at didactic lessons and it was shown that the learning by the direct involvement in the project was very fruitful.

In addition, the science teacher trainees participated at one of the most important project dissemination event - the public seminar held in February 2016. The students could benefit from the programme that primarily targeted teachers of science subjects from lower and upper secondary schools that were not involved in the project. The purpose was to provide the teachers with information about inquiry-based methods and approaches used in Chain Reaction, to introduce examples of activities that could be effectively used in the classrooms at physics, chemistry, biology and geography lessons. The participants were invited to contribute to the programme with sharing their experience in the field of inquiry-based science education and this way they educated the students as well.

Besides those special events the project approaches were continually incorporated into syllabuses of several courses within Masters science teacher training programme. For example, the content of Didactics of Physics I was enriched in a theme "Inquiry methods and student activities", Didactics of Physics III contains now a theme "Inquiry activities with various levels of student autonomy". Those obvious improvements of science teacher training curricula were accompanied with less visible changes of the work style of teacher trainers who have started to use inquiry methods in their everyday teaching of purely scientific subjects as Mechanics, Practicum of Optics etc. Thus teacher trainees could develop both the theoretical knowledge and the practical skills with the use of inquiry approaches at science lessons.

Also, as a result of our involvement in Chain Reaction, a new course, Active Learning Strategies in Physics, was introduced within the Master's study programme. The course takes two hours weekly in one semester and is compulsory for all future physics teachers. Other teacher trainees may opt for it as well. Inquiry-based methods and activities form a core part of the syllabus of the course and will serve as a vehicle for spreading the IBSE approaches to the next generations of science teachers.

CONCLUSION AND FUTURE DIRECTION

The stated results of the three-year implementation of inquiry-based methods in science subjects at Slovak secondary schools have proved the benefits of these approaches for both teachers and students. We consider that the most important outcome is the increased motivation of students towards participation and engagement in science and school science. Teacher involvement in Chain Reaction has resulted in increased enthusiasm for science teaching, improvements in classroom practice and positive changes in attitudes towards engaging in professional development of this nature. Most of the participating teachers spontaneously expressed their intention to continue using inquiry approaches where appropriate. The participating teachers found the provided materials and EUPRBs ideas were meaningful, interesting and motivating to students as well as helping them develop a range of skills.

The benefits of the new teaching and learning styles of the project participants were proven for example in a national competition *Mission to Mars* in 2016. Student teams from schools across Slovakia competed to design a project that could be used by an international research team at the simulated Mars station in the USA. The three best teams in this competition came from Chain Reaction partner schools and were led by teachers that had participated in the project (Sutaz Misia Mars, 2016). These schools have embedded the Chain Reaction model in their schemes of work and continue to deploy student inquiry-based projects, organizing school conferences and disseminating the project approaches within the teaching community.

Our experience with Chain Reaction resulted in the development of the in-service teacher training course "Implementation of inquiry-based activities into educational process" that serves as a tool for sustainability of the project approaches in the future. In addition, we have brought our experience to a consortium of the new national project (IT-Akademia, 2017) that will further promote inquiry-based science education supported with digital technologies at lower and upper secondary schools through the whole country.

REFERENCES

Centrum vedecko-technickych informacii Slovenskej republiky. (2012 – 2016). *Statistika prijimacieho konania na vysoke skoly SR*. Retrieved August 15, 2017 from http://www.cvtisr.sk/cvti-sr-vedecka-kniznica/informacie-o-skolstve/statistiky/statistika-prijimacieho-konania-na-vysoke-skoly-sr.html?page_id=9723

Establish: European Science and Technology in Action: Building Links with Industry, Schools and Home. (2010). Retrieved from http://www.establish-fp7.eu

Fibonacci: Disseminating Inquiry-based Science and Mathematics Education in Europe. (2010). Retrieved from http://www.fibonacci-project.eu

IT Akademia – vzdelavanie pre 21. Storocie. (2017). Retrieved August 15, 2017, from http://itakademia.sk/sk/domov/

Jeskova, Z., Kires, M., McLoughlin, E., Finlayson, O., Ottander, Ch. & Ekborg, M. (2016). In-service and pre-service teacher education in IBSE: The ESTABLISH approach. *Il Nuovo Cimento C, 38*(3). doi:10.1393/ncc/i2015-15099-8

Kires, M. (2016). *Badatelske aktivity v prirodovednom vzdelavani*. Bratislava: Statny pedagogicky ustav.

Lapitkova, V. (2008, February 5). Reforma vzdelavania – informacie, vychodiska a trendy pre fyzikalne vzdelavanie. Presentation given at the videoconference. *Reforma fyzikalneho vzdelavania*.

Organisation for Economic Co-operation and Development. (2006). *Evolution of Student Interest in Science and Technology Studies*. Retrieved August 15, 2017, from http://www.oecd.org/science/sci-tech/36645825.pdf

Polakova, Z. (2016). *Badatelsky orientovane vyucovanie prirodovednych predmetov v podmienkach Gymnazia Andreja Sladkovica*. A talk given at the seminar "Inquiry-based education in science subjects". Banská Bystrica.

Primas: Promoting Inquiry in Mathematics and Science Education Across Europe. (2010). Retrieved from http://www.primas-project.eu

SAILS – Strategies for Assessment of Inquiry Learning in Science. (2015). Retrieved from http://www.sails-project.eu/

Statny pedagogicky ustav. (2015a). *Fyzika – nizsie stredne vzdelavanie.* Retrieved August 15, 2017, from http://www.statpedu.sk/files/articles/dokumenty/inovovany-statny-vzdelavaci-program/fyzika_nsv_2014-12-03.pdf

Statny pedagogicky ustav. (2015b). *Statny vzdelavaci program pre nizsie stredne vzdelavanie – 2. stupen ZS.* Retrieved August 15, 2017, from http://www.statpedu.sk/files/articles/dokumenty/inovovany-statny-vzdelavaci-program/svp_nsv_6_2_2015.pdf

Sutaz Misia Mars pozna svojich vitazov. (2016). Retrieved August 15, 2017 from https://www.seas.sk/clanok/sutaz-misia-mars-pozna-svojich-vitazov/273

Trna, J., & Trnova, E. (2016). Inquiry-based Science Education Modules and Their Effects on Teacher Education. In Education Applications & Developments II (31-46). Lisboa, Portugal: Science Press.

Trna, J., Trnova, E., & Sibor, J. (2012). Implementation of inquiry-based science education in science teacher training. *Journal of Instructional and Educational Studies in the World, 2*(4).

Zelina, M. (2007, December 5). A talk at the meeting of the Curricular Council, Casta, Pila.

Section 3

Sharing the Experience of the Teacher and Project Management

This section details the experience of Chain Reaction from the perspective of a teacher involved in the project and those involved in management of the project.

Chapter 11
Experiences of IBSE and Chain Reaction:
Reflections on Evolution of Practice and Curriculum

David King
Junior Cycle for Teachers, Ireland

ABSTRACT

This chapter considers the impact of professional development on IBSE in the Chain Reaction (CR) project, Ireland. The author offers reflections, based on his own experiences of CR, suffused with the theme of "evolution." Evolution experienced includes that in thinking, practice, curriculum, and teacher engagement during the project and follow-on professional development endeavors. He speaks as a practicing teacher who, during the lifetime of his engagement with CR, took up a national leadership role in supporting Science Education change in Ireland. He offers that professional development that seeks to enhance social capital creates opportunities for teachers to develop their capacity as science curriculum developers. This has been a guiding precept for CR Ireland and broader professional development endeavors across Ireland in recent years. Evolutions in science education across Ireland have allowed for the integration of IBSE opportunities afforded by the introduction of national policy and the promotion of social capital at CPD to compliment policy intentions.

INTRODUCTION

Reflecting on this as a teacher, I can draw parallels between my experiences in CR, and indeed beyond since the years of the project's inception, progression and completion. I offer in this chapter a teachers' perspective on engaging with the CR project, and its impact on the adaptability to change in IBSE classroom practice. The project, and my involvement with it, happened in a climate of national curriculum change in Irish Science education. Through a professional learning community (PLC), the CR Ireland team set out to 'disturb practice'. I contend it is our ability to adapt to this disturbance in a supportive

DOI: 10.4018/978-1-5225-5439-4.ch011

community, connected to the classroom, that influences our achievement of agency in engaging with IBSE. This has implications for emerging programmes of continuous professional development in Irish education and beyond.

TEACHER CONTEXT

At the time of writing this chapter, I have been working in Irish Science education for 17 years. I have taught Junior Cycle Science (lower secondary, typically 13-15 years old) and Biology, Physics and Agricultural Science to Senior Cycle Science students (upper secondary, typically 16-18 years old). I worked within a school culture that supported innovation and change, which was reflected in our Science students varied and successful endeavours in Science fairs at a local, national and international level. My motivations to engage with CR Ireland were in part response to my perceptions of curriculum at the time. In my early years as a novice teacher my classroom work was, in many ways, defined by coverage of content. I demonstrated excellence in the delivery of subject matter. I developed student competency in practical science through recipe-style scientific investigations. The expectation from parents and students was high grades in the Junior and Leaving Certificate exams, and I measured myself by these expectations. My engagement with my subject was characteristic of the technical-rational interest that has been so pervasive in Irish curriculum discourse (Gleeson, 2010, p.122), and market-driven expectations of performativity (Ball, 2003). I was conflicted by the quality of learning experiences I was providing for my students. I was a master of 'drilling content', but this did not help to develop students as learners and to instil the love of education I had experienced as a child. Through mentoring students for science fairs, I saw their passion for learning develop as they explored their interests as real scientists. This was disconnected from the reality of 'doing science' in the classroom. I felt the 'goodness' of my subject was being outsourced to extracurricular activities such as Science fairs, disconnected from the Science classroom. Rather than become dispirited, I turned my energies to experimenting with my practice (Huberman, 1989; 1993). I began to introduce varying IBSE experiences into classroom investigative work, and invited students to begin developing their own approaches rather than following a set procedure to pre-defined knowledge. I created conditions for students to review and support each other's work, activating them as learning resources for one another in the classroom and the digital space through virtual learning environments. I found through creating the space for authentic IBSE experiences in the classroom, students began to flourish as Science learners. This was evidenced by their successful engagement with Science fairs and improved learner outcomes in the classroom.

This was not a straightforward experience and took time and the resilience to get up and learn after multiple failures. In my desire to progress my own practice in IBSE and to support fellow teaching colleagues, I began to look beyond the walls of my own classroom and school. I worked part-time in teacher CPD with Science teachers and facilitated professional development in workshops around IBSE at a Junior and Senior Cycle level. At this point, I was invited onto the Technical Board (TB) for CR Ireland.

BEING A CR TEACHER: VALUES, PURPOSE, AND ACTIONS

One of the ways CR Ireland aimed to support teachers in their enactment of IBSE was to provide a forum in a professional learning community (PLC) for reflection and discussion of experiences in the classroom. Through this approach, we endeavoured to raise the social capital of the teachers involved in the group. That is, "the resources teachers can access through peer collaboration to support their ongoing learning" (Johnson et al., 2011, p.1). The social capital model of professional development contends that incremental steps lead to changes in teachers' practice, as they consider new ideas and try out new approaches. At its core, the social capital model centralises collaboration to the process of changing practice. Applying this approach to the CR PLC, we wanted to consider ideas around using IBSE in our classrooms, articulate and commit to new approaches, experience them with our students and through collaboration reflect upon the process.

At the start of the project, I joined CR solely as a TB member, and in 2014 worked with the inaugural cohort of teachers in developing our understanding of IBSE through the PLC. From the outset, we (TB members and participating teachers) entered into the PLC with a commitment to learning from and supporting each other. We committed to this onsite, in the online space and in the crucible of practice in classrooms. We engaged in dialogue on what learning would look like as part of the PLC (Fig. 1). Interestingly, and resonate with my own experiences, there was a dominant view that we should step out of our comfort zones with our students through not jumping in with the 'right answer' to expedite the learning in a lesson, only to find later that the desired learning did not actually take place. We committed to making sense of our experiences as a community of learners. Through this iterative and supportive process which usefully aligned with school self-evaluation goals for individual, department and school improvement, we endeavoured to understand how we *enacted* IBSE in our classrooms, rather than conform to perceived models of delivery and implementation in our existing curricula.

We wanted to embrace curriculum as a social practice, wherein meaning was constructed as a product of student, teacher and PLC interaction (Snyder, Bolin, & Zumwalt,1992). As part of these social interactions, we acknowledged our professional concerns around teaching Science through inquiry (Fig.

Figure 1. Developing a sense of purpose for CR Ireland: who we are and why we are here

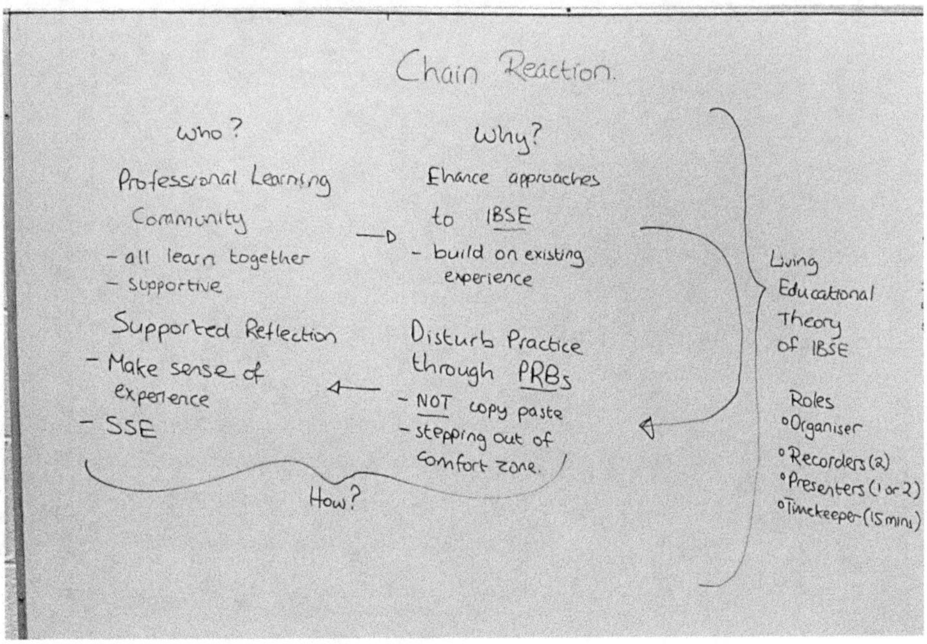

2). These concerns included school contexts such as professional culture, resources and time constraints. They also included system contexts such as exam pressures driving teaching and learning. Finally, personal and professional contexts were raised. These included our own self-efficacy and the need to develop a base of knowledge around inquiry and practical approaches to introduce it to our students, cognisant of the previously outlined contextual conditions. There was a spirit of deliberative reflection, self-confrontation and sharing within in the group, the types of behaviours which according to Day (1987, p.215) result in teacher change. Our intention was to, as much as possible, encourage a "flattening of hierarchies" (Drew et al., 2016, p.9) in the PLC as we developed our understanding of the use of IBSE in our practice and the affordances this offered us and our students. In doing so, it was our ambition to, as Whitehead (1989) puts it, construct a living educational theory in which students, teachers and the PLC learned from and with each other.

In their study of social capital models of professional development, Johnson et al. (2011, p.9) found that resources which could improve social capital included "access to expert knowledge, emotional support from shared commitments, and the ability to analyse instructional practices".

Through our engagement with the PLC, there were a number of resources available to facilitate our ongoing learning. In this context, the resources we drew upon included

- Expert knowledge on IBSE from our colleagues in the University of Limerick (UL) and from "role model" scientists
- Suggested approaches to lessons as provided in the Pupil Research Briefs (PRBs)
- Space through post lesson conversations and on PLC professional development days to reflect upon and analyse our evolving IBSE practices

Figure 2. Professional concerns about teaching science through inquiry

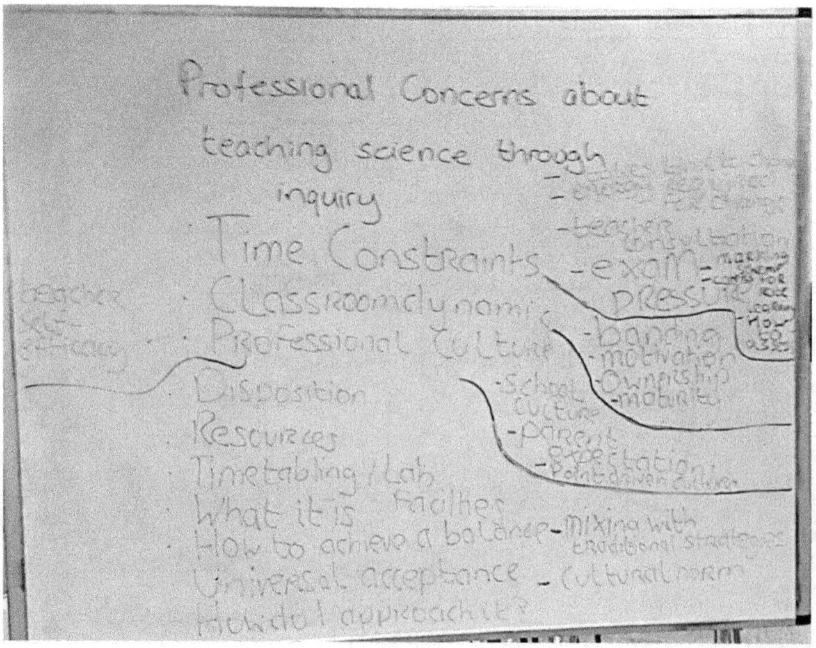

In the next section, I offer reflections upon the experiences of CR with my Science students. I consider how my engagement with the suggested approaches to IBSE in the PRBs and the role model experience were positive resources in supporting my ongoing professional learning and served to enhance my social capital in realising aspirations for IBSE in my classroom.

THE EXPERIENCE OF CR IN MY CLASSROOM

The CR experience in my school was related to scientific communication and the responsibilities of scientists. It can be divided up into three key areas:

- Enacting the PRB
- Post-lesson discussion
- Role model experience

Each of these areas shall be addressed below.

Enacting the PRB

Early on, it became clear to the TB that we would struggle to retain Science teachers in our PLC[1]. Conscious of this, and in the spirit of the community, I requested to be allowed to participate as a teacher in the inaugural year of the project, as well as a TB member. I wanted to understand the process teachers were going through in the classroom to inform my own practice and my dual contributions to the PLC and TB. I chose to write my own Pupil Research Brief (PRB) based on the style and templates of the existing 'off the shelf' PRBs. I made this decision for I wanted to engage students with learning that had relevance to where we were at in our current curriculum plan, but also brought us somewhat away from the regularly worn track of classroom content. Further, I wanted to gain first-hand experience in producing curricular content pertaining to Science concepts which were contextualised in student experience (Wallace & Priestley, 2016). Having recently learned about food and the digestive system, I chose to write a PRB titled "Tackling Obesity in Ireland" (See Appendix 1). This was a paper-based, non-experimental PRB. It employed the jigsaw method to get students working in teams to construct knowledge around scientific communication about obesity in Irish society. Students were tasked with producing an advertisement for the Irish public on the dangers of obesity in Ireland. As part of the PRB, students engaged in four specific areas pertaining to obesity:

1. Definitions and causes of obesity, including calculation of BMI values
2. Occurrence and statistics of obesity with a specific focus on Ireland
3. Impacts of obesity
4. Prevention and treatment of obesity

The levels of inquiry involved in this learning experience fall between structured and guided (Banchi & Bell, 2008). Students were invited to generate explanations supported by the evidence they collected from information provided by the teacher (structured). They were also invited to design a presentation to the Irish pubic on the dangers of obesity, having been given this as a brief by the teacher (guided).

The lesson was enacted with a class of mixed ability first year students in January 2014. Students broke up into expert teams to engage with obesity occurrence and statistics, impact and health problems, prevention and treatment, and factual information, including calculation of BMIs for sample clients. In their home groups, they were then invited to come together and advise the 'health trust' through producing a scientifically accurate advertisement on the dangers of obesity and ways to tackle it. Following the activity, students expressed a desire to produce different types of artefacts for their advertisement – e.g. posters, leaflets. To ensure a common standard, we co-constructed success criteria for the production of the advertisement artefact.

In the following sections, I will share reflections and testimony from a rich post-lesson discussion and the 'role-model' experience.

Post-Lesson Discussion

Following the lesson, I engaged students in conversations concerning the 'bigger picture' of how we worked to develop new understandings, and how this work parallels with that of real scientists particularly in the context of how to frame and communicate scientific arguments suitable for purpose and audience. Through a guided class discussion, we arrived at a number of key points concerning the work of scientists in society.

In the initial conversation, students discussed the **purpose** of the information they had put forward in their advertisements:

TEACHER – what do we want this advertisement to do ultimately?

Mike: Change people's minds about losing weight

TEACHER: I think you've hit a really key point there Mike, I'm actually going to write that down [Writes 'change people's minds' on board – so this is a bit bigger than people's weight, isn't it? The role of Science in changing people's minds. Whether it's about obesity, nuclear power, building electricity pylons in our community – how important a role do you think Science plays in changing people's minds?

Aine: I think the facts can kind of shock you so then you'd want to change your mind

Michelle: You kind of want people to follow the leaders

Next, we delved deeper to consider the **arguments** being put forward and the importance of validity and accuracy in our arguments:

TEACHER: Let's look at the information side of it – was that good? Remember your role, what are you trying to be here?

Class: Scientists

TEACHER: Exactly. So when you're communicating Science to the public – yes, we want to change people's minds, but also – what about the information? What do we need that information to be?

Mike: Clear.

Sophia: Scientific

TEACHER: Good, so this brings me onto something else I wanted to talk to you about, and I'm going to pop up a few of my own questions here: So the first thing – is our argument a strong one? Do you think we made a strong argument?

Eoghan: Yeah, because we showed what could happen if you were obese and how more and more people are dying each year

TEACHER: Good, good. Michelle?

Michelle: it's like getting the information into their heads, it's so shocking that it would make them change.

TEACHER: OK, very good. Mark?

Mark: Arguments need to be strong because if they're not strong people are going to go "ah sure this is just a daft ad, we'll just move on". So it needs to be strong to actually change people's minds.

TEACHER: That's a very good point - that our arguments need to be strong to be taken seriously, but also because – who are we? What are we?

Mark: Scientists

TEACHER: Exactly. And we put forward not just a strong argument, but a *correct* argument.

Mike: If you gave false information, it'd be like a teacher teaching a child something wrong

TEACHER: Go on..

Mike: So if a teacher taught a child something wrong, they'd think like that's the right thing and then they'd do that, so if you're telling the public the wrong thing they'll go out doing the wrong thing.

TEACHER: That's a very good point you're making there. But also, if we don't communicate the stuff right and it isn't factual, and we're supposed to be the experts – will we be believed?

Class: No

TEACHER: So in a sense – what do we have as scientists, in terms of our duties to these people? Do you think we have a responsibility? [class agrees]

Upon further discussion, students were invited to interrogate the arguments they had framed about obesity, in the context of communicating Science. From the dialogue, it is clear that students are beginning to weigh up the strength of the argument with the purpose of the communication in changing people's minds. Some suggest there should be a positive spin to win people over, whilst others argued that issues such as obesity demand a serious tone of communication. They also begin to **critically interrogate** their arguments in the context of current media portrayals of other scientific issues, as well as limitations in the information they were given:

TEACHER: So - is our argument missing anything?

Mark: We didn't put in like the fun parts about it

Mike: We didn't put in like the positives, we just put in the negative stuff like heart disease and all that. We could have put in like – if you lose weight, like, you will feel better

TEACHER: Now, that's actually a good point – did any of us put in any positives?

Class: No

TEACHER: Why didn't we put in any positives?

Jessie: There's not really many positives about being overweight…

Michelle: Yeah, but there's positives about losing weight!

TEACHER: OK, when we made this advertisement, why didn't we think to put in any positives or fun facts?

Michaela: Because we were only thinking of the negatives

TEACHER: And why was that?

Marie: Because that was kind of the title of the thing, about obesity and the negatives…

Mark: Because the negative points of obesity stand out…

Tom: The sheets you gave out were mostly negative!

TEACHER: So there's an interesting point – So if I could sum up what we've been discussing here – I would say that you produced mostly negatives because you worked with what you were given, yes? [Class agrees]. Now, there's nothing wrong with that at all, you did the task you were asked, but it kind of explains where the negatives have come from yeah?

Fiona: If we looked for ourselves on the internet we might end up finding some positives.

Aine: I think it was important to keep it [the advertisement] serious though, to keep the science serious.

TEACHER: OK, fair point – can you give me some examples of other science adverts you see on the TV that touch on the serious side of things?

Mike: Ads about, like, safe foods, and when your child comes home from school, make them drink water, don't let them have sugary foods and drinks because that could cause obesity if they have them everyday

TEACHER: OK, cool. Aine?

Aine: Smoking adverts

What started as a discussion about making an argument through the student advertisements turned into a deeper conversation which scratched at the surface of the need to be critical when sourcing information and framing an argument. Students began to realise how you communicate your argument depends on the purpose and audience of the communication. Next, we spoke about the **responsibilities of scientists** as demonstrated through the arguments they make:

TEACHER: Excellent. OK, so are there any experiments we could carry out that might strengthen the argument we've put forward? Based on the arguments you've put forward?

Eoghan: We could get kids at a birthday party and see which foods they'll go for and which ones they won't, and whether the obese ones will go for the sugary and unhealthy foods.

TEACHER: OK so let me get this straight. What you are saying is have a kid's party, with a mix of some kids who are obese and some who aren't. Have healthy and unhealthy foods and see which ones go for which…anybody see a problem with that?

Fiona: Because, like, it's a kid's party – you're hardly gonna want to be eating lettuce and carrots!

TEACHER:OK, good point! [laughs from the class]. But let me just ask another question on that experiment – would it be right to do it?

Class: No

TEACHER: Why not?

Olivia: Because you could be encouraging the kids to get fatter with the experiment

Fiona: And you can't just, like, test ideas on children [class laughs]

TEACHER: OK, right, so fancy word time – have any of you heard this word before? [teacher writes 'ethics' on the board]. Any ideas? Anybody heard this used before? [after long pause…] Ethics means we ask ourselves if what we are doing is right, if it's morally OK to do the thing you are suggesting. So, going back to your original idea Eoghan, and after the discussion – do you think we should do it?

Eoghan: No, because we have a responsibility

TEACHER: Great stuff. So, who do we have a responsibility to?

Aine: The public

TEACHER: Good, who else do we have a responsibility to?

Mike: The science…the scientific world…like, if one scientist gets a bad reputation from doing something that the public doesn't like, the public might think like if one scientist is like this, the others might be the same.

TEACHER: Good point Mike. Would everybody else agree? [class agrees]. Now beyond obesity, let's think about like other issues like, for example, the pylons thing that is going on[2], that's a massive debate isn't it? So when things like that occur, it's important to think about our responsibility to the public but also to the science community and to consider all the evidence isn't it? [class agrees, bell rings].

By the end of the post-lesson discussion, students had developed a strong sense about the importance of scientific communication as appropriate for purpose and audience, as well as the roles and responsibilities of scientists in making claims based on valid evidence. This experience solidifies the implications of previous research (Erduran et al., 2004) that argumentation is a form of discourse which, in order to promote acquisition of knowledge and the development of scientific habits of mind, needs to be taught explicitly in schools.

Role Model Experience

To connect the students' learning with that of real scientists in a further and visible way, I then invited a role model to visit the group. I elected to choose my own role model and former student, Dr. Elaine McCarthy, who completed her PhD in Nutritional Science, to come and talk to the students. I chose Elaine both for her exemplary work as a scientist, but also as a source of inspiration for my students who could, if so desired, walk a similar path to her as a graduate of our school and become an actual scientist. Elaine drew upon the experiences of the classes previous and married them with her work as a scientist. She reinforced key ideas the students would have engaged with, such as the multitude of diverse ways to communicate Science. She stressed the need for integrity and honesty in engaging with Science in the laboratory and in sharing new knowledge with wider audiences. She gave interesting examples of how she communicated Science to young children partaking in longitudinal research and how she felt ethically responsible for the accuracy and clarity of this communication. This was an inspiring message for students to hear and gave validity to the work they had completed in their previous lessons. The role model experience was extremely positive in not only reinforcing the work of students, but also in connecting the real world of Science with the experiences of students in schools. This experience validates the case for periodically introducing scientists to students as a means of positively reinforcing their attitudes to Science and, in this case, women in Science (Smith & Erb, 1986). Elaine's post – lesson reflections offer further value to this assertion:

One thing I did think important about the role model experience was that it had the potential to highlight how diverse "science" is. Sometimes, I think the public perception of science is skewed and this is something that I think turns people off of science, especially girls. Science covers a very broad spectrum of topics, including my own field of nutrition, which is very vast and highly applicable to "real-life". I think "science" is essentially a way of doing things in a very logical, clear and transparent manner, that can be applied to lots of different fields like nutrition and chemistry.

Figures 3 to 5 below are examples of messages Elaine conveyed to students. The experience of a role model provides positive opportunities for students to engage with Science as a living and interconnected discipline, of which their work is seen to play an important part. This aspiration is congruent with recent STEM reform intentions nationally, as adumbrated in the Irish STEM Education Policy Statement (Department of Education and Skills, 2017, p.8). The policy acknowledges that STEM learning takes place in formal and informal settings across a multi-stakeholder, multi-programme landscape and seeks to bring cohesiveness to the STEM learning experience for each learner through connecting these experiences. Role model engagement with students in the classroom is one measure that could help to enhance this connectivity.

Based on my students' engagement with the PRB and subsequent role model experience, I contend this was a valid experience of IBSE, insofar as students had the opportunity to consider evidence, use logic and reason to extract salient points of information to communicate and defend their argument (Bell, 2010). Whilst there was an overarching focus on scientific communication through a structured/ guided inquiry experience, I do lament the loss of opportunity to address some misconceptions which arose during the learning experiences. For example, I was hoping for a more robust discussion on the differences between Body Mass Index (BMI) and Basal Metabolic Rate (BMR) as valid measures of human health, however this did not happen. In future experiences, I would deem it important to address such issues to ensure better student attainment of Science concepts.

Engagement with IBSE in this hands-on and minds-on way has the potential to realise exciting aspirations for Science education. These include an enhanced understanding of scientific communication, the responsibility and contributions of Science to society and vice versa, and a lived experience of the work of scientists authentically connected with the classroom. However, aspirations require an appropriate curricular structure to support their realisation (Priestley & Humes, 2010). In a poignant moment following Elaine's presentation, a student asked her "*why did you want to become a scientist?*", to which she replied "*it was because I did Science Fairs with Mr. King. I really loved Science because of that*".

Figure 3. Aspects of 'being a scientist' shared by role model with students

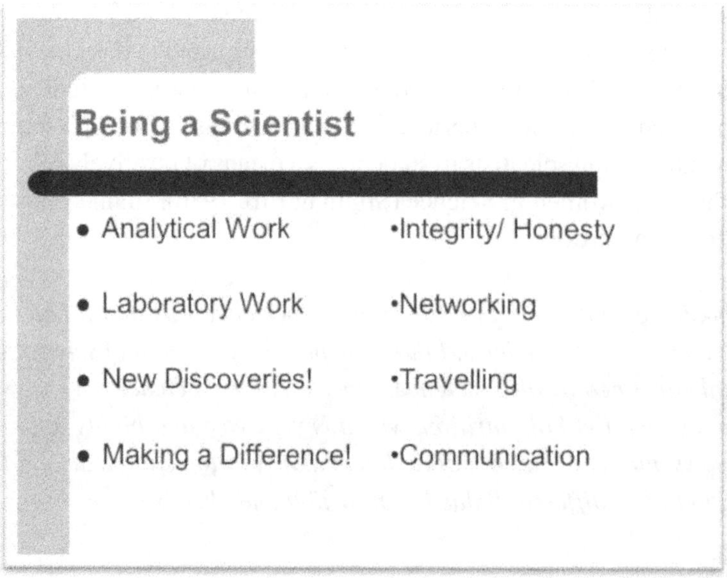

Figure 4. Role model communicating science to a young child partaking in research

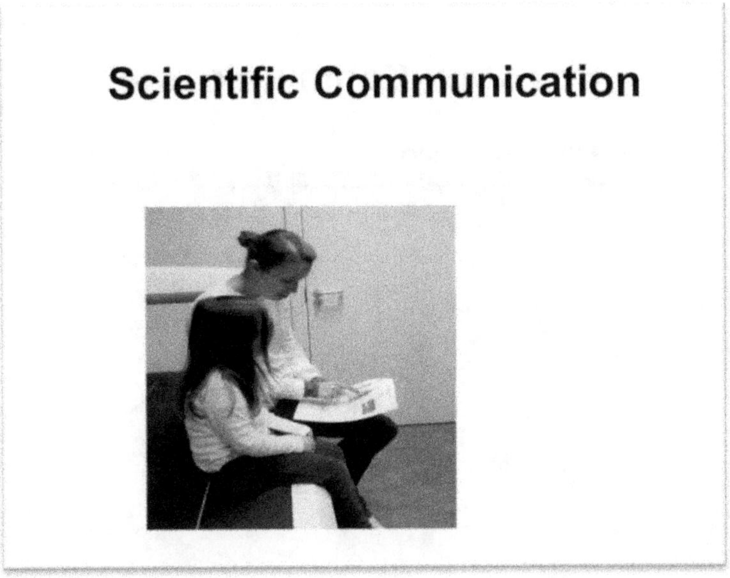

Whilst this answer was heartfelt, it was also stark in its absence of reference to the study of Science as a subject within the curriculum. It seems through engagement with real scientific work for extra-curricular Science fairs, Elaine had cultivated an enthusiasm for Science, contributing to her aspirations to become a scientist. Whilst the previous Science Syllabus (2003) had some lofty aims concerning the acquisition of Science knowledge and skills, as the promotion of responsible citizenry and the study of Science subjects at upper secondary level (Department of Education, 2003, p.3), it did not explicitly aim to support students to engage with the living field of Science, working as scientists within society. Next, I outline how the curriculum has evolved to support these aspirations.

CURRICULUM CHANGE IN IRELAND

From Specifications to Learning Outcomes

In 2014, the Department of Education and Skills launched a *Framework for Junior Cycle*. An aspiration of the *Framework* was to allow for greater deregulation of the curriculum, so that schools could develop a responsive Junior Cycle (JC) curriculum narrative appropriate to their context, student needs and passions. The *Framework* also aspired to centralise teachers in the assessment process, wherein previous JC summative assessment for certification was conducted entirely by an independent body, the State Examinations Commission (SEC). Following a significant period of political unrest, changes were made to the policy document and intentions, leading to a rewriting of the *Framework* in 2015. In *Framework '15*, teachers would not assess their students for certification purposes, this would still be conducted by the SEC. However, teachers are now explicitly recognised as central to guiding students through ongoing teaching, learning and assessment, with responsibility for overseeing and assessing

Figure 5. Examples of scientific communication shared with students

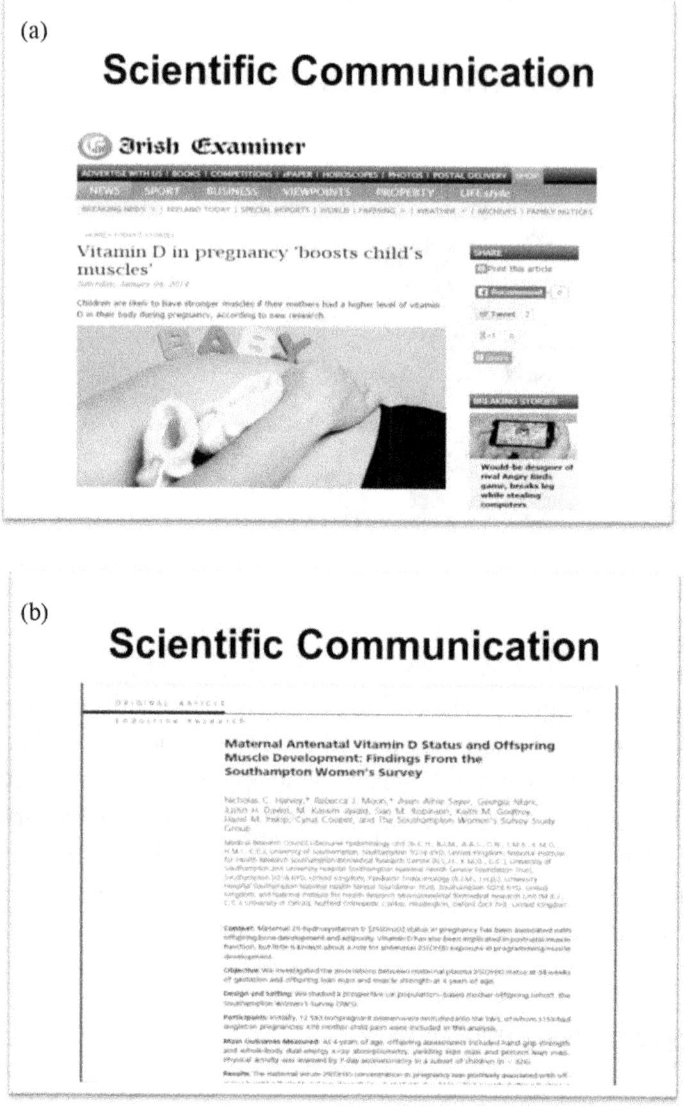

two classroom-based assessments (CBAs). Students achievements in these CBAs are documented and reported upon as part of a Junior Cycle Profile of Achievement (JCPA).

A number of key structural changes to the curriculum were conducive to supporting the types of connectivity which were deemed beneficial in the CR community. Learning in JC is now organised across a number of key areas, including:

- Subjects, e.g. Science which are taught for a minimum of 200 hours.
- Short courses: Schools can opt to run short courses that met the needs of the students and school and may draw from a range of disciplines, e.g. STEM, caring for animals, computer coding. A short course is taught in approximately 100 hours.

- Other areas of learning: These can include social, cultural, pastoral, scientific, entrepreneurial and other activities, e.g. if students enter a science quiz or compete at a science fair.

In keeping with the motivation for curriculum deregulation, our subject documents have changed to what are known as *curriculum specifications*. Our previous curriculum documents were known as syllabi. In the case of the 2003 Science Syllabus, "topics and sub-topics are described, together with associated learning outcomes. In general, these are presented in an increasing order of difficulty—later investigations and experiments build upon and extend the knowledge and skills developed earlier." (p.5). With a curriculum specification, we are presented with a document that describes what students should be able to do at the end of a unit of learning, with the possibility of flexible teaching and learning approaches to reach those end points (Walshe, 2015, pp. 99-100). The Science Specification shows "how junior cycle science is linked to central features of learning and teaching outlined in the *Framework for Junior Cycle*." (2016, p.6). The Science Specification outlines examples of how the learning in Junior Cycle Science is linked to particular statements of learning and key skills, which informed the writing of 46 learning outcomes, i.e. "Statements that describe the understanding, skills and values students should be able to demonstrate after a period of learning". It provides a rationale and aims for Junior Cycle Science, as well as a description of each of the strands and its learning outcomes. An example of an open learning outcome in JC Science which speak to the aspirations I held for my own students' learning is:

Students Should Be Able to Evaluate Media-Based Arguments Concerning Science and Technology

As can be seen, there are multiple entry and exit points to this type of student learning, which provides rich opportunities for student and teacher exploration. These types of outcomes invite a construction of meaning of curriculum as a social practice, much in keeping with the values and desires of teachers involved in CR Ireland. With open learning outcomes, an opportunity exists for teachers and students to develop a curriculum narrative in which the end point may be defined, but flexible routes to this point can be determined according to the school context. This structural change to how we frame learning shows great promise in allowing authentic inquiry experiences to flourish in classrooms and an increased ownership of the curriculum by teachers and students.

A curriculum specification describes expectations for students, which "links learning outcomes with annotated examples of student work" (NCCA, 2016, p.15). These examples (See Fig. 6 for example) are developed by teachers and students in Irish Science classrooms, in collaboration with the National Council for Curriculum and Assessment (NCCA) and are updated on an ongoing basis and shared with teachers online. The purpose of the examples of student work is "to show the extent to which the learning outcomes are being realised in actual cases" (ibid, p.15). In this regard, a curriculum specification and its associated assessment guidelines, along with annotated examples of student work, are intended to be a set of interactive, online resources, accessible by teachers, students and parents. This ensures a degree of curricular coherence without mandating an overarching level of curriculum control. Further, it allows for evolution in curricular engagement and understanding, as new examples of the curriculum in action are generated on an ongoing basis.

Figure 6. An annotated example of student work from NCCA
Source: www.curriculumonline.ie

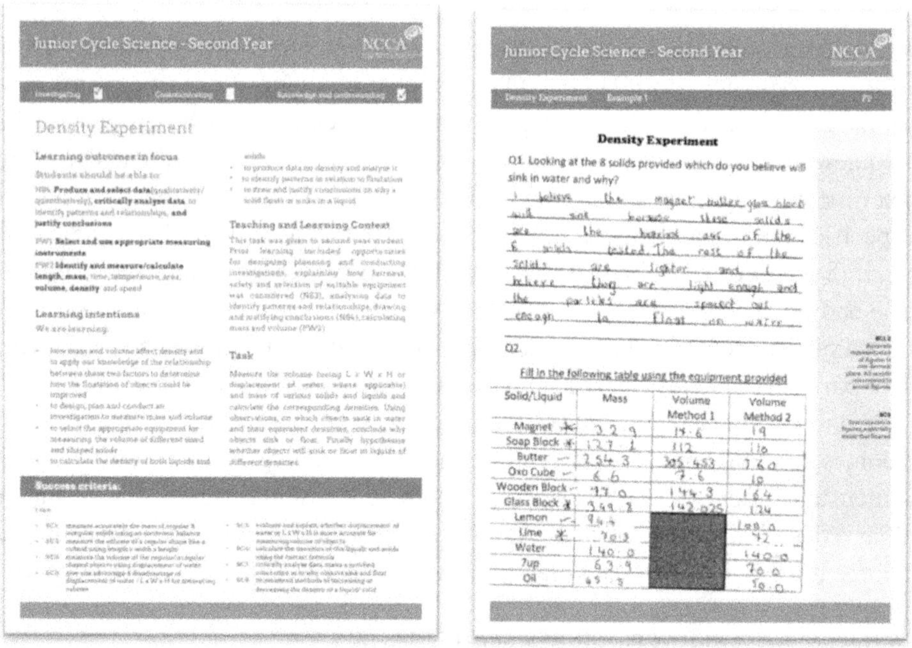

A Closer Look at Science: Aims, Rationale, and Learning Outcomes

The Curriculum Specification for Junior Cycle Science frames a rationale and aims for the teaching of Science in Junior Cycle (NCCA, 2016, pp.4-5). In this, one can find rich similarities between the policy intentions of the Specification and many of the desired aspirations of the teachers involved in CR Ireland, which now may be rendered possible thanks to appropriate curricular structures. These include:

- A desire for Science to be a collaborative, creative human endeavour, stemming from children's natural curiosity in thoughtful and deliberate action
- An acknowledgement that Science is not a tidy package of knowledge
- A desire "to develop students' evidence-based understanding of the natural world and their ability to gather and evaluate evidence: to consolidate and deepen their skills of working scientifically; to make them more self-aware as learners and become competent and confident in their ability to use and apply science in their everyday lives"
- An explicit aim of JC Science in developing a scientific habit of mind and inquiry orientation

The aims and rationale for JC Science are realised through the learning outcomes of the Specification, 36 of which are organised across four contextual strands providing the subject specific context for student learning. These strands are called Earth and Space, Chemical World, Physical World and Biological World. All strands are divided into elements, which are called:

- Building blocks

- Systems and interactions
- Energy
- Sustainability

Through the elements, teachers are invited to see the connecting content, thinking and cross-cutting concepts interwoven through the world of Science, as opposed to a narrow viewing of biology content, for example, as separate from matters pertaining to chemistry and physics. This further reinforces the intention that Science is not a neat and tidy package of knowledge, an intention which the CR project aimed to represent in the cross-disciplinary learning of the PRBs.

Underpinning and suffusing the learning within the contextual strands is the Nature of Science strand, which seeks to provide opportunities for students to learn about, with and through Science in every class they experience. There are 10 learning outcomes in the Nature of Science strand, divided across a number of elements unique to this strand, which are focused on

- Understanding about Science
- Investigating in Science
- Communicating in Science
- Science in Society

Figure 7. The learning outcomes of the JC science specification
Source – www.jct.ie

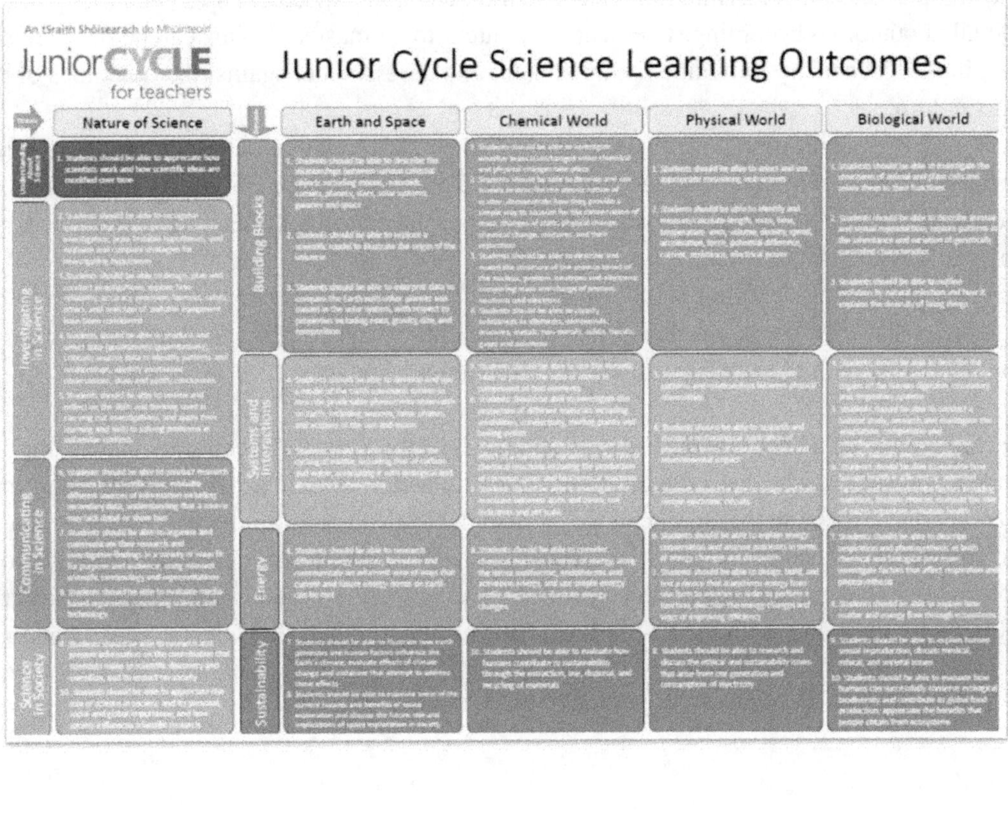

It is in the Nature of Science strand that we see the true potential of IBSE to flourish. The Specification identifies “a strong focus on scientific inquiry” (NCCA, 2016, p.11) as central to realising the learning within this strand. It is envisaged that students would experience this learning in every Science class as the learning from the Nature of Science strand is experienced through the content and activities of the contextual strands. This sits appropriately with the way learning was framed in the PRBs and with my own students. Through engagement with content related to obesity and health, they learned how to produce and select data, organise and communicate findings appropriate to a specific audience and engaged in dialogue which led to a more practical understanding of how scientists work. These aspirations for student learning are aligned with the learning outcomes of the Nature of Science strand. I have experienced first-hand the affordances of an open learning outcomes approach to my practice in terms of evolution in my students’ engagement with Science and evolution in the power dynamic within the classroom. Through adopting an open learning outcomes approach, a more holistic and coherent practice emerged from the classroom experience (Drew & Priestley, 2014).

Whilst an open learning outcomes approach may seem laudable, we must also remember the concerns raised by CR Ireland teachers, and also reflected by Irish teachers more broadly, around the need for structure and balance in the curriculum, particularly in light of perceived expectations of parents and assessment requirements. In this context, a shift to open learning outcomes can pose challenges to the system. However, Priestley (2017) counters the concerns over open learning outcomes with a summation of the challenges posed by learning outcomes which are over-specified. These include:

- A narrowing of the curriculum and a reduction of learning to an assessment performance.
- Over-specified outcomes risk becoming ‘tick-box’ in nature leading to teaching to the test, strategic compliance and performing to the letter of curriculum expectations
- Detailed outcomes becoming statements of content to be mastered. This can be particularly challenging if teachers try to audit existing content and assessments against the outcomes of the new curriculum.
- A “spiral of specification” (Wolf, 1995, in Priestley, 2017), where outcomes are translated into a large set of unwieldy standards and benchmarks. This has been the case in other countries such as Scotland, where over-specifying added unnecessary layers of complexity and bureaucracy, making courses difficult to teach.
- An assumption that learning is a linear process. The learning outcomes in Junior Cycle Science celebrate that learning is non-linear and provide autonomy for teachers and students to develop and guide the path of learning according to their own contexts.

The changes brought about by the learning outcomes of the Science Specification have implications for teachers and students in terms of classroom planning and in teaching, learning and assessment. To support teachers in this time of major change, a formidable CPD programme was required at a subject specific and whole school level. Next, the approaches to CPD adopted by the system are considered in the context of challenges and opportunities for IBSE to embed into classroom practice, both through initiatives such as CR and as aligned with the curricular intentions of the Science Specification.

CPD BY TEACHERS, FOR TEACHERS

In 2014, the DES launched a new professional support service dedicated to support teachers, schools and school leaders in the enactment of JC reform. The Junior Cycle for Teachers (JCT) Support Service exists to "inspire, support and empower teachers in the transformation of junior cycle education in Ireland" (JCT, 2018, p.18). To realise this vision, JCT offers CPD support across a range of areas reflective of the school programme and curricular aspirations. During my time involved with CR Ireland, I was privileged to be seconded to the JCT Science Team[3]. Our team supports the professional learning of over 5,500 Science teachers across Ireland through the design and delivery of high quality, research-informed CPD that is onsite, online and ongoing. Our approach to professional development seeks to place high trust in teachers as curriculum developers. On JCT Science CPD days, we aim to support teachers' professional learning in a number of areas, including

- Developing pedagogical practices
- Developing understanding of outcomes-based education
- Developing assessment practices
- Challenging assumptions concerning the teaching of Science
- Promoting the centrality of the Nature of Science in the curriculum
- Promoting student voice and a student-centred curriculum
- Giving value and validity to teachers' professional judgements

Central to all of this is the promotion of collaborative practices within and across subject departments. We endeavour to create space on CPD days for the promotion of professional conversations in which curricular sense making occurs in a collaborative environment. This is similar to some of the approaches to professional development adopted in the CR Ireland PLC and is reflective of the desire to enhance, through an ongoing approach to CPD, the social capital amongst Science teachers:

What is crucial is what happens between workshops. Who tries things out? Who supports you? Who gives you feedback? Who picks you up when you make a mistake the first time? [..] Learning is the work, and social capital is the fuel. If social capital is weak, everything is destined for failure. (Hargreaves & Fullan, 2012, p.92)

Curriculum alignment ensures coherence between the intended curriculum, the enacted curriculum and the assessed curriculum (Porter, 2007; Webb, 1990; Ziebell, Ong & Clarke, 2017). I contend that promotion of social capital is one measure that promotes fidelity to the intended curriculum. Drew, Priestley and Michael (2016) have reported on the issue of the "implementation gap between policy intention and classroom practice". By understanding the contextual factors influencing the intended, enacted and assessed curriculum from the teachers' perspective, we create an opportunity to narrow the gap and enable this through CPD that promotes social capital.

This opportunity was optimised in how the DES and JCT organised CPD arrangements for schools. Prior to the introduction of JC reform, it was common practice for subject teachers to convene at local education centres. Due to the demands of school supervision and class contact time, this meant teachers from the same subject department would normally attend different CPD days. Now, in a first for Irish education, schools across the country were clustered together for subject-specific JC CPD from December 2016. This 'cluster model' saw schools in a locality (typically 6-10 schools) close on the same day to facilitate CPD provision. From a subject perspective, this meant whole subject departments from neighbouring schools could attend the same CPD day. This has the potential to improve social capital through weaving together cognitive and social aspects of teacher learning. (Johnson et al, 2011, p.12). Pedagogical concepts combined with structured conversations and reflections can lead to resource building that supports ongoing learning about the curriculum. This further leads to the building of resources that can improve social capital within subject departments and across schools in regions.

In my intimate experience of leading CPD for Science, I found space could be created for collaborative opportunities within and across schools in localities to be explored. With whole subject departments and neighbouring schools in attendance, the possibility existed to inculcate a rich, context-specific dialogue around the enactment of the curriculum, and to look at possibilities for these types of practices to embed into classroom and school culture, much like the rich conversations within the CR Ireland PLC. These conversations on CPD days may serve as a catalyst for the promotion of social capital. It has always been the intention of JCT to continue this support between CPD days as well, cognisant that the day in and of itself is not enough to sustain a change in practice. Hence, conversations on CPD days are always steered towards actionable change in schools and classrooms, and online CPD supports from JCT Science serve to guide teachers as they take these actions. These onsite and online measures combined may serve to support an achievement of agency amongst teachers in the ways advocated by Priestley et al. (2015, p.31) – through a focus on capacity building, within and across schools, to develop agents of change and professional developers of the curriculum. However, the supportive and dialogic nature of this professional development, as manifest in a local/regional context, may also help to develop the right types of structural, cultural and relational conditions to enable achievement of teacher agency to engage with educational purpose and principles of curriculum development. Hopefully these local onsite and continuous online CPD possibilities will help to reduce uncertainty and fear of change as teachers engage in the collaborative process of curricular sense-making.

In my experience of Science CPD across a number of years, I contend the importance of professional conversations grows with time. The potency of professional development to promote meaningful and lasting change is rooted in a number of factors, including a change in classroom practices, changes in attitudes and beliefs and an improvement in learner outcomes. According to Guskey (2002, p.383), significant changes in teacher attitudes and beliefs occur after they have seen a change in learner outcomes. Initially, JC Science CPD was being offered a year prior to subject introduction, so teachers had no classroom experience of the new curriculum to speak to. Over the years, as the CPD programme progressed, teachers began to gain experience of teaching the JC Science curriculum. Through a cluster model of CPD, improved learner outcomes and the contexts which led to them can now be discussed amongst teacher colleagues. Reflecting upon classroom experiences at CPD creates a resource for social capital, given that reflection is both a social and individual activity (Johnson et al., 2011, p.10). This can have a rippling effect on other teachers within and across schools, encouraging them to take on new approaches, see improved learner outcomes for their own students, change their attitudes and beliefs and evolve their own practice.

CONCLUSION

The CR Ireland PLC allowed me a freedom and space to engage with the principles and purpose of IBSE with fellow colleagues, and through classroom experiences my students and I developed aspirations for IBSE in our everyday experiences. Through expert knowledge supports, the shared commitments of my fellow CR teachers and the collaborative and reflective spaces afforded by the PLC, resources were made available that served to enhance our social capital and support ongoing learning around IBSE. Since the dissolution of CR Ireland, the changed curricular structures in JC reform and ongoing CPD supports show promise in continuing to promote the types of aspirations being sought, and at times realised, in the CR initiative.

We are living through an exciting time of change in Irish Science education. National review of STEM education, a number of active FP7 projects[4] situated in Ireland over recent years, and the ongoing enactment of *Framework '15*, have contributed to the evolution of Science curriculum and practices. The lessons from JC Science CPD speak to initiatives such as CR which sought to promote a sustainable change in IBSE amongst the teachers within its community. Sustainability refers to meeting the needs of today without jeopardising the future. In seeking a sustainable approach to IBSE, we need to support professional learning in ways that is rooted in understanding current practice but also has an eye on the future; a future in which teachers and students have the confidence, competence and capability to engage with IBSE as a norm and in school contexts which enable this, cognisant that curriculum enactment is mediated by, with and through context (King, 2017, p.168).

Social capital recognises that teachers play a more autonomous and conscious role in their own development. Professional development may increase their access to social capital, but teachers decide for themselves if and how this capital is used (Johnson et al., 2011, p.3). This speaks to the sustainability of approaches to changing practice in projects such as CR Ireland and also in curricular changes such as JC reform. Professional development can support increased access to social capital, however it is up to teachers to decide the degree to which social capital is used. One can only hope that through seeing changes in learner outcomes and the benefits of this for their practice and schools, along with the value of collaborative practices within subject departments, teachers will express increased desires to use social capital as a powerful force to grow and sustain their engagement with IBSE and curriculum change. Evolution takes time but it does happen - if the conditions for change are right.

REFERENCES

Ball, S. J. (2003). The teacher's soul and the terrors of performativity. *Journal of Education Policy*, *18*(2), 215–228. doi:10.1080/0268093022000043065

Bell, R.L. (2010). Teaching about scientific inquiry and the nature of science: Toward a more complete view of science. *The Journal of Mathematics and Science: Collaborative Explorations*, *13*, 5-25.

Day, C. (1987). Professional learning through collaborative in-service activity. In J. Smyth (Ed.), *Educating teachers: Changing the nature of pedagogical knowledge* (pp. 207–222). New York: Falmer Press.

Department of Education and Skills. (2015). *A Framework for Junior Cycle*. STEM Education Policy Statement.

Donnelly, D., McGarr, O., & O'Reilly, J. (2014). "Just be quiet and listen to exactly what he's saying": Conceptualizing power relations in inquiry-oriented classrooms. *International Journal of Science Education, 36*(12), 2029–2054. doi:10.1080/09500693.2014.889867

Drew, V., Priestley, M., & Michael, M. K. (2016). Curriculum Development Through Critical Collaborative Professional Enquiry. *Journal of Professional Capital and Community, 1*(1), 92–106. doi:10.1108/JPCC-09-2015-0006

Erduran, S., Simon, S., & Osborne, J. (2004). TAPping into argumentation: Developments in the application of Toulmin's argument pattern for studying science discourse. *Science Education, 88*(6), 915–933. doi:10.1002ce.20012

Gleeson, J. (2010). *Curriculum in Context: Partnership, Power and praxis in Ireland*. Bern: Peter Lang.

Guskey, T. R. (2002). Professional Development and Teacher Change. *Teachers and Teaching, 8*(3/4), 2002.

Hargreaves, A., & Fullan, M. (2012). *Professional Capital: Transforming Teaching in Every School*. London: Routledge.

(1989). How do i improve my practice? *Cambridge Journal of Education, 19*(1), 41–52.

Huberman, M. A. (1989). The professional life cycle of teachers. *Teachers College Record, 91*(1), 31–57.

Huberman, M. A. (1993). *The Lives of Teachers*. New York: Teachers College Press.

Johnson, W., Lustick, D., & Kim, M. (2011). Teacher Professional Learning as the Growth of Social Capital. Current Issues in Education, 14(3).

Junior Cycle for Teachers. (2018). *JCT Development Strategy – Building on our Achievements*. Author.

King, D. (2017). *Junior Cycle Reform: Why Context Matters. A context-centric analysis of curriculum reform in lower Irish secondary education*. Doctoral Thesis.

Leat, D., Livingston, K., & Priestley, M. (2013). Curriculum deregulation in England and Scotland - Different directions of travel? In W. Kuiper & J. Berkvens (Eds.), *Balancing Curriculum Regulation and Freedom across Europe. CIDREE Yearbook, 2013* (pp. 229–248). Enschede, The Netherlands: SLO Netherlands Institute for Curriculum Development.

National Council for Curriculum and Assessment. (2003). *Junior Certificate Science Syllabus*. Ordinary Level and Higher Level.

National Council for Curriculum and Assessment (2016). *Curriculum Specification for Junior Cycle Science*. Author.

National Council for Curriculum and Assessment. (2016). *Guidelines the Classroom Based Assessments and Assessment Task (Assessment Guidelines)* (1st ed.). Science.

Porter, A. C., Smithson, J., Blank, R., & Zeider, T. (2007). Alignment as a teacher variable. *Applied Measurement in Education, 20*(1), 27–51. doi:10.1080/08957340709336729

Priestley, M. (2017). *Approaches to specifying curriculum areas of learning*. Stirling: The Stirling Network for Curriculum Studies.

Priestley, M., Biesta, G., & Robinson, S. (2015). *Teacher Agency: An Ecological Approach.* London: Bloomsbury. doi:10.4324/9781315678573-15

Priestley, M., & Humes, W. (2010). The development of Scotland's Curriculum for Excellence: Amnesia and déjà vu. *Oxford Review of Education, 36*(3), 345–361. doi:10.1080/03054980903518951

Smith, W. S., & Erb, T. O. (1986). Effect of women science career role models on early adolescents' attitudes toward scientists and women in science. *Journal of Research in Science Teaching, 23*(8), 667–676. doi:10.1002/tea.3660230802

Wallace, C., & Priestley, M. (2016). Secondary Science Teachers as Curriculum Makers. *Mapping and Designing Scotland,* 1.

Walshe, A. (2015). *From Curriculum to Classroom in Upper Second Level Science.* Doctoral Thesis.

Webb, N. (1999). *Research monograph No. 18. Alignment of science and mathematics standards and assessments in four states.* Washington, DC: Council of Chief State School.

Whitehead, J. (1989). Curriculum for Excellence. *Journal of Research in Science Teaching.*

Ziebell, N., Ong, A., & Clarke, D. (2017). Aligning curriculum, instruction and assessment. In T. Bentley & G. Savage (Eds.), *Educating Australia: Challenges for the decade ahead* (pp. 255–274). Melbourne: Melbourne University Publishing.

ENDNOTES

1 Based on our experiences in year one, we chose to actively 'head hunt' teachers for the PLC in year two. Whilst we got a full cohort of positive teachers, with little attrition in this year, one could argue against the potency of initiatives such as CR to support sustainable approaches to IBSE if 'running with the runners' is the *modus operandi* which leads to optimal success.

2 At the time of filming the class, the local school communities were in conflict with the national electricity supply groups in Ireland over plans to erect large pylons in the area. Following this class experience, we went on to do a separate PRB concerning the arguments for and against the erection of pylons in the area and had a class debate on whether it should take place or not.

3 At the time of writing this chapter, I am currently Team Leader for Science CPD

4 Example projects include CR Ireland, and the Establish and SAILS projects from Dublin City University

APPENDIX: OBESITY PRB

Tackling Obesity in Ireland

You are a team of researchers in Ireland acting as consultants to the Health Trust, a committee whose mission it is to promote healthier living amongst Irish people. Your brief (i.e. your job described) is to come up with an advertisement to inform the Irish public about the dangers of obesity and the ways they can tackle it. You will present information on specific aspects of obesity, including causes, Irish statistics on the matter and measures people can take to prevent it

Science you should know before you start this activity:

- The functions of different food groups in the body
- How to have a balanced diet based on the types and portions of foods in the food pyramid
- The role of food additives and the types of food additives

o Science you will learn about in this activity:

- The basal metabolic rate (BMR) of the human body
- Definitions and causes of obesity
- What a BMI score is and how to calculate it
- How to visually represent statistics using bar charts and/or line graphs

Checking your learning:

You will produce a newspaper advertisement informing the public about the dangers of obesity in Ireland and the ways this can be tackled. You should make sure you produce the following items as you work through the activity:

Item 1

- Information on obesity– definition and causes
- How to calculate a Body Mass Index (BMI) value for an average adult male/female

Item 2

- Obesity in Ireland – Occurrence and statistics in Ireland

Item 3

- Impacts of obesity – associated health problems

Item 4

- Prevention and treatment of obesity

Table 1

The Health trust

An Irish healthy living committee — Meeting Agenda

Table 2.

Item	Description	Suggested times	Presenter
1	Captain's introduction **in your 'HOME' group** SPLIT INTO EXPERT GROUPS AFTER CAPTAIN'S INTRODUCTION. Then **spend 10 minutes in your expert groups**. After this – **RETURN FROM EXPERT GROUPS, HOME GROUPS FORM AGAIN Items 2-5 below: in home groups: each expert has 5 minutes to present their expert knowledge to the home group**	5 mins 10 mins	
2	What is obesity?	5 mins	
3	Obesity in Ireland	5 mins	
4	Impacts of obesity	5 mins	
5	Prevention and treatment of obesity	5 mins	
6	Discussion about newspaper advertisement (class 1)	30-35 mins	
7	Presentation of the final draft of newspaper advertisement (Class 2)	5 mins	

Committee Meeting Notes for the Captain

Item 1 Starting the Meeting: Introduction

- Welcome the members who will be acting as consultants to the committee
- Outline the purpose of the meeting:

The Health Trust is a committee whose mission it is to promote healthier living amongst Irish people. Your brief (i.e. your job described) is to come up with an advertisement to inform the Irish public about the dangers of obesity and the ways they can tackle it. You will present information on specific aspects of obesity, including causes, Irish statistics on the matter and measures people can take to prevent it

- Read out the items on the agenda
- BREAK UP THE TEAM MEMBERS INTO THEIR EXPERT GROUPS. THEN GO TO YOUR OWN EXPERT GROUP
- **PEOPLE STAY IN EXPERT GROUPS FOR 10 MINUTES, GAINING THEIR EXPERT KNOWLEDGE TO BRING BACK TO THE HOME GROUP**
- **AFTER TEN MINUTES**, EVERYONE RETURNS TO THE HOME GROUP FROM THEIR EXPERT GROUPS

Items 2 – 5: In the Home Group

- Invite each member to give her/his presentation in turn based on their expert knowledge
- After each presentation ask if anyone has a question
- **Try to make sure each item lasts no more than the 5 MINUTES of time allotted on the agenda**

Items 6 - 7

- Chair the discussion about the style and content of the newspaper advertisement (**it should be no more than 80 words and include an eye-catching graphic**)
- Oversee the final draft of the advertisement.
- Decide on presenters of different sections – THIS SHOULD BE A TEAM PRESENTATION WITH THE CAPTAIN AS THE LEADER

Notes for experts
Please read this before proceeding:
You are going to give a brief presentation on this topic in your home group (approx. 5 minutes)

- Prepare yourself by studying the information in box 1.
- Use the pictures and graphics provided to help you
- It might be useful to calculate the BMI values for the example male and female adult provided

Eg – an average adult male/female– if they are 5 ft 7 inches tall (1.7 metres tall) and weigh 14 stone (90kg) – what would their BMI be? Based on their BMI score – how would we classify them? What would their optimum weight need to be for a healthy BMI?

Item 3 Obesity in Ireland: Occurrence and Statistics

Notes for experts
Please read this before proceeding
You are going to give a brief presentation on this topic in your home group (approx. 5 minutes)

- Prepare yourself by studying the information in box 2.
- Use the facts and figures provided to help you

Box 1.

In rich, industrialised countries like Britain and USA, more people suffer from overeating than from eating too little. If a person takes in more energy in food than they use up, then the excess is stored as fat. Taken to extremes, people can put on so much weight that they become **obese**.

Major causes of obesity include:

- High intake of fatty foods and refined foods containing a lot of added sugar
- Too little exercise
- Social and emotional stress, leading to 'comfort eating'

The overwhelming influences of obesity in 99% of the population are environmental and include marketing, advertising, increasing portion sizes, accessibility and availability of foods and facilities, increased automation and increased car use among other factors

There are two ways in which a person can identify being obese:

- ⇨ Being 20% above the recommended weight for his or her height
- ⇨ Having a **Body Mass Index (BMI)** greater than 30
    - The **body mass index (BMI)** is a measure for human body shape based on an individual's mass and height.
    - Devised between 1830 and 1850 by the Belgian polymath Adolphe Quetelet, it is defined as the individual's body mass divided by the square of their height – with the value universally being given in units of kg/m^2.
    - An adult with a BMI of less than 20 is underweight, between 25 and 30 is overweight and more than 30 is obese.
    - If an adult has a BMI of 20-24 they are at an optimum (ideal) weight

Figure 8.

$$\text{Body Mass Index} = \frac{\text{Weight (in kg)}}{\text{Height}^2 \text{ (in m)}}$$

- It might be useful to create bar charts or line graphs to visually represent the information given below

Notes for experts
Please read this before proceeding
You are going to give a brief presentation on this topic in your home group (approx. 5 minutes)

- Prepare yourself by studying the information in box 3

Box 2.

Obesity is the most common nutritional disorder in the world. The prevalence of overweight and obesity has increased with alarming speed over the past twenty years. It has recently been described by the World Health Organisation as a 'global epidemic'. In the year 2000 more than 300 million people worldwide were obese and it is now projected that by 2025 up to half the population of the United States will be obese if current trends are maintained. The disease is now a major public health problem throughout Europe.

In Ireland at the present time 39% of adults are overweight and 18% are obese. Of these, slightly more men than women are obese and there is a higher incidence of the disease in lower socio-economic groups.

Statistics related to obesity for Irish adults (source - Irish Heart foundation):

Irish adults (18 years and older)

Two out of five Irish adults - 39% (45% of men; 33% of women) are overweight

• One in four - 25% (24% of men; 26% of women) is obese.

• The trend towards obesity in Ireland is increasing. Between 1990 and 2000 the prevalence of obesity increased by 67% overall, up 1.25 fold in women (from 13%) and up 2.5 fold in men (from 8%)

• Ireland has the fourth highest prevalence of overweight and obesity in men in the EU and the seventh highest prevalence among women

Irish Teenagers (13-17 years)

• One in five teenagers is overweight or obese (11% overweight and 8% obese)

• There has been a significant increase in teenage obesity since 1990 with an 8-fold increase in males (1% to 8%) and a 2-fold increase in females (3% to 6%)

Economic Burden

• Costs for treating obesity in Ireland is estimated at €0.4 billion. The number of premature deaths annually attributable to obesity currently approximates to 2,000

- It might be useful to make a bullet-pointed summary of these different conditions and to rank them according to seriousness

Item 5 Prevention and Treatment of Obesity

Notes for experts

Please read this before proceeding

You are going to give a brief presentation on this topic in your home group (approx. 5 minutes)

Box 3.

Obesity has a far-ranging negative effect on health. Each year obesity-related conditions cost over 150 billion dollars and cause an estimated 300,000 premature deaths in the US. The health effects associated with obesity include, but are not limited to, the following:

- ⇨ **high blood pressure** - Additional fat tissue in the body needs oxygen and nutrients in order to live, which requires the blood vessels to circulate more blood to the fat tissue. This increases the workload of the heart because it must pump more blood through additional blood vessels. More circulating blood also means more pressure on the artery walls. Higher pressure on the artery walls increases the blood pressure. In addition, extra weight can raise the heart rate and reduce the body's ability to transport blood through the vessels.
- ⇨ **diabetes** - Obesity is the major cause of type 2 diabetes. This type of diabetes usually begins in adulthood but, is now actually occurring in children. Obesity can cause resistance to insulin, the hormone that regulates blood sugar. When obesity causes insulin resistance, the blood sugar becomes elevated. Even moderate obesity dramatically increases the risk of diabetes.
- ⇨ **heart disease** - Atherosclerosis (hardening of the arteries) is present 10 times more often in obese people compared to those who are not obese. Coronary artery disease is also more prevalent because fatty deposits build up in arteries that supply the heart. Narrowed arteries and reduced blood flow to the heart can cause chest pain (angina) or a heart attack. Blood clots can also form in narrowed arteries and cause a stroke.
- ⇨ **joint problems, including osteoarthritis** - Obesity can affect the knees and hips because of the stress placed on the joints by extra weight. Joint replacement surgery, while commonly performed on damaged joints, may not be an advisable option for an obese person because the artificial joint has a higher risk of loosening and causing further damage.
- ⇨ **sleep apnea and respiratory problems** - Sleep apnea, which causes people to stop breathing for brief periods, interrupts sleep throughout the night and causes sleepiness during the day. It also causes heavy snoring. Respiratory problems associated with obesity occur when added weight of the chest wall squeezes the lungs and causes restricted breathing. Sleep apnea is also associated with high blood pressure.
- ⇨ **psychosocial effects** - In a culture where often the ideal of physical attractiveness is to be overly thin, people who are overweight or obese frequently suffer disadvantages. Overweight and obese persons are often blamed for their condition and may be considered to be lazy or weak-willed. It is not uncommon for overweight or obese conditions to result in persons having lower incomes or having fewer or no romantic relationships. Disapproval of overweight persons expressed by some individuals may progress to bias, discrimination, and even torment.

- Prepare yourself by studying the information in box 4
- It might be a good idea to compile a list of do's and don'ts in terms of food and activities to prevent and treat obesity

Box 4.

The food you eat in a day should provide you with enough energy to get through that day, to keep your heart beating, your lungs working, your body temperature and all the chemical reactions in your body going. The 'ticking over' speed at which our bodies work is called the **basal metabolic rate (BMR).** The BMR varies from person to person, but roughly it uses 7000kJ per day. So even if you lie around doing nothing you still need this amount of energy. People with a low BMR are more likely to get overweight.

So, how do we prevent this? : A balance of food intake and physical activity is necessary for a healthy weight. The foods we individually consume and our participation in physical activity are the result of a complex supply and production system. **The growing research evidence that energy dense foods promote obesity is impressive and convincing.** These are the foods that are high in fat, sugar and starch. Of these potentially the most significant promoter of weight gain is fat and foods from the top shelf of the food pyramid including spreads (butter and margarine), cakes and biscuits, and confectionery, when combined are the greatest contributors to fat intake in the Irish diet.

In company with their adult counterparts **Irish children are also consuming large amounts of energy dense foods outside the home.** A recent survey revealed that slightly over half of these children ate sweets at least once a day and roughly a third of them had fizzy drinks and crisps with the same regularity. Sugar sweetened carbonated drinks are thought to contribute to obesity and for this reason the World Health Organisation has expressed serious concerns at the high and increasing consumption of these drinks by children.

Physical activity is an important determinant of body weight. It is now widely accepted that adults should be involved in 45-60 minutes, and children should be involved in at least 60 minutes per day of moderate physical activity in order to prevent excess weight gain.

Treating obesity?

Obesity is caused by lifestyle choices, and these habits need to be broken if an obese person is to stand a good chance of losing weight. There are no quick fixes. Your GP can refer you to a dietician or a nutritionist who will draw up a personal diet plan to help you lose weight. There are a number of **drugs** which can cause rapid weight loss results, often by stimulating the body's metabolism while suppressing the appetite. However, taking a drug for weight loss is a short-term treatment, as it does not confront the underlying behaviour that leads to gaining weight. **Surgical treatment** for obesity is not pleasant and is only considered for people who have a BMI level well over 30, and for whom diet and exercise has proved unsuccessful in shifting the weight. Options include stapling the stomach or wiring the jaw shut to reduce the potential for food intake by the obese person

Chapter 12
Reflections on Inquiry

Stuart Bevins
Sheffield Hallam University, UK

Louise Lehane
St. Angela's College, Ireland

Josephine Booth
Sheffield Hallam University, UK

ABSTRACT

This chapter provides concluding comments and reflections on the Chain Reaction project as well as inquiry approaches in general. The authors also include three reflective stories based on their involvement in the program. These stories attempt to bring differing roles and aspects of the project to life while identifying successes and barriers within the project. They also discuss implications for the future of science education across Europe and summarize their outcomes from the Chain Reaction project.

INTRODUCTION

In this chapter we reflect on our experiences of Chain Reaction from our roles as managers of this large and complex collaborative project, the main purpose being to provide a more personal account of the project from three key project team members in contrast to the previous chapters which are purely objective in their reporting. We will highlight lessons learned and identify some of the problems and complexities of such a large education project which impacts a diverse group of stakeholders, be it teachers, students, teacher educators or policy makers, and share our experiences of three different roles within the project consortium. Firstly, we set out our position regarding inquiry as an approach to teaching and learning in the science classroom, giving the reasons why we focused on Inquiry Based Science Education (IBSE) and highlighting key literature.

DOI: 10.4018/978-1-5225-5439-4.ch012

A BELIEF THAT INQUIRY IS EFFECTIVE

Our belief that inquiry approaches can add significant value to students' learning experiences in the science classroom strongly influenced the design of the Chain Reaction project and this belief has deep roots over many years of experience and scholarship in the field of science education. Historically, the idea of inquiry dates back to the time of Greek philosophers (Shields, 2014), with a more recent attempt to encourage science learning through inquiry originating with Dewey (1938). Since this point, the roots of scientific inquiry have been fertilised over the years by the work of scholars such as Herron (1971), Driver (1986; 1994), Osborne (1998), Harwood (2004), Bybee (2006a) and Crawford (2000). Our belief in inquiry is underpinned by a desire to support an approach that situates the student as an agent within the inquiry process and provides greater opportunities for encouraging a more positive attitude to the study of science and personal growth by encouraging student ownership of their own learning. We are not alone in this, indeed, the effectiveness of inquiry in science teaching and learning is supported by a large pool of scholars whose empirical work reports positive learning outcomes for students in terms of achievement, enthusiasm, ownership and scientific skills development (Minner, Levy, & Century, 2010; Minstrell & van Zee, 2000) as they practice science which mirrors, to some extent, authentic processes outside of the classroom. This work signifies continuing attention on inquiry teaching in science education and suggests a perceived importance placed on inquiry by a good portion of the science education community which we will now reflect upon. The following section will explore literature which expresses specific positive outcomes from engaging in the practice of inquiry.

IMPROVING STUDENT ATTITUDES

Crawford (2000) undertook a yearlong observational activity whereby she reviewed the teaching of a particular teacher who claimed to teach through inquiry. During this time the researcher attempted to understand students' attitudes towards learning science in this manner. She found that the authentic nature of the inquiry projects, which they engaged in throughout the year, motivated the students and made them consider the importance of doing science in this way for future college studies. Additionally, some students revealed the respect they had developed for their teacher because he included them as necessary contributors to the classroom 'research team' (Crawford, 2000). This implies that inquiry practices can help to foster positive relationships between teacher and student, a phenomenon reported by many teachers and students in the Chain Reaction project. Crawford (2000) did note however, that such positive interactions were not always met as some students felt that the teacher's high expectations of them had a negative impact. That said, the majority of students valued the rigor and effort needed from them (Crawford, 2000). Rees, Pardo and Parker (2013) endorsed Crawford's finding by suggesting that when using a supportive framework during an open inquiry activity, participating students can be highly engaged in the task.

INCREASING ACHIEVEMENT LEVELS

Mao and Chang (1998) investigated the effect of inquiry-oriented instruction on the achievement of students through earth science using a comparative study method. The results showed that students who were taught through inquiry scored significantly higher than those taught using a more traditional method. Wilson, Taylor, Kowalski and Carlson (2010) studied two groups of students both of whom had the same learning outcomes to reach. One group were taught through inquiry-based instruction and the other group through the traditional, more common, didactic styles of teaching. The results of their study showed that there were significantly higher achievement levels for students who were subject to the inquiry-based approach both immediately after instruction and four weeks post-instruction. Smolleck and Yoder, (2008) venture that better learning outcomes like the ones found above are achieved because an inquiry approach is more suited to encouraging the development of higher order thinking. Higher order cognitive thinking is best explored through Bloom's taxonomy of cognitive learning. Historically the work of Benjamin Bloom has been used to examine the different domains of learning, cognitive, affective and psychomotor. The cognitive domain involves knowledge and development of intellectual skills (Bloom, 1956). The domain has since been revised by Anderson, Krathwohl, Airasian, Cruikshank, Mayer, Pintrich, Raths, and Wittrock (2001). The revised cognitive domain of learning consists of 6 levels which are (1) remembering, (2) understanding, (3) analysing, (4) applying, (5) evaluating and (6) creating. As one moves up these levels (from 1 up to 6) thinking becomes more higher order, requiring students to move beyond remembering and understanding content to being able to apply, analyse, evaluate and create knowledge. As Smolleck and Yoder (2008) contend, engaging in scientific inquiry practices leads to students engaging in higher order thinking by analysing, applying, evaluating and creating knowledge which in turn leads to a greater conceptual understanding of the content being explored.

Blanchard, Southerland, Osborne, Sampson, Annetta and Granger (2010) also engaged in research to determine the effectiveness of inquiry-based instruction over traditional didactic instruction and revealed that students who engaged in inquiry-based learning activities had higher knowledge retention rates than the other students. They do stipulate however, that it is dependent on the level of inquiry. They found that when students were engaged through a guided inquiry approach they developed a much stronger understanding of subject matter than when taught by a more 'cookbook' approach to science (Blanchard *et al.,* 2010). Colburn (2000b) undertook a comprehensive literature review of studies which demonstrated increased student learning from an inquiry-based approach from the 1960's to the 1990's. He reported that existing literature indicates that inquiry instruction is equal to or superior to traditional instructional strategies in aiding students' achievement levels. Brickman, Gormally, Armstrong and Hallar (2009) found that students increased their scientific literacy significantly when engaged in an inquiry approach in comparison with traditional approaches. Moreover, their process skills also improved during investigations reflective of the scientific inquiry approach (Brickman *et al.,* 2009).

INCREASING STUDENT INTEREST

Palmer (2009) undertook a study with the objective of uncovering whether or not student interest levels increased during an inquiry approach. His primary objective was to understand teaching strategies which could enhance student motivation, focussing on short term situational interest, which is defined as interest present as a result of a specific situation. Research has concentrated on this type of interest as it centres on classroom events and their immediate impact on students, as opposed to personal interest (Palmer, 2009). The latter type of interest is difficult in lesson planning as it is near impossible to consider all students personal interests when planning a lesson (Palmer, 2009). This research found that students' situational interest was apparent during inquiry-based activities but that it did fluctuate with the type of activity students were engaged in. He found that one of the main sources of interest was the novelty factor of doing something different. He also found that giving students a choice in terms of how they approached solving a problem increased their interest levels which suggests these students enjoyed having autonomy and responsibility to design their own investigations. Other factors which influenced their interest levels were (1) that they were learning and (2) they were engaged in social interactions as they learned (Palmer, 2009).

Of course there are those who have argued that inquiry is not the 'holy grail' some laud it to be and suggest that there is empirical evidence to show that minimally guided instruction is less effective than instructional approaches that place emphasis on the teacher teaching. There are also critics of overly scaffolded models of inquiry which are thought to reduce inquiry to a series of procedures (Bevins & Price, 2016). For example, Kirschner, Sweller and Clark (2006) argue that instructional strategies that have minimal guidance from the teacher and therefore, ignore the limits of the working memory, are unlikely to be effective. They also state that there is no evidence to suggest that presenting learners with partial information enhances their ability to learn more than if they have been given full information (Kirschner *et al.* 2006). Moreover, Johnston (2008) emphasises that there may not be enough evidence currently to support the usefulness of inquiry solely for the purpose of refining student achievement on standardized measures. Other studies such as that carried out by Gormally *et al.* (2009) reveal that students resented activities that were inquiry-oriented because of the increased workload which they had to contend with in terms of engaging in critical thinking. This enhanced experience in critical thinking is not something that they necessary engage in when experiencing traditional didactic methods. Klahr and Nigam (2004) also show opposition to the use of discovery learning, which similar to scientific inquiry, models a constructivist approach to learning. However, it is important to note that Blanchard *et al.* (2010) examined the studies undertaken by Klahr and Nigam (2004) and Kirschner *et al.* (2006) and concluded that the discovery approach of Klahr and Nigam (2004) and the inquiry approach used by Kirschner *et al.* (2006) resembled open inquiry.

However, it is not the purpose of this book to delve deeply into arguments for and against 'inquiry' as a pedagogy for effective science education. We have already made our belief clear—inquiry is an effective approach to teaching and learning science, and probably many other subjects, and can produce benefits for teachers and students during their subject learning that more structured approaches do not.

CHAIN REACTION

In recognising the above mentioned benefits, the overall aims of the Chain Reaction project were to make a contribution towards the development, delivery and sustainability of 3 key strands of activity which are:

- Inquiry Based Science Education (IBSE)
- Teacher professional development
- European teacher network

The first strand looked at the focus of IBSE in the development of student intrinsic interest, motivation and participation in science. Effectively, the project leaders wished to provide opportunities for students within the consortium countries to experience science that, as much as possible, would reflect the spirit of professional science.

The second strand related to teacher professional development. How this varied within the different countries has been discussed already in previous chapters but in effect, our professional development offer focused on how they approached the teaching of science; how they understood inquiry; and how they provided a hook for students to engage their critical thinking minds.

Our final strand intended to develop not just a locally situated teacher network, but to develop a European-wide teacher network. Our reading and thinking inspired us to encourage participating countries to engage with each other as much as possible to share knowledge, understanding, ideas, problems and ask questions of each other. Essentially, we felt that providing a platform for the development of a community of science teachers would establish an effective support mechanism which would be owned, solely, by the teachers.

It is also the case that each partner country had additional aims to suit their own educational context. For example, in Ireland, as discussed in chapter 4, the project took place during a period of curricular change, therefore one of the aims of the project was to use the experiences from the Chain Reaction project to partly influence the design of the science specification within the new Junior Cycle curriculum and vice versa. Effectively the new science specification, focused on nature of science and scientific inquiry, served as an intrinsic hook to recruit teachers willing to engage in the project. Likewise in Georgia, a new inquiry based curriculum meant that teachers were keen to take part in the project.

The underpinning philosophy of Chain Reaction was based on an intention to bring about a cultural change in terms of how science is taught—regardless of national policies and local contexts. We will go on to reflect on the project philosophy, individual strands and outcomes and discuss the deployment of the project within that context, identifying whether or not we can consider the project a success. However first, we'll discuss the importance of reflexivity in order to identify a reflexive frame for this chapter.

REFLECTING ON THE PROJECT AS TEACHER EDUCATORS AND AS PROJECT COORDINATORS

We felt that offering our individual reflections on the project would provide a more personal perspective than the previous chapters which seek to provide objective and descriptive accounts of inquiry and the project. Reflection requires us to consider our processes, practices, roles and identities, very often, within a specific setting or situation. Refection has become an important philosophy and tool in educational research (Moon, 2013) and is also used frequently within teacher development contexts. In other words, it offers the opportunity to reflect on whether or not there was shift in teachers' thinking and which triggered a change in their practice and, subsequently, students' learning.

We also reflect on the relationship between the project leaders (academic researchers and teacher educators) and the teachers as asymmetrical and speculate on the ways the academic-teacher interaction may have been exacerbated by decisions or presumptions arising from our project decisions, or more subtle cues such as socio-economic contexts or cultural background. Through this we may be able to honestly identify successes and failures while providing experience-based reasoning.

Teacher educators can often be so consumed by their day to day tasks and roles within their institution that often time may not allow for them to consider the impact they are having on pre-service teachers and in-service teachers. This can be especially true at the end of a complex, highly demanding and pressured project where both motivational levels and shear ability to step back and reflect on the strengths, weaknesses and overall professional learning experiences being involved in such a project had for those at the forefront of both the management and delivery of the project.

As teacher educators and/or academic researchers involved in any project we need to think about how we have learned from being involved. This requires pulling back from the situation and thinking critically about the project. We will now illustrate this through three brief reflective accounts from key members of the project team. These reflective accounts are presented from each individual's perspective and intend to highlight our role within the project and to reflect on how we engaged in the project and/or key issues we experienced from perspectives of both a management and delivery.

Louise: Ireland Project Manager

I was a student who was taught through inquiry when I was in primary school and understood the impact that this had on me not only in terms of my understanding of the content but in how confident I felt in being able to share my new-found knowledge with my classmates. It is my belief that using scientific inquiry practices in the classroom can not only help to develop conceptual and procedural knowledge but it can help to develop key life skills such as the ability to communicate, work with others and express creativity. However, this experience of inquiry was not consistent as I progressed through compulsory education and left me frustrated as I knew what it felt like to learn science more authentically. I felt passionately about learning science and wanted to bring my earlier school experiences to students and so decided to train to become a science teacher where I was determined to, despite the many constraints that act as a barrier to inquiry, provide students with that "eureka' moment I enjoyed so much when I was practicing inquiry-based science. Through my training I was provided with the knowledge and support to engage in inquiry practices when on teaching practicum experiences. However, this support was because I was fortunate to work with a teacher educator who was passionate about inquiry, therefore it was not something that all student teachers I knew experienced during initial teacher education courses. I embarked on a PhD directly after qualifying to become a teacher as I ultimately wanted to develop further as a professional educator. My PhD focused on the use of a professional learning community to develop pre-service science teachers understanding of scientific inquiry using a pedagogical content knowledge lens.

I was involved in the Chain Reaction project as I was completing the PhD. My role within the Chain Reaction project was diverse but in effect I had three key roles:

- *Teacher training*
- *Research*
- *Project management*

My role in terms of teacher training was to facilitate the recruitment of teachers and engage in continuous communication with them, to organise the workshops and national conferences, to help facilitate the workshops and to support the teachers and students in the in-school delivery of the EUPRBs.

While Chain Reaction was a Knowledge Exchange (KE) project and not a research project, deliverable 8 of the Impact and Delivery section of the proposal allowed for space to conduct research crucial to the progressive refinement of the project. The research focused on the teachers but also the student voice was critical to understanding the impact of the project.

From a project management side, I completed the necessary reports needed and collated the information, such as timesheets, required for these reports. I attended consortium project management meetings and managed the budget within my institution.

The above summaries identify the different 'hats' I, as a novice, had during the project. There were significant learning moments as this was the first major project that I had been involved in on so many different and complex levels. My overall impression was that of the unique learning experience this project had given me in such an early stage of my professional career.

Because I was a novice in terms of teaching experience, I did at times, feel an imposter in the sense that I was involved in providing training to teachers who had much more teaching experience and realistic understanding of life in the classroom than I did. While I considered myself an expert in IBSE research, mainly through my PhD work, I did not have the realistic application of same in the classroom. This was a tentative string to pull at times for me. This uncertainty was amplified when having professional dialogue with friends of mine who I trained with. I would often hear of their negative attitudes towards working with teacher educators who have not had the classroom experience necessary to 'preach' about how it should be done, and they ultimately felt that we are unrealistic in our blue sky approaches to science teaching and learning. While the term "preach" may sound like a strong word to iterate here, it is directly the kind of language used as I engaged in dialogue with friends. This made me very uncertain when I began working with teachers on the project, indeed teachers who had vastly more experience than that of my friends. It made me think, if my friends, who were essentially in the induction stage of teaching, held these views, what kinds of beliefs would teachers in the in-career stage of the continuum of professional development hold?

It can be argued that this experience is not unique to my situation but is something that many early career researchers and teacher educators may find. I believe that my honest approach is what enabled the teachers in the project respond well to me. I never claimed to 'know more' or 'have all the answers'. I understood the rigid nature of the classroom, due in part to curricular constraints. I used workshops and visits to schools as a chance to learn myself and in turn I vocalised this with teachers. Chapter **x** discussed the idea of disturbing practice to get to the root of what it means to be a professional and how my practice can be enhanced. I went through this during the project. At the beginning of the project, I viewed myself as 'David' and the teachers and other teacher educators as 'Goliath' but as the project progressed, I begun to see my influence on the teachers and how they responded positively to me. In effect they saw my vulnerable nature and realised that vulnerability is OK to acknowledge and this in turn helped them to engage effectively while having such feelings as they delved into their own practice.

Josephine: Overall Project Manager

As a project manager it is very easy, in fact too easy at times to get caught up in the minutiae of a project—the targets, reporting and the finances and, therefore miss the bigger picture of what is achieved as a whole. However, from a personal perspective the most impressive part of the project for me was watching young people present their work, both at the UK national student research conferences and at the international student research conferences. These were a core part of the Chain Reaction project, making the 'real life' science of the project's IBSE ethos come alive with the conferences intending to imitate that of professional scientific conferences attended by academics, with poster sessions, presentations, and at the international conferences, eminent key note speakers. This sharing of research outcomes is of course a key part of working in science and vital to an inquiry approach.

For me, it was important to remember that the majority of the students presenting at the national conferences were very young, just 13 or 14 years old, and also the vast majority had not been engaged in project based work before, nor had they presented to a large number of their peers. In addition they were doing this at a university, where given their ages, few of them had ever been, so it would have been unsurprising had they been intimidated by the surroundings. However, in both the poster and verbal presentations the work presented was of an excellent standard and students were animated and passionate when talking about their work individually and as a group.

The teachers at the national conferences were also often slightly nervous about this part of the project in terms of expectations for the day and also of their students, and seeing how they would fare in the academic environment and in presenting their work. This, in particular, was seen by the teachers as a huge challenge for their students in terms of confidence and presentation skills. I found teachers' reactions quite profound and moving when they realised their students could achieve high levels of success. They were incredibly proud of their students and all they had accomplished.

Turning to the international conferences, the work presented by students was inevitably of a high standard, given that these were identified as the best projects from each country. Across the project more widely the way in which students had been introduced to the Chain Reaction project work varied according to their educational contexts, with some taking place at after school science clubs and societies, while in the UK for example, some of the teachers tied the topics into classroom curriculum time. Perhaps as a result of the freedom given by carrying out the work after formal school time had finished or a more flexible curriculum some students had been able to take their work that bit further; for example in canvassing opinions from the public, and lobbying public officials with their research findings and building prototypes, which was very impressive. Students were able to demonstrate a deep understanding of their research, as well as a broader understanding of the societal issues in which these were rooted, as well as future implications and potential risks. The pride that they took in their work and the amount of time and effort which they committed was wonderful to see. Group working, thinking and discussion in the projects, again an important element of the IBSE approach, was also evident from the work presented, with individuals taking turns to present different facets of their work.

Another personal highlight was the international teacher workshop at the end of the project and seeing groups of teachers from different countries working together as learners on IBSE style projects. While the project had some weaker points, overall it was a success, and everyone involved, teachers, pupils and partners learned from their involvement. I can honestly say I enjoyed it, and am very proud of the work that we accomplished as a consortium. I learned a huge amount as a result of the project, from working with teachers, pupils and partners, in terms of IBSE and how to work with the EC and as a project manager.

Stuart: Overall Project Coordinator

When I look back now, the complexity of the project astounds me. The incorporation of three significant strands of work all of which were intended to bring about a sea change in teachers' thinking and practice while they were engaged in the usual milieu of the day to day science classroom was rather ambitious to say the least. Having the role of overall project coordinator brought added pressures of responsibility but also offered crucial licence to pull back from micro issues and reflect on some of the subtleties and issues of personal research of interest to me, an example of which I will share with you through this account.

One of the most personally and professionally satisfying aspects of the project was being able to talk to teachers. Through support visits to schools and student research conferences I was fascinated by their stories regarding student participation and their own approaches and arising issues. The most common issue mentioned by teachers, by far, was 'time.' Unfortunately this was discussed in a negative sense as the teachers reported that time, or lack of it, was their biggest enemy in using an inquiry approach. Talking with the participating teachers inevitably led to their assertion that what is practised in classrooms often lacks parity with academic research and policy directives which places them awkwardly between a high stakes testing regime which requires total curriculum coverage and curricular policy and research acclaiming inquiry as the most effective way to teach science. The frustration here is that the teachers recognise inquiry as being a very effective way of teaching and learning science, where appropriate, but cannot allocate the time needed to the approach which frequently involves lengthy investigations and project-based work which attempts to reflect professional science activity. This kind of inquiry approach tends to focus on developing conceptual understanding through exploration and knowledge construction, not merely delivering subject content to be crammed in lieu of examinations. It would appear that high stakes testing has vastly increased pressure on teachers to ensure that their students are exposed to as much curriculum content as possible in order for them to recall facts during examinations to obtain the highest grades possible and, therefore, keep the education department wolves away from the school doors.

No wonder the teachers I spoke with described complete dissatisfaction with the education system (across many countries involved). So, how did the participating teachers deal with a lack of time? Well, many of the teachers delivered the project through after school science clubs or students continued project work on their own time out of school. Of those teachers that delivered the project through science clubs many expressed disappointment with that approach as it restricted the activities to students that already had an interest in science and prevented students who may have benefitted from interested science contexts and activity from engagement.

This issue became a major barrier for a large majority of the teachers across the project with none offering a satisfactory solution. This left me in a somewhat confusing position of leading a successful project but having to contemplate a certain amount of failure regarding the project when there was so much to applaud. However, the teachers' contextual dissatisfaction— teacher's assessment of contextual characteristics of their work (Gess-Newsome et al., 2003) strongly indicates that policy factors (high stakes testing environment) are the restricting factor and not the project philosophy or aims. While the opportunity to attach blame to the relevant countries education departments is appealing, it offers no comfort.

So, we find ourselves post-project in the situation of having delivered a highly successful project in terms of positively impacting large numbers of teachers and students, developing an approach to guided inquiry which is effective in terms of inspiring students, and providing stimulating professional development for teachers, but having to accept that many teachers will not be able to embed the approach in their schemes of work due to high stakes testing regimes. I will end this brief reflective account with an interesting note. Finland, whom score extremely highly in PISA rankings, generally adhere to minimal testing and, more importantly, minimal interference from government (Berliner, 2011). I will leave you to make of this final point what you will.

The above three accounts attempt to reflect on our personal perspectives regarding our roles, delivery of the project, the different aspects and strands of Chain Reaction, our interests, and/or learning. We will now go on to discuss key lessons learned from both the project delivery and its outcomes.

KEY LESSONS LEARNED

At the time of writing it has been two years since the project ended and all consortium partners have claimed that it is something which positively impacted our thinking and practice in terms of what we have learned going forward as science teacher educators and academic researchers.

Overall, the main lesson learned was that change cannot be forced upon professionals and that a bottom up approach to change is critical to altering habits of mind and practice. Ultimately, a teacher is trying to do their best for students and if they believe that a certain approach to teaching serves their student's best, then it can be difficult to tap into these assumptions and offer alternatives. Therefore, they need to be involved from the start in any professional development opportunity and a constant cycle of reflection should be established and include those training the teachers in order to understand what works and what does not. This reflective examination can often be difficult as it can lead to vulnerability about one's own practice, but it is necessary. As teacher educators engaged in professional development, we need to consistently and critically reflect on our own practice too and how this can be improved.

Each member of the consortium has reflected on the barriers that emerged during project delivery and considered whether or not such barriers negated our chance to make IBSE sustainable in the teachers' classrooms. By far, the largest barrier identified has been time —actually classroom time to engage in meaningful inquiry. The large majority of participating teachers stated that deploying an inquiry approach can be extremely time consuming and because each country practices within an examination driven culture covering curriculum content is the most important priority. While the teachers acknowledge that inquiry approaches would also enable them to address curriculum content it is simpler and more efficient in terms of time to use more traditional methods of teaching.

The participating teachers also complained of a distinct lack of professional development opportunities which could be seen as another indictment of education systems and policies which undervalue teachers' professional status. The importance of professional development is widely accepted (e.g. Guskey, 2003) but seems to fall off school and policy agendas regularly—most likely related to cost. However, it is obvious that for teachers to develop their practice effectively and innovatively, professional development is essential.

Finally, the importance of maintaining high quality project management of a complex education programme of this nature is clear. Regular, honest and open discussion and communication between and within partners was crucial, alongside a progress/problem reporting structure that was frequent enough to be of benefit, but light touch enough not to be regarded as onerous. It was also essential that the expectations relating to this and other aspects of the project were set out very early on in the project including the role and responsibilities of each partner in terms of the management of the project.

We will now consider science education in Europe and ask whether or not reform in science education needed.

Science Education in Europe: Is Reform Needed?

The European Commission Rocard Report - '*Science Education Now: a new pedagogy for the Future of Europe*' (2007) highlighted the decline in numbers of young people interested in pursuing study and careers in science and mathematics. This report, together with a range of literature, bemoaned the existing state of science education across Europe warning that the consequence of failure to address the issue would leave Europe in a vulnerable position in terms of global competition in innovation and development (European Commission, 2004; 2007). These concerns about a deficit of suitably qualified STEM graduates have been consistently voiced for at least the last 70 years and have resulted in many, often expensive, initiatives to encourage more young people to study and participate in science.

It is believed by some that a significant reason for young people's lack of interest and participation in science is the 'prevailing orientation' of science education which favours more traditional approaches to teaching and learning (Kampanadze & Eilks, 2014; Bevins and Price, 2016) and educational reform in science education throughout many European countries has appeared inevitable for some time (Janiuk & Mazur, 2010). However, rarely does reform come cheap and the cost to the European Commission alone has been approximately 50 million Euros to fund projects focused on developing inquiry based science education between 2010 - 2017 as part of the FP6 and FP7 funding calls.

So, the desire and financial investment exists but even now after several years of significant funding in Europe and countries beyond there seems to be little change in the way school students experience science in the classroom, particularly in terms of students' scientific literacy development (PISA, 2015). So, why are many European schools still practising teacher-centred pedagogies?

This book has identified a number of reasons that schools find it challenging to include inquiry approaches in their diet of teaching methods but none more important than the allocation of 'time' to engage their students in an inquiry. Evaluation data shows us that the large majority of teachers from the Chain Reaction project have all bemoaned the excessive time, they believe, it takes to deploy an effective inquiry. However, we suggest that this belief is strongly influenced by teachers' lack of familiarity with setting up and managing an inquiry approach and the additional levels of involvement required for an effective inquiry approach. In her study of a biology classroom Crawford (2000) suggested that during an inquiry session teachers are required to take on many roles which are *much more demanding than have been commonly depicted* (p. 933). Her study showed that teaching science through inquiry means that it is necessary for teachers to be guides, research collaborators, modellers of techniques such as analysis of data, and advocators of science communication. All of these roles demand so much more than just the transmission of knowledge. Shifting between these roles consistently, forces a complex mode of teacher participation which many teachers are not prepared for through initial teacher education programmes. These shifting roles can allow for the provision of what Crawford calls a *holistic experience* (p. 934) as opposed to a sequential, reductionist approach evidenced by more traditional types of teaching. It is this complex teacher participation, and not merely an excess of time needed to engage students through inquiry, that we believe is the real inhibitor of change in the science classroom. This has obvious and significant implications for current initial teacher education and professional development programmes that do not recognise the complexities involved in using inquiry for teaching and learning science.

Another point to note, and literature does suggest this also, the experiences of us as teacher educators has been that those choosing to enter Initial Teacher Education programmes often have clouded orientations (beliefs and values) to science teaching where they were "successful' in studying science through a rote-learning, product-oriented approach to learning. If it worked for them, why would it not work for others, is often the rhetoric expressed by these student teachers. We would suggest that it is a difficult task to be able to tap into and alter student teachers notions of perceived best practice and therefore the cycle of rote-learning continues with the next generation of teachers.

Interestingly, findings from recent research conducted in the UK suggest that STEM graduates are being attracted to other areas or failing to find work in STEM roles and what could be conceived as the final insult to STEM is that the majority of these graduates are not going into highly paid sectors such as banking and finance but into education and lower-level management roles (Smith & White, 2018).

That education seems to be an option for STEM graduates may offer some hope for the field. That said, even though teaching may be the most common occupational destination for STEM graduates, at least in the UK, many degrees offer nothing in the way of subject-focused pedagogy or delineate inquiry to any great depth.

While this research uses data from the UK it is possible that similar results would be found across Europe if the study were replicated and the final point is of particular significance given that the recruitment of science teachers, across Europe, has traditionally been difficult (REF).

The question of whether or not reform is needed in science education across Europe is probably the wrong question to ask. A better question might be 'where is reform needed most?' If the career opportunities are limited for students in science disciplines, there is need to investigate why, particularly in the light of so many claims from industry and policy makers that we are in a recruitment crisis, and address subsequent findings. Alternatively, if a lack of interest is causing issues of recruitment and, therefore, causing a reduction in discipline provision then the wealth of existing studies which have investigated this need to be synthesised and actions assessed appropriately. Thus, identifying the correct question to be investigated is paramount. There is evidence that indicates wider socio-cultural issues impact interest in science education and science and studies which indicate science education is itself inadequate.

However, there is need for much more research that focuses on these issues and more—curricular issues and recruitment of science teachers to name but two. What is not in doubt is that science education should provide young people with the knowledge and understanding of the main themes that science offers and develop their ability to engage critically with science in their future lives as European citizens. Unfortunately, it would seem that this practical subject area and one which is so important to lives of everyone globally has managed to switch young people off, particularly at secondary school level. If this is not addressed fully the impact on societies may well be significant.

REFERENCES

Anderson, L. W., Krathwohl, D. R., Airasian, P. W., Cruikshank, K. A., Mayer, R. E., Pintrich, P. R., ... Wittrock, M. C. (2001). *A Taxonomy for Learning, Teaching, and Assessing: A revision of Bloom's Taxonomy of Educational Objectives*. New York: Pearson, Allyn & Bacon.

Berliner, D. (2011). Rational responses to high stakes testing: The case of curriculum narrowing and the harm that follows. *Cambridge Journal of Education*, *41*(3), 287–302. doi:10.1080/0305764X.2011.607151

Bevins, S., & Price, G. (2016). Reconceptualising inquiry in science education. *International Journal of Science Education*, *38*(1), 17–29. doi:10.1080/09500693.2015.1124300

Blanchard, M. R., Southerland, S. A., & Granger, E. M. (2009). No silver bullet for inquiry: Making sense of teacher change following an inquiry-based research experience for teachers. *Science Education*, 321–360.

Blanchard, M. R., Southerland, S. A., Osborne, J. W., Sampson, V. D., Annetta, L. A., & Granger, E. M. (2010). Is inquiry possible in light of accountability?: A quantitative comparison of the relative effectiveness of guided inquiry and verification laboratory instruction. *Science Education*, *94*(4), 577–616. doi:10.1002ce.20390

Bloom, B. S. (Ed.). (1956). Taxonomy of Educational Objectives, Handbook I: The Cognitive Domain. New York: David McKay Co Inc.

Braun, V., & Clarke, V. (2013). *Successful Qualitative Research: A Practical Guide for Beginners*. Sage.

Brickman, P., Gormally, C., Armstrong, N., & Hallar, B. (2009). Effects of inquiry-based learning on students' science literacy skills and confidence. *International Journal for the Scholarship of Teaching and Learning*, *3*(2), 1–22.

Bybee, R. W. (2006a). *The BSCS 5E model Instructional Model Origins and Effectiveness*. Colorado Springs, CO: BSCS.

Colburn, A. (2000b). An inquiry primer. *Science Scope*, *23*(6), 42–44.

Crawford, B. A. (2000). Embracing the essence of inquiry: New roles for science teachers. *Journal of Research in Science Teaching: The Official Journal of the National Association for Research in Science Teaching*, *37*(9), 916–937. doi:10.1002/1098-2736(200011)37:9<916::AID-TEA4>3.0.CO;2-2

Dewey, J. (1991). Logic: The theory of inquiry. In J. A. Boydston (Ed.), *John Dewey: The Later Works, 1925–1953* (Vol. 12, pp. 1–5). Carbondale, IL: SIU Press. (Original work published 1938)

European Commission. (2004). *Europe needs more Scientists*. Brussels: European Commission.

European Commission. (2007). *Science Education now: A renewed Pedagogy for the future of Europe*. Brussels: Directorate General for Research, Science and Economy.

Gess-Newsome, J., Southerland, S. A., Johnston, A., & Woodbury, S. (2003). Educational reform, personal practical theories, and dissatisfaction: The anatomy of change in college science teaching. *American Educational Research Journal*, *40*(3), 731–767. doi:10.3102/00028312040003731

Gormally, C., Brickman, P., Hallar, B., & Armstrong, N. (2009). Effects of inquiry-based learning on students' science literacy skills and confidence. *International Journal for the Scholarship of Teaching and Learning*, *3*(2), 16. doi:10.20429/ijsotl.2009.030216

Guskey, T. R. (2003). What makes professional development effective? *Phi Delta Kappan*, *84*(10), 748–750. doi:10.1177/003172170308401007

Harwood, W. (2004). An activity model for scientific inquiry. *Science Teacher (Normal, Ill.)*, *71*(1), 44–46.

Herron, M. D. (1971). The nature of scientific enquiry. *The School Review*, *79*(2), 171–212. doi:10.1086/442968

Janiuk, R. M., & Mazur, P. (2010). Basic Structure of the Educational System. *Teaching Chemistry Around the World*, 291.

Johnston, A. (2008). Demythologizing or dehumanizing? A response to Settlage and the ideals of open inquiry. *Journal of Science Teacher Education*, *19*(1), 11–13. doi:10.100710972-007-9079-y

Kampanadze, M., & Eilks, I. (2014). Supporting reform in science education in central and Eastern Europe: Reflections and perspectives from the project TEMPUS-SALiS. *Eurasia Journal of Mathematics, Science & Technology Education*, *10*(1), 47-58.

Kirschner, P. A., Sweller, J., & Clark, R. E. (2006). Why minimal guidance during instruction does not work: An analysis of the failure of constructivist, discovery, problem-based, experiential, and inquiry-based teaching. *Educational Psychologist*, *41*(2), 75–86. doi:10.120715326985ep4102_1

Klahr, D., & Nigam, M. (2004). The equivalence of learning paths in early science instruction: Effects of direct instruction and discovery learning. *Psychological Science*, *15*(10), 661–667. doi:10.1111/j.0956-7976.2004.00737.x PMID:15447636

Mao, S., & Chang, C. (1998). Impacts of an Inquiry Teaching Method on Earth Science Students' Learning Outcomes and Attitudes at the Secondary School Level. *Proc Natl Sci Counc*, *8*(3), 93–101.

Minner, D. D., Levy, A. J., & Century, J. (2010). Inquiry-based science instruction—What is it and does it matter? Results from a research synthesis years 1984 to 2002. *Journal of Research in Science Teaching*, *47*(4), 474–496. doi:10.1002/tea.20347

Minstrell, J., & van Zee, E. H. (Eds.). (2000). *Inquiry in to Inquiry Learning and Teaching in Science*. Washington, DC: American Association for the Advancement of Science.

Moon, J. A. (2013). *Reflection in learning and professional development: Theory and practice*. Routledge. doi:10.4324/9780203822296

Palmer, D. H. (2009). Student interest generated during an inquiry skills lesson. *Journal of Research in Science Teaching*, *46*(2), 147–165. doi:10.1002/tea.20263

PISA. (2015). *Excellence and Equity in Education Volume 1*. Available from https://read.oecd-ilibrary.org/education/pisa-2015-results-volume-i_9789264266490-en#

Rees, C., Pardo, R., & Parker, J. (2013). Steps to opening scientific inquiry: Pre-service teachers' practicum experiences with a new support framework. *Journal of Science Teacher Education*, *24*(3), 475–496. doi:10.100710972-012-9315-y

Shields, C. (2014). Aristotle. *The Stanford Encyclopedia of Philosophy*. Available from http://plato.stanford.edu/cgi-bin/encyclopedia/archinfo.cgi?entry=aristotle

Smith, E., & White, P. (2018). Where Do All the STEM Graduates Go? Higher Education, the Labour Market and Career Trajectories in the UK. *Journal of Science Education and Technology*, 1–15.

Smolleck, L. A., & Yoder, E. P. (2008). Further development and validation of the teaching science as inquiry (TSI) instrument. *School Science and Mathematics*, *108*(7), 291–297. doi:10.1111/j.1949-8594.2008.tb17842.x

Wilson, C. D., Taylor, J. A., Kowalski, S. M., & Carlson, J. (2010). The relative effects and equity of inquiry-based and commonplace science teaching on students' knowledge, reasoning, and argumentation. *Journal of Research in Science Education*, *47*(3), 26–30.1.

Appendix

APPENDIX 1: INTRODUCTION TO THE CHAIN REACTION RESOURCES

Throughout this book, reference has been made to the 11 EUPRBs; alongside a set of teacher's notes, these were the materials upon which the school-based implementation of Chain Reaction centred. These were originally written as part of the Pupils Research Initiative (PRI) described in Chapter 2 and were updated and rebranded for the Chain Reaction project. These were then adapted and translated for use by partner countries.

Inevitably some EUPRBS were more popular than others across the consortium; Table 1 gives the eight most popular sets of materials and the areas that they cover.

At the time of writing all of 11 of the original EUPRBS can be found in English on the project website (http://www.chreact.eu/), and can be accessed translated into German (http://chreact.de/), Greek (http://www.chreact.gr/), Slovenia (http://www.chreact.si/), Slovakian (http://www.chreact.umb.sk/), Georgian (https://chain-reaction.iliauni.edu.ge/), and Turkish (http://www.chreact.org.tr/). Teacher Guidance notes that should be used and considered alongside them can be found in Appendix 2 and are translated on the websites previously mentioned.

It is inevitable that teachers will wish to adapt, change and modify material; including these EUPRBs to suit their students, context, resources and curriculum. In doing this you may wish to reduce the length of introductory texts, change the emphasis of an experiment, the equipment used or re-frame the inquiry in a way that suits a different topic area or learning outcome. The EUPRBs can be thought of as a jumping off point for exploring IBSE in the classroom or trying out a new approach to learning with students.

The remaining appendices (3-8) look at the Chain Reaction "toolkit" which was developed by the consortium to help teachers with their understanding and implementation of IBSE in the classroom. Based around the Plants in Space EUPRB, the toolkit gives advice on adapting and using EUPRBs in the classroom and communicating science, as well as an introduction to IBSE.

Table 1. most popular EUPRBs

EUPRB	Science	Engineering	Technology	Other
Collision Course	Physics Cosmology Astronomy Geology	Telescope engineering Computer engineering Computer imaging	Information technology specialisms	Science journalism/writing
Out of site out of mind	Environmental/Earth science Chemistry geology	Waste management engineering	Environmental management	Scientific writing
Mars-ology	Vulcanology Astronomy Geology	Fluid engineering Space hardware design/ engineering	Information technology specialisms	
Green Heating	Physics	Environmental engineering Heating engineering	Solar panel design	
Green light	Physics Environmental science	Electrical/electronic engineering Environmental engineering	Light bulb design	Scientific writing/ journalism
PHEPPs	Physics	Electrical/electronic engineering		
Feed the World	Chemistry Biochemistry Biology/plant biology	Environmental engineering	Laboratory technical work	Scientific writing
Plants in Space	Plant biology Biochemistry Human biology/medicine Space science	Space survival engineering Design engineering	Life support computer technology	Scientific writing

APPENDIX 2: TEACHER NOTES FOR IMPLEMENTING AN INQUIRY CLASSROOM

Introduction

Inquiry-based instruction is an important approach to science teaching that involves students in investigating questions and using data as evidence to answer those questions (Crawford, 2000). Teaching through inquiry is thought to promote scientific literacy and has the potential to improve both student understanding of science and participation and engagement in science and classroom science. Moreover, inquiry-based instruction provides a context to begin learning about the nature of inquiry and the nature of scientific knowledge. Unfortunately, many teachers have limited experience with scientific inquiry and hold naïve conceptions of the process by which scientific knowledge is generated. Lack of knowledge and experience with inquiry is thought to act as a barrier for teaching science in this way (Blanchard, Southerland, & Granger, 2009). This lack of knowledge and experience likely puts serious limitations on teachers' ability to plan and implement lessons that will help their students develop an image of science beyond the familiar body of knowledge (Capps & Crawford, 2012).

Inquiry can be thought of as a content area and in this sense learners begin to understand how scientists do their work. For example, students should understand that scientists ask questions, perform different types of investigations and produce explanations based on their observations. These understandings about inquiry reflect the philosophical nature of scientific inquiry. Another element of classroom inquiry is a student's ability to scientific inquiry. Abilities to do inquiry include asking and identifying questions, planning and designing experiments, collecting data, using data, and connecting data as evidence with explanations. Also, classroom inquiry can be viewed as a kind of pedagogy, or a teacher's ability to employ inquiry-based instruction in the classroom in order to address key science principles and concepts.

Teaching students to do scientific inquiry involves teachers engaging their students in the practices of science. These practices include the various activities and processes carried out by scientists to answer questions and develop explanations and models using logic and critical thinking. As they engage in scientific practices, both students and scientists use observations and inferences to develop conclusions and evidence-based explanations.

The EUPRBs are resources designed to support the teaching and learning of science through an inquiry-based approach. The EUPRBs provide opportunities for investigative work, with some offering the opportunity for practical investigations.

Although the EUPRBs show a wide range of opportunities, a unifying feature is that each one presents teachers and students with opportunities to experience, as far as is reasonably possible, the ways that science and engineering researchers think and work.

Inquiry Based Methods in Science Education (IBSE): Developing Key Skills

The EUPRBs are a way to develop inquiry-based learning in science. In particular the EUPRBs encourage the development of the skills shown below. The Teachers' Notes for each Brief indicate which of these skills apply to that activity. But, in general, they focus on how science works. The necessary skills that can be developed in science are as follows:

Data, evidence, theories and explanations:

1. Students will learn:
   a. How scientific data can be collected and analysed
   b. How the interpretation of data, using creative thought, provides evidence to test ideas and develop theories
   c. How explanations of many phenomena can be developed using scientific theories, models and ideas
   d. That there are some questions that science cannot currently answer, and some that science cannot address.

Practical and Inquiry skills

2. Students will learn to:
   a. Plan to test scientific ideas, answer scientific questions, or solve scientific problems
   b. Collect data from primary or secondary sources, including ICT sources and tools
   c. Work accurately and safely, individually and with others, when collecting first-hand data
   d. Evaluate methods of collection of data and consider their validity and reliability as evidence.

Communication skills

3. Students will learn how to:
   a. Recall, analyse, interpret, apply and question scientific information or ideas
   b. Use both qualitative and quantitative approaches
   c. Present information, develop and argument and draw a conclusion, using scientific, technical and mathematical language, conventions and symbols and ICT tools.

Applications and implications of science

4. Students will learn:
   a. About the use of contemporary scientific and technological developments and their benefits, drawbacks and risks
   b. To consider how and why decisions about science and technology are made, including those that raise ethical issues, and about the social, economic and environmental effects of such decisions
   c. How uncertainties in scientific knowledge and scientific ideas change over time and about the role of the scientific community in validating these changes

The EUPRB Model

This has been achieved by adopting a simple yet powerful model in the design of the EUPRBs (see Fig 1). This model describes a set of categories into which the varied activities carried out by researchers can be placed. It is not an attempt to describe the scientific method, since most people would agree that no single method can be identified. The model is a way of representing the activities involved in research in a way which allows them to be translated into structured curriculum materials.

Figure 1 The EUPRB model

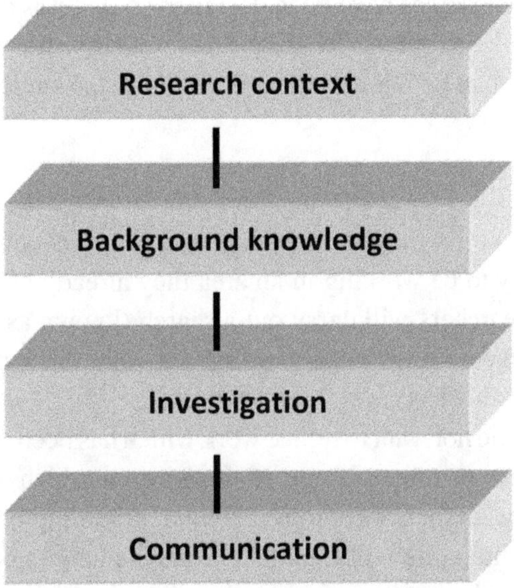

Science and engineering research is not a linear process. A research project can have numerous false starts and blind alleys, and work can be carried out along several lines of inquiry at once. Activities which tend to come only at the end of a school science investigation (such as report writing or communicating findings) tend to occur at all stages in a research project. Discussions with science and engineering researchers showed that all four categories of activity could be going on simultaneously within any one research project.

The Four Stages of the EUPRB Model

Context Setting

All research projects exist for a reason. The reason could be the personal interests of the researcher, or it could be the need to address a particular issue/question raised by the research community. Research can be directly linked to a commercial enterprise. It could be more speculative and lacking obvious immediate application – so-called *blue-sky* research. Much of the research into particle physics and astronomy could be said to have little industrial application. It takes place to push forward the frontiers of knowledge, so that we gain an increased understanding of the Universe around us. However, the construction of the machines needed to carry out this research (including particle accelerators, rockets and telescopes) and the associated computer systems represent some of the most sophisticated engineering research. There can be spin-offs for other research and industrial applications not envisaged by the people doing the research.

In attempting to simulate this professional world of research, it is important to construct curriculum scenarios where the purpose of the investigation is made clear to the students. The EUPRBS have been constructed so that the students are led in by means of one or two activities which establish the context for what is to come. This is brought about by a number of types of documents – emails, press releases, newspaper or journal articles, letters, etc. These have been designed to simulate 'reality'.

All student documents have been written without stepping outside the 'reality' created by these documents. The documents themselves have been designed so that it is obvious what the students must do having read and discussed their content. In some cases, 'hand-written' notes have been added to the documents so that instructions can be made clearer, or to direct the students' attention to salient points.

Background Knowledge

All research projects are based on a form understanding of the relevant science. Researchers embarking on a new venture are likely to be working in an area they already know a lot about. Even so, at the start of a new project, the researchers will carry out a search (known as a *literature review*) of what is currently known about the subject. They will try to find out what relevant work is going on in research institutions around the world. This has several purposes.

Firstly, in general scientists do not want to repeat work which has been carried out elsewhere (although successfully repeating the work of others is one of the ways in which scientists validate each other's work). Scientists will usually want their work to be original, providing new knowledge and understanding. A thorough search of the literature will establish if this is likely to occur.

Secondly, reading through relevant papers and talking to other scientists working in the same field will mean that researchers are learning more and more about the subject under study. This may help them to frame the right questions, construct the most promising hypotheses, or make the most telling predictions to guide them in their future work.

Thirdly, the research literature will contain a range of investigatory techniques that could be used in the research project about to be embarked upon. In this case, researchers may indeed repeat the work of others in order to become familiar with a technique that will be useful for their own particular work. Throughout the time of the project, researchers will keep up to date with current thinking, so that inconclusive reporting in journal articles, for example, can be avoided, or new knowledge gained by other researchers can be incorporated into their plan.

The EUPRBs attempt to represent this range of activities by means of 'research papers' specially written for the purpose, reports, meeting summaries, extracts from books and notes. All, written at a language level the students can understand. An EUPRB can often start with an email which introduces the subsequent papers, articles or notes. In some cases, extra guidance may be needed, and so briefing sheets are provided. Most of the research papers students must deal with contain 'hand-written' annotations, as if written by the research supervisor. The documents used in the context setting and background knowledge categories of the EUPRBs involve students in a wide range of active learning approaches (see later) such as small group discussion, problem solving, active reading and brainstorming. These activities will help them to think their way into the investigation which is the focus of the EUPRB.

The first two phases of the EUPRB help to create the appropriate 'mind set', where the purpose of the investigative work is clear to the students, and where they have some stake in its successful outcome.

Investigation

In the world of science and engineering, researchers design experiments and investigations to answer questions and to test hypotheses and predictions. In many cases scientists might actually carry out a set of procedures based on little more than a 'hunch'. "I wonder what would happen if I..." has probably been the true starting point for more lines of inquiry than researchers would like to admit. This sort of approach is acceptable as part of a range of strategies scientists will bring to the question they are trying to answer. It is a way of eliminating factors affecting the particular phenomenon under investigation, or removing 'suspects' from the list of possible solutions to a tough problem.

The EUPRBs involve students in carrying out one or more investigations. In some cases, this is a whole investigation. The pupils begin by framing the question, hypothesis or prediction, or taking an idea presented to them and planning a procedure for the investigation. After carrying out their investigation they analyse their findings, draw conclusions and evaluate the outcomes.

Communication

Researchers tell each other about their work. The traditional method of doing this is to publish papers in academic journals. Researchers also make presentations to each other in departmental seminars or at conferences. They inform the wider public through popular science journals, through newspapers and magazines and through TV and radio, through the internet (electronic versions of journals, teleconferencing, etc) including social networking sites.

The EUPRBs involve students in communicating science in a variety of ways, reflecting the diversity of media and methods used in the professional world. Some require students to write up a standard laboratory report. Others involve writing proposals for funding. The students can write articles for journals, construct websites or produce and exhibit and/or presentation for a national and international conference. There are also opportunities for presenting their work to others in their class through seminars and plenary sessions. They will be able to publish their work on the Chain Reaction website so that students and teachers from other countries can see what they have been doing.

Teaching and Learning Approaches

One of the aims of the Chain Reaction project is to actively involve students in their science studies and enable them to experience the excitement and challenges of experimental and investigative science. Active involvement in learning leads to greater motivation and achievement. However, active involvement does not mean an over-emphasis on certain learning strategies to the exclusion of others. There needs to be diversity in teaching and learning approaches in science lessons and a selection of activities that best suit the circumstances. 'Fitness for purpose' should be an important criterion when selecting strategies from the wide choice available. Variety and balance in the menu of strategies is important. Students are likely to lose interest if they are continually presented with the same type of activity in lessons. Lack of motivation can sometimes lead to underachievement and a 'switch off' to science. There are a range of teaching and learning techniques, coupled with more traditional methods, that will allow the curriculum to be covered and which provide variety and interest for the student.

Another important reason for introducing a varied diet of approaches is that many of the 'active learning' techniques encourage the development of a diversity of skills. Important in personal development, these skills include problem solving, communication, organisational and investigative capabilities. Student-centred approaches often provide opportunities for students to make decisions, be creative, take more responsibility for their own work and practise working as a member of a team.

The student-centred approaches used in the EUPRBs create valuable time for the teacher to work with individuals and small groups of students. This results from the change in role of the teacher from director of the whole group to manager of the learning experience. Time spent with individuals or small groups can be used to offer support for those with learning difficulties, provide other challenges for students needing additional stimulation and 'encourage' those students who are not working to their full potential.

Another important factor when promoting active involvement in learning is the benefit of providing realistic and relevant contexts in which to present the learning. Wherever possible it is important to relate the learning context to the students' everyday lives or top situations that have some meaning or consequence for them.

The EUPRBs are designed to encourage active learning and small group collaborative work. As well as providing opportunities for the teacher-led activities and teacher input, the EUPRBs include a range of active learning strategies. It is not possible here to describe all of the strategies in details but the ones below are the most commonly adopted in the EUPRBs.

Small Group Discussion (SGD)

SGD is probably the most valuable and commonly used of all the strategies. It is central to most other strategies. Many opportunities are provided within the EUPRBs for students to engage in SGD activities. Two SGD techniques that are particularly useful for supporting effective discussion are described below:

1. Step-by-step discussion (agenda building)

This type of group work allows the teacher to maintain close control and is particularly useful for pupils not yet skilled in open, free ranging discussion work. The key preparation step is to create an agenda for discussion. This is most effective when the first items on the agenda are concrete (e.g. *list five reasons why...think of four types of...*) and later items become more abstract (e.g. *why do you think...? try to decide which...*). The agenda is prepared and moves in steps from the factual or simple recall to the application of knowledge and speculation. This approach could be used in many of the EUPRBs.

The discussion can be widened to the whole class by encouraging interchange between groups, with the teacher orchestrating the discussion. Alternatively groups can be asked to present their findings. This could be in the form of a poster, a paper or a short presentation.

2. Jigsaw method

This method encourages group discussion and depends on collaboration between group members and/ or groups. Individual group members or groups are given responsibility for researching a particular topic. They then report back to their respective groups, or the groups report back to the whole class. Rather than have every small group cover all the topics, a great deal of time can be saved if each group covers a different topic. It is important that sufficient time is allowed for each group to report back. Students should be given time to question each other in order to clarify their understanding. It is important that they have a written record of the key points of the topic they did not research themselves.

Supporting successful group discussion and team work

- Create the right climate - agree some ground rules:
    - Don't interrupt other members
    - Listen when others are talking
    - Respect each others' opinions
    - You don't have to say anything if you don't wish to
    - No silly or unpleasant remarks
    - Share out the jobs and allocate roles and responsibilities
    - Help others if you can but don't interfere
- Encourage active listening and give practice in listening skills
- Set clear tasks and goals with timescales, and make sure that the students have sufficient background information and knowledge
- Thinks about how groups are assembled:
    - Groups of 3 or 4 are normal and manageable
    - Groups can be self-selective but reserve the right to change them
    - Allow sufficient time to develop group working relationships, e.g. half a term

Problem Solving/Investigative Work in Teams

Once teams are formed, they will work more effectively on their task if the team:

- Is clear about the purpose and has a clear understanding of the task
- Is able to identify and describe the criteria for success
- Is able to generate ideas and select the most promising
- Draws up an action plan and allocates tasks to members with timescales
- Monitors the action plan and takes any necessary steps to keep to the plan
- Reviews progress continuously and evaluates the outcomes.

A help sheet is shown below in Table 2 which has been developed to help students tackle a problem or investigation using a more structured approach:

Simulations

Most EUPRBs are simulation activities, which involve students working as realistically as possible within a real-life scenario. Information is supplied which allows students to operate within the simulation. The students are not role playing. Role playing can be described a 'person-centred' in that issues and concepts are explored by asking students to imagine they are in someone else's shoes. Simulations are 'job-centred' as the emphasis is on carrying out a certain task or tasks and students behave according to the task in hand.

Simulation provides a framework and philosophy in which students are not only active participants but are given considerable responsibility for their actions and decisions. The technique encourages a cooperative approach, provides a stimulating and creative atmosphere and develops problem solving and communication skills, as well, as knowledge and understanding of scientific concepts.

Table 2. How to be a successful problem solver

Key question…		These suggestions will help you answer the questions
Purpose	**What is the real problem?**	Be sure that you understand the problem. It might help to write it out in your own words, sketch it or describe it to someone else.
Review	**What are you really trying to achieve?**	You may need more information so check
Criteria for success	**What are the different things that the problem depends on and what do you need to find out about them?** **What information and materials are available to help you?** **How will you know when you have been successful?**	The problem depends on… I need to find out about… I can find out from… Materials available are… People available are… Time available is… Essential criteria (the solution must contain these).A successful solution must… Desirable criteria (these can give a better solution, e.g. design, colour, etc). A successful solution may also…
Ideas	**What ideas have you got for solving the problem?**	Brainstorm ideas, be imaginative.
Review	**Are your ideas suitable?**	Check that you are using your criteria.
Choose	**How can you decide which idea seems most promising?**	Make a shortlist of your best ideas. You could use a scoring system (advantages +; disadvantages -) to help you decide. The advantages of each idea are… The disadvantages of each idea are…
Review	**Which idea best meets your criteria?**	Check using your criteria.
Do it	**What is your chosen plan?** **How will you put it into practice?**	Draw up an action plan. The material and information needed are… A checklist and schedule for carrying out the plan is… Who does what, when and how?
Monitoring & control	**How will you check that things are going according to your plan?**	Keep checking your action plan. Keep checking your criteria for success.
Review it	**Have you succeeded?** **Could you have done it better and how?**	Match the solution against your criteria for success.

Within the context of the EUPRBs, simulation enables the students to gain insight into, and awareness of, the roles and activities carried out by science and engineering researchers. A number of helpful pointers for running a simulation activity are given below:

Five stages of a simulation:

1. **Preparation**: Check that you are fully familiar with the aims, content, activities and organisation of the simulation. This will involve sorting out groupings for each stage, student materials, apparatus, what you will be doing and when. This is where the EUPRB Teachers' Notes will be very helpful.
2. **Briefing**: Introduce the Brief, setting the context for the activity and stimulating student interest and 'readiness' to get involved. You can explain the structure of the activity and give out the Student Activity Sheets. You may in fact take on the role of a research director or supervisor - simply giving an overview of what is expected and creating a sense of realism and purpose. If there is additional

information about the topic it can be found in the background information for teachers, at the end of the Teachers' Notes for each EUPRB.

3. **Action**: In most cases you will step back and observe. The students will generally take responsibility for working through the activity. However, you will be required to coordinate changes of activity such as arranging groups for investigation tasks, reporting back, etc. You will also want to facilitate by sitting in on a discussion, helping move things forward if necessary and giving help and advice when appropriate. However, try to do this in a way that will maintain a sense of realism, for example by using questions such as, 'how do you think this would be done in a research laboratory?', 'what information do you need in order to carry out this method?', 'where would a researcher get this information from?'.
4. **Debriefing**: Group report-backs can lead into discussion of the activity - either in small groups or as a whole class. Important learning points are drawn out and misconceptions are challenged and corrected sensitively. An overview of the Brief is created and students are reminded of the outcomes and what needs to be completed.
5. **Follow-up**: Following completion of reports, presentation plans, etc more formal presentations or displays can take place. Links with future work can be made and the experience can be evaluated.

Remember to:

- Give careful thought to groupings, roles and functions
- Encourage participation and cooperation
- Set timescales and keep things moving along briskly
- Take the opportunity to observe and record progress.

Active Reading Techniques (Directed Activities Related to Text: DARTs)

Written materials in science are often demanding and there are a number of techniques which enable students to interrogate the text so that they read for understanding. DART activities help students to see underlying ideas in the text, compare them with those they already hold, pick out essential information and process it in a form they can use and understand. The following are some of the different types of DART in use.

- **Cloze**: words are deleted from a section of text by the teacher at regular or irregular intervals. Students then have to complete the text by filling in the spaces with the appropriate words. Removing words at irregular intervals is particularly effective since key terms and words can be selected by the teacher. Although cloze activities are not used in the EUPRBs, since this would detract from the realism of the scenarios, you may wish to use this technique in follow up written materials.
- **Sequencing**: paragraphs and sections of a text are placed out of the correct sequence and students have to re-assemble it into the correct order. This activity encourages group discussion and gets students to really think about the context. You could use this technique wherever an experimental procedure is described in a Brief. Students could reconstruct the procedure before carrying it out.

- **Emphasis**: this technique requires students to highlight single words, sentences, data or phrases from the text. By underlining, highlighting or circling key words, students have to make decisions about what they are reading. This helps to develop skills required for selecting important details from a mass of information, e.g. laws, equations, explanation, etc. These highlighted words or phrases can be discussed.
- **Labelling**: labelling of diagrams can be used to encourage reading to gather specific information. Students can find appropriate labels from a section of text. This ensures that students are considering ideas hidden in the text and makes them annotate diagrams rather than simply copy them down. This could be particularly useful when students are producing presentations. You could supply students with some diagrams needed for the presentation. Students would then need to interrogate the relevant text in the Brief, their notes, or secondary sources to label the diagram(s). You may want to support some students in this activity with additional written prompts and questions.
- **Devising questions and/or identifying key learning points**: students are given a text and asked to identify key learning points and devise a series of questions. These questions can be given to another group to answer. In this way the text is interrogated at least twice - once when the questions are being devised and again when the answers from another group are being checked.
- **Editing/correcting the work of others**: this requires careful concentration and scrutiny of text and encourages active reading.

Active Writing, Presentations and Displays

The *Communication* stage in the EUPRB model provides realistic context for students to engage in active writing, presentations and display work and will be useful when preparing for the *Express Yourself* conference.

Active writing enables students to write for a variety of audiences and purposes and includes, for example, writing in the form of:

- Diaries
- Investigation reports
- Newspaper, magazine and journal articles
- Scripts for TV and radio interviews
- Research proposals and applications
- Consumer reports
- Letters, poetry, plays
- Presentations and displays

Active writing should be fostered so that students can be asked to clarify and order scientific ideas and information for a real purpose - and in their own words. It teaches students to:

- Take responsibility for their own writing
- Clarify and express their own ideas
- Communicate their findings to others
- Express personal feelings and reactions to scientific ideas and theories.

It also helps maintain the sense of realism established in the EUPRBs with students being provided with opportunities to communicate to audiences other than their teacher and each other, such as the *Express Yourself* conferences.

Creative writing is particularly valuable in helping to raise student interest and enthusiasm in science - it is also an enjoyable and pleasant change from the formal notes and report writing experience offered in some science lessons. Student diaries provide an important mechanism for developing skills of critical reflection, as well as giving teachers valuable insights into student understanding. This approach could be used as the students work through a Brief.

Some EUPRBs provide students with opportunities to present reports from group investigations to the rest of the class. These provide valuable active writing experience. Posters, leaflets and PowerPoint presentations, etc can be made to help communicate ideas to others.

Examples of active writing can produce a stimulating display for a school laboratory/classroom, for public areas and/or for an *Express Yourself* conference. Students can share responsibility for preparing and monitoring displays, which will help to promote their own self-esteem and contribute to the class, department and school ethos.

Students need considerable encouragement and practice in preparing and making presentations and displays, particularly for events such as the *Express Yourself* conferences. It is a great advantage if students can develop the skills and confidence for presentations while at school. It helps prepare for late life. A booklet called 'The Young Scientist's Guide to Giving Talks' is available for use with students but a few tips can also be found below:

Know your objectives

- Why am I giving this presentation?
- What is its point?
- What am I trying to tell people?
- What science is involved?
- What information do I need to find out?

Know the situation

- Who is in the audience?
- What will the room be like?
- What resources are available?
- What do they already know?
- How long will the presentation last?

Know your subject

- Have I fully researched the subject?
- What questions might be asked?
- Will I be able to answer them?
- Should I include a demonstration?

Write the presentation

- What structure will the presentation take?
- Is the sequence logical?
- How will I introduce the topic?
- Have I written it clearly?
- Have I included all the important points?
- Is the purpose of the presentation clear?

Rules for presentation
DO:

- Speak clearly and at a steady pace
- Try to get rid of 'ers' and 'ums'
- Control your mannerisms - don't fiddle with clothes, pens, etc
- Vary the volume and pitch of your voice
- Be yourself (there is no need to act)
- Be enthusiastic and smile.

DON'T:

- Be tempted to race on (use pauses)
- Speak to the screen, whiteboard, poster, etc. (look at the audience)
- Read from your notes (speak naturally).

Visitors

Using visitors as an active learning experience for the students can have many benefits. Visitors to the classroom, such as Role Models, can be very motivating for students, not least because they are 'somebody different'. Also they are often 'experts' in their field who can bring a 'real world' perspective into the classroom. Often the work done in organising a visitor and managing the visit is done by the teacher. However, students can take an active role in the planning and managing of the visit. Preparation is vital. The opportunity should be provided for a 'rehearsal' prior to the visitor's arrival. The following checklist may be helpful.

Preparing for a visitor

- What is the purpose/objective of the visit?
- Who shall we invite? What is her/his background?
- Who makes the invitation? In writing or in person?
- What do we want to ask? What scientific questions should we raise?
- Can we ask personal questions?
- How can we encourage and help everyone to say something?
- How can we prevent individuals taking over?

- Who will ask what and in what order?
- Where is the best place to meet?
- How do we greet and welcome the visitor?
- Where will the visitor sit? How will the room be arranged?
- Who has a few prompt questions if things dry up?
- How do we bring the session to a close?
- Who will give the vote of thanks?
- Who will escort the visitor to and from reception?
- How do we review the visit?
- Who will write the letter of thanks?

This requires preparation time, but it is very worthwhile and it almost guarantees a successful visit. It is also an excellent learning experience for the students and most visitors experiencing this type of visit have found it very rewarding.

Experimental and Investigative Science

A key aim of the EUPRBs is to offer support for experimental and investigative work in the form of interesting and realistic contexts which link the learning of scientific knowledge and skills with investigations and problem solving.

An investigation allows students to use their skills, knowledge and experience in a flexible and creative way. Working productively in small groups, with the opportunity to plan their own work, gives many students a tremendous stimulus and insight into science as a human activity. No one person learns in exactly the same ways, most do not learn linearly, and investigations provide a major opportunity to follow different pathways.

Careers Education Opportunities

The supply and demand issues relating to graduates in Science, Technology, Engineering and mathematics (STEM) disciplines have been a matter of concern since numbers began to drop over a decade ago. Uptake of the physical sciences, chemistry and mathematics continues to raise concern.

Young people have broadly positive views about the role of science and technology in society even if this does not always translate into the post-16 study of scientific subjects. They largely agree that science and technology is important for society; that it can contribute to curing diseases; that it can create greater opportunities for future generations; and that it can make everyday life healthier, easier and more comfortable. Girls generally exhibit less confidence and lower levels of optimism than boys about the contribution of science and technology to society (Jenkins, 2006). A majority of 11-14 year olds agree that science has a positive influence on society and that it is important for a country to have well qualified scientists, while around half think that science makes an important contribution to the wealth of a nation (National Science Learning Centre, 2007).

Careers aspirations have been shown to be one element influencing further study. Murphy and Whitelegg (2006) found that the perceived strategic usefulness of physics is a significant predictor of the choice of physics for both boys and girls. Students want careers advice well in advance of being 16 years old so 13-14 years old is a good time as it is often the time when these students are starting to consider their post 16 options.

Context-based and Science-Technology-Society (STS) approaches to teaching have been found to produce significantly more positive attitudes to science in both boys and girls than their gender peers in classes using a traditional approach (EPPI, 2005). This study also shows that boys and girls in classes using this approach gain significantly better conceptual understanding of science than their male and female peers using a traditional approach. This method also narrows the gap between boys and girls in their attitude to science. Girls in classes using the context-based/STS approach developed a significantly more positive attitude towards taking a science career compared with boys in these classes (EPPI, 2005).

Another factor influencing the participation of females in STEM subjects and careers is their perception of the STEM workforce. Murphy and Whitelegg (2006) found evidence that as women progress from secondary to higher education they 'increasingly see no future in a career in the sciences' (p 4). This is an issue that can begin to be addressed by female students coming into contact with female role models from what used to be considered non-traditionally female careers.

The EUPRBs highlight a whole range of careers from the obvious such as becoming a scientist, engineer of technologist to the less obvious such as becoming a scientific journalist. These can be highlighted as each Brief is tackled perhaps in a discussion after completing the work.

Some of the misconceptions about what scientists and engineers do can be discussed and combined with work with a role model, some myths around the stereotypes can be dispelled and maybe interest raised in the students.

Table 3 highlights some of the obvious and less obvious career areas that can be discussed around each of the EUPRBs. It should be emphasised that these career areas can range from the scientists and engineers with higher degrees to highly qualified technical support roles, to trainee technicians, etc. This is not an exhaustive list but it may form a starting point for any students who wish to learn more about careers in the STEM subjects.

Discussion with career advice specialists, visits from role models and visits to specialist centres, universities and research centres all enhance understanding of career opportunities within the fields of the Earth and Universe.

Health and Safety

Guidance on health and safety matters is given in the teachers' notes for each EUPRB, according to best practice at the time. However, no written text can fully anticipate the laboratories, equipment, resources, and materials available to the teacher, nor the abilities, background or temperament of the students with whom they are working. Also, one of the aims of the EUPRBs is to encourage flexible and creative scientific activity. **Therefore, Chain Reaction cannot accept any responsibility for actions taken by teachers or their students.**

A risk assessment needs to be carried out before any activity that may present a hazard. It is the duty of employers to provide the information needed for professional teachers and technicians to make day-to-day judgements on what is necessary to enable practical activities to go ahead. Certainly the guidance given in each EUPRB can be construed as helping to supply that information, but more is needed. It is assumed that employers (teachers and technicians) have available standard health and safety references.

Table 3. Possible career choices around each EUPRB

EUPRB	Science	Engineering	Technology	Other
Collision Course	Physics Cosmology Astronomy Geology	Telescope engineering Computer engineering Computer imaging	Information technology specialisms	Science journalism/ writing
Out of site out of mind	Environmental/Earth science Chemistry geology	Waste management engineering	Environmental management	Scientific writing
Mars-ology	Vulcanology Astronomy Geology	Fluid engineering Space hardware design/ engineering	Information technology specialisms	
Green Heating	Physics	Environmental engineering Heating engineering	Solar panel design	
Ozone Conference	Environmental science Atmospheric sciences Meteorology	Atmospheric engineering	Measuring emissions technology	Weather forecasting
Green light	Physics Environmental science	Electrical/electronic engineering Environmental engineering	Light bulb design	Scientific writing/ journalism
PHEPPs	Physics	Electrical/electronic engineering		
Feed the World	Chemistry Biochemistry Biology/plant biology	Environmental engineering	Laboratory technical work	Scientific writing
Plants in Space	Plant biology Biochemistry Human biology/medicine Space science	Space survival engineering Design engineering	Life support computer technology	Scientific writing

Where information is given in an EUPRB it is assumed that practical work is conducted in a properly equipped and maintained laboratory. It is assumed that mains-operated equipment is properly maintained, care is taken with normal laboratory operations (such as heating and handling heavy objects) and that good laboratory practice is observed when chemicals are handled, prepared and disposed.

Open-ended, investigative science or project work need to be carefully planned in respect of health and safety. In the first instance students should be encouraged, as part of the planning exercise, to assess risks and devise suitable precautions. However, since teachers cannot rely on students having carried this out adequately, they must always check their plans, whether written or otherwise. Sometimes, the implications of the plans may not be clear, either to the student or to the teacher and in any case, plans may change as the investigation or project proceeds. Teachers' supervisory skills will be tested to the full when all the students in the class are all working on different investigations. It is therefore best to set investigations in the context of relatively safe activities, avoiding, for example, the use of potentially hazardous chemicals or dangerous equipment. It is intended that students progressively learn to make risk assessments themselves, with their teachers monitoring the process and taking final responsibility. This planning and checking of plans, with overall monitoring, creates an excellent opportunity for formative assessment of learning.

This is also an excellent example of education for citizenship and industrial awareness in action, with students looking out for one's own health and safety and that of others and making risk assessments, just as they will have to do in their working lives.

REFERENCES

Blanchard, M. R., Southerland, S. A. & Granger, E. M. (2008). No silver bullet for inquiry: making sense of teacher change following an inquiry-based research experience for teachers. *Science Teacher Education.,* 93 (2), pp 322 – 360.

Capps, D. K. & Crawford, B.A. (2012). Inquiry-based Instruction and teaching about nature of science: Are they happening? *Journal of Science Teacher Education*. Published on-line 26.08.12. Springer.

Crawford, B. A. (2000). Embracing the essence of inquiry: new roles for science teachers. *Journal of research in Science education*.37 (9), pp 916 – 937.

EPPI, (2005). *Review: The effects of context-based and Science-Technology-Society (STS) approaches in the teaching of secondary science on boys and girls, and on lower-ability pupils.*

Jenkins, E. (2006). Student opinion in England about science and technology. *Research in Science and Technological Education*. 24 (1), pp 59 – 68.

Murphy, P. & Whitelegg, E (2006). *Girls in the Physics Classroom: a Review of the Research on the Participation of Girls in Physics*. Institute of Physics and Open University.

National Science Learning Centre (2007). *Annual National Survey of Year 9 Students' Attitudes to Science.*

APPENDIX 3: TOOLKIT ON SCIENTIFIC INQUIRY- UNDERSTANDING THE USE OF PUPIL RESEARCH BRIEFS (PRBS)

Most PRBs are simulation activities, which involve students working as realistically as possible within a real-life scenario.

Information is supplied which allows students to operate within the simulation. Simulations are 'job-centered' as the emphasis is on carrying out a certain task or tasks and students behave according to the task at hand.

Simulations provide a framework and philosophy in which students are not only active participants but are given considerable responsibility for their actions and decisions.

The technique encourages a cooperative approach, provides a stimulating and creative atmosphere and develops problem solving and communication skills, as well, as knowledge and understanding of scientific concepts.

Within the context of the PRBs, simulation enables the students to gain insight into, and awareness of, the roles and activities carried out by science and engineering researchers.

First Stage of a Simulation: Preparation

Check that you are fully familiar with the aims, content, activities and organisation of the simulation. This will involve sorting out groupings for each stage, student materials, apparatus, what you will be doing and when. This is where the PRB Teachers' Notes will be very helpful.

Second Stages of a Simulation: Briefing

Introduce the Brief, setting the context for the activity and stimulating student interest and 'readiness' to get involved. You can explain the structure of the activity and give out the Student Activity Sheets. You may in fact take on the role of a research director or supervisor - simply giving an overview of what is expected and creating a sense of realism and purpose. If there is additional information about the topic it can be found in the background information for teachers, at the end of the Teachers' Notes for each PRB.

Third Stages of a Simulation: Action

In most cases you will step back and observe. The students will generally take responsibility for working through the activity. However, you will be required to coordinate changes during an activity such as arranging groups for investigation tasks, reporting back, etc. You will also want to facilitate by sitting in on a discussion, helping move things forward if necessary and giving help and advice when appropriate.

However, try to do this in a way that will maintain a sense of realism, for example by using questions such as, *'how do you think this would be done in a research laboratory?', 'what information do you need in order to carry out this method?', 'where would a researcher get this information from?'.*

Fourth Stages of a Simulation: Debriefing

Group report- either in small groups or as a whole class. Important learning points are drawn out and misconceptions are challenged and corrected sensitively. An overview of the Brief is created and students are reminded of the outcomes and what needs to be completed.

Fifth Stages of a Simulation: Follow Up

Following completion of reports, presentation plans, etc. more formal presentations or displays can take place. Links with future work can be made and the experience can be evaluated.

Remember to:

- Give careful thought to groupings, roles and functions
- Encourage participation and cooperation
- Set timescales and keep things moving along briskly
- Take the opportunity to observe and record progress.

Recommendations for a Successful Group Discussion

- Create the right climate - agree some ground rules:
    - Don't interrupt other members
    - Listen when others are talking
    - Respect each other's' opinions
    - You don't have to say anything if you don't wish to
    - No silly or unpleasant remarks
    - Share out the jobs and allocate roles and responsibilities
    - Help others if you can but don't interfere.
- Encourage active listening and give practice in listening skills.
- Set clear tasks and goals with timescales, and make sure that the students have sufficient background information and knowledge.
- Thinks about how groups are assembled:
    - Groups of 3 or 4 are normal and manageable
    - Groups can be self-selective but reserve the right to change them
    - Allow sufficient time to develop group working relationships, e.g. half a term.

Background Knowledge

All research projects are based on a basic understanding of the relevant science. Researchers embarking on a new venture are likely to be working in an area they already know something about. Even so, at the start of a new project, the researchers will carry out a search (known as a *literature review*) of what is currently known about the subject.

Why do you think scientists do this?

Research Papers Used in the PRBs

The PRBs attempt to represent the range of literature review activities by means of 'research papers' specially written for the purpose and can be in the form of reports, meeting summaries, extracts from books and notes etc.

All are written at a language level that students can understand.

A PRB can often start with an email which introduces the subsequent papers, articles or notes. In some cases extra guidance may be needed, and so briefing sheets are provided. Most of the research papers students have to deal with contain 'hand-written' annotations, as if written by the research supervisor.

The first two phases of the PRB help to create the appropriate 'mind set', where the purpose of the investigative work is clear to the students, and where they have some stake in its successful outcome.

APPENDIX 4: TOOLKIT ON SCIENTIFIC INQUIRY-EXAMPLE OF A PRB

Name of PRB: Plants in Space

Teachers' Notes

Knowledge and Understanding

- Green plants photosynthesise when it is light.
- During photosynthesis light energy is absorbed by chlorophyll, found in chloroplasts in some plant cells.
- This light energy is used to convert carbon dioxide and water into sugar (glucose).
- Oxygen is released as a by product.
- The rate of photosynthesis is limited by low temperatures, shortage of carbon dioxide and shortage of light.
- For healthy growth plants need a range of mineral ions.
- The body needs a balanced diet to remain healthy.
- The more active people are, the more carbohydrate and fat they need for energy.

Introduction

In this Brief the students take on the role of research scientists in a university department of bio-astronomy. They are asked to carry out some preliminary work to use in a competition to design a life-sustaining unit for use in a crewed flight to Mars. They are asked to carry out investigations into various aspects of photosynthesis, to work out what plants could best be used in the life-support system, and into human physiology to work out what needs a crew would have for oxygen and food on the journey.

IBSE Links

- Using scientific ideas and models to explain phenomena and developing them creatively to generate and test theories.
- Critically analysing and evaluating evidence from observations and experiments.
- Using a range of scientific methods and techniques to develop and test ideas and explanations.
- Assessing risk and working safely in the laboratory.
- Planning and carrying out practical and investigative activities.
- Obtaining, recording and analysing data from a wide range of primary and secondary sources and using their findings to provide evidence for scientific explanations.
- Using appropriate methods, including ICT, to communicate scientific information and contributing to discussions about scientific issues.

This activity also provides opportunities for students to

1. Research, experiment, discuss and develop arguments
2. Use real-life examples as a basis for finding out more about science.

Figure 2. Route Through This BRIEF

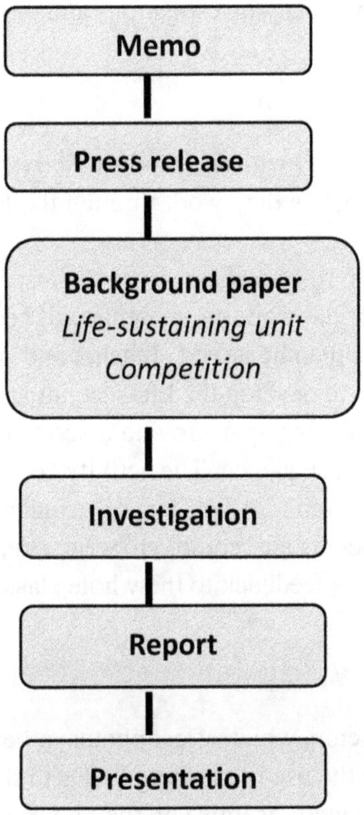

Prior Knowledge

Before attempting this Brief, students should have covered some basic work on photosynthesis and methods for measuring the rate of photosynthesis, but this can be taught during the activity.

Running the Activity

1. Students could work in a number of groupings during this activity.
2. Timing

Table 4. Suggested grouping formats

• Initial briefing, email & press release -	• Whole class; teacher introduces topic & sets the context for the activities
• Background paper *Life-sustaining -Unit Competition*	• Whole class or small group
• Carrying out investigations & other -research	• Pairs, threes or fours (depending on availability of equipment)
• Report -	• Individuals or small groups
• Presentation -	• Each group presents their results to the whole class

This Brief involves students carrying out a number of investigations. This work could take 3 to 5 hours of classroom time, depending on how many investigations and research tasks are done.

3. Activities

The teacher should issue the students with the Student Activity Sheet which provides them with a summary of what they should produce as they work through the Brief. The checklist lets each student monitor his/her progress.

The Director of the research team (the teacher) issues the researchers (students) with the email and press release. These set the scene for the work the students will be doing. The background paper 'Life-sustaining Unit Competition' should then be issued. Teacher and students go through this paper, which gives details of the work being done to develop the life-sustaining unit for a crewed Mars mission. The handwritten notes in the margins of this paper provide a series of investigations and other activities which all students should carry out in sequence. The activity covers a number of key investigations on the topic of photosynthesis, which, taken together, form a stimulating way to tackle parts of this topic. Following their investigation(s), students are required to write a report of their own work. A nominated member of each group should provide feedback to the whole class during the plenary session.

4. Investigation details

A key investigation is into the factors which affect photosynthesis. The background paper mentions research which is trying to develop the use of flashing LEDs to illuminate plants. Students should investigate the effect of colour (wavelength of light) on the rate of photosynthesis. Since LEDs come in red, green, blue and yellow, these colours should be chosen for the investigation. Students do not need to use LEDs for the work. They could illuminate pondweed (*Elodea canadensis*) with light fitted with a filter. Different student groups could investigate different colours. Before carrying out these experiments the students should design a standard method, so that the work of different groups can be compared. An important factor to control is the light intensity for each of the colours of light being tested.

The other tasks and investigations mentioned are:

- *A dietary analysis for eight people on a space mission.*
- This would be based on researching into secondary sources.
- *Methods of recycling water.*
- Students could investigate ways to evaporate and condense water. Although not strictly a biological investigation, it provides a context-based approach to the water cycle.
- *How physical activity affects oxygen consumption and carbon dioxide production.* This could involve using a spirometer, or calibrated breathing bag, to measure breathing rate and tidal volumes when a student is at rest and during different levels of activity.
- *The combination of nutrients needed to be supplied for healthy plant growth.* This provides a context for carrying out hydroponics experiments.
- *The effect of light intensity on the rate of photosynthesis.* Pondweed under different light regimes.

- *The effect of light and dark periods on plant development.* Research into secondary sources on the need for light and dark periods to cause the onset of flowering. The need for dark periods will affect the potential for oxygen production.
- *The effect of CO2 concentration and temperature on plant growth.* Pondweed in water containing various amounts of sodium hydrogen carbonate, which provides plants with CO2. Temperature can be varied using ice baths, room temperature and a range of temperatures created using water baths.
- *The range of plants needed for a balance between protein production and efficient photosynthesis.*
- This is a link back to the activity to design a diet for long space travel. If the spacecraft is taking food plants along to supply the crew, what are the best plants to produce a balanced diet? This is likely to be a secondary source search, most likely involving diet tables.
- A note at the end of the paper asks for any further suggestions for investigative work. Students may suggest testing a range of food plants to see which ones are the most efficient at photosynthesising. This may not produce conclusive results, but it could be a worthwhile experiment to try – if only to show that sometimes scientists do not get definite answers. One possible approach would be to illuminate various food plant leaves (as fresh as possible) in a closed environment with a CO2 indicator.

5. Safety issues

N.B. It is also important that you prepare your own risk assessment for the practical work in this Brief in the usual way.

Many suggestions for independent student investigations are contained in this Brief.

All student plans should be subjected to check and risk assessment by the teacher before being allowed to proceed.

Student Activity Sheets

Setting the Scene

You will be working as a researcher in a university department of bio-astronomy. You are part of a team that is putting together a proposal to win a contract to design a life-sustaining unit for a spaceflight to Mars. You and your colleagues will carry out some research into photosynthesis and seek out other information about dietary needs and energy requirements for astronauts during their mission. You will then write a report of your findings to present at a seminar.

Science you will learn about in this activity:

- Green plants photosynthesise when it is light.
- During photosynthesis light energy is absorbed by chlorophyll, found in chloroplasts in some plant cells.
- This light energy is used to convert carbon dioxide and water into sugar (glucose).
- Oxygen is released as a by product.

- The rate of photosynthesis is limited by low temperatures, shortage of carbon dioxide and shortage of light.
- For healthy growth plants need a range of mineral ions.
- The body needs a balanced diet to remain healthy.
- The more active people are, the more carbohydrate and fat they need for energy.

Outcome Checklist

You will carry out one or more investigations or written tasks and produce a written report of your findings. This will be presented to the rest of the class.

You should make sure that you produce the following as you work through the activity:

- Investigation plan
- Investigation report
- Presentation

Item 1: Background Paper

Email from: Daphne Jefferson
To: Research teams
Subject: Space Agency 'Life-sustaining Unit' competition

I have just got the details of a new Space Agency project to develop plans for a Life-sustaining Unit for use on possible future space flights. It seems that the Agency has decided to hold a world-wide competition to produce the best design for a life support system for use on long space flights. Astronauts on such flights will need food, water and oxygen supplies for the journeys of 2 years or more. The winning team will get the job of designing the system and developing the technology to go with it. I think that we should have a go – we have some good staff who are imaginative and hard working. I am sure we would have a good chance of getting the final contract. Your job will be to provide advice and ideas about some of the biological issues, e.g. getting maximum oxygen and food production from the plants taken on the spacecraft and relating this to the needs of the crew. The Space agency project was announced by a Press Release on the internet. I have attached this along with further details about the competition. I also downloaded a background paper (see attached).

I have looked through the paper and made some notes in the margins about possible investigations and background research we could carry out. Doing this work will help us get to grips with some of the problems we will need to consider when we begin drawing up the design proposal. Perhaps these could be shared out amongst the team. You will need to:

- Read through the press release and background paper
- Select one or more of the issues I have highlighted and carry out any necessary research or investigations
- Produce a report of your findings
- Attend a meeting where the results of the various investigations can be discussed.

At this meeting we will start thinking about putting our ideas together to design the Life-sustaining Unit.

Item 2: Press Release

Space Agency Calls for Design Proposals to Develop a Life-Sustaining Unit for Use on Crewed Space Flight to Mars and Beyond

The Space Agency is pleased to announce a competition, open to all institutions to develop a Life-sustaining Unit for use on crewed space flights of over two years duration. The competition will result in a short-list of 10 designs, which will form the main feature of a major public exhibition to be held in one year's time to celebrate the imagination, creativity and capability of the world's scientists and engineers. The winning design will be chosen by a team of international scientists and engineers, and announced at a press conference during the final day of the exhibition. The winning team will be awarded a contract to carry out further research, and to produce a working prototype of the Life-sustaining Unit.

The design of the life-sustaining unit will be the first step on the road to the achievement of a space flight to Mars carrying astronauts. Following the first crewed space flight in 1958, and then the first crewed Moon landing in 1969, the International Space Station, the successful landing on Mars of the Rover Curiosity in 2012, man going to Mars represents a major milestone in the history of human exploration of space.

Producing the most effective design for a Life-sustaining Unit for use on crewed space flights to Mars and beyond. At the moment plans for a mission to Mars are at the planning stage. These include the European Space Agency long-term vision of sending a human mission to Mars in 2033 and NASA considering 2037.

This short paper sets out some of the key issues in the design of a life-sustaining unit.

Item 3: Creating a Life-Sustaining Unit for Astronauts on a Mission to Mars

Introduction

We need a full dietary-needs analysis for 8 individuals on an 800 day journey in space – total amounts of nutrients, energy needed, etc. – list the factors which could affect the totals, and HOW they would affect them.

In every space mission so far carried out all materials needed to sustain the astronauts (food, water, oxygen) have been sent up with the crew. Getting large masses out of the earth's gravity is very expensive. The longer the mission, the bulkier the payload. There comes a point when the cost of sending up all the food and oxygen needed (water is recycled) is greater than the cost of installing a self-sustaining biological supply system on board the spacecraft. It is calculated that the break-even point is two years. For flights less than this time, it is cheaper to send the food and oxygen up with the astronauts.

You could come up with and try out various methods for recycling water.

For flights longer than this it is cheaper to build a supply system on the space craft. Space missions to Mars will take over two years to complete, and so the Space Agency is seriously considering the possibility of putting a balanced ecosystem, or Life-sustaining Unit, into a Mars-bound space craft, to provide the crew with all the food and oxygen they will need for the flight. It is vitally important to control such an ecosystem precisely, supplying energy in finely judged amounts so that as little as possible is wasted, and using plants that will provide the highest possible yield of food and oxygen from the minimum amount of light and nutrients.

Item 4: Growing Plants in Space

Growing Medium

Can we produce data on how physical activity affects oxygen consumption and CO2 production – what, apart from physical activity, will affect these levels?

An adult human requires, on average, 602 litres of oxygen a day. So, plants used on the space craft must be as efficient as possible at photosynthesis in order to supply the oxygen needed by the crew of the space craft. The plants will not be grown in soil, but instead a soil substitute will be used. The material contains all the nutrients

plants require and it releases them in controlled amounts, thus making it much more efficient at supplying minerals than ordinary soil.

What combination of nutrients needs to be supplied to plants for healthy growth? You could also look at the types of growth mediums which could be used.

Light

The greatest technical problem is supplying plants with enough light. The Life-sustaining Unit will be powered by solar cells which will produce electricity. Some of the electricity will power banks of electric lights which illuminate the plants. Plants will use the light energy to photosynthesise, making both food and oxygen for the astronauts.

How does light intensity affect the rate of photosynthesis?

Each stage of this chain of energy transfers is inefficient. Light is first converted to electricity by the solar cell. Then it is converted back into light by the electric lights. Finally, light is converted into chemical energy in plants during photosynthesis. Even under greenhouse conditions, only about 7% of the light arriving at a plant leaf will be used in photosynthesis. This is partly because chlorophyll molecules in the leaf's cells can only process one photon of light at a time. Any other photons arriving at chlorophyll molecule when it is busy dealing with the last one will be wasted.

Do plants need a period of dark each 24 hour period in order to develop? If they do, what are the implications for oxygen supply to the crew?

Scientists at the European Space Agency have suggested using lights that flash on and off very quickly, between 2000 and 12000 times a second. This leaves a gap between the photons arriving at chlorophyll molecule, allowing enough time for the molecule to be ready to receive its next packet of energy. A light source which can switch on and off this quickly is a light emitting diode (an LED). LEDs have the advantage of being very long lasting and using very small amounts of electrical energy. LEDs emit red, blue, green or yellow light, depending on the material used when they are made.

What colour of the spectrum will produce most photosynthesis? And could plants survive on Mars where the light is mainly at the red end of the spectrum?

Type of Plant

What are the optimum CO2 and temperature levels for plant growth?

The plants in the Life-sustaining Unit need to be both nutritious, providing the right nutrients to keep the crew healthy, and efficient at photosynthesis_so the maximum food production occurs. One square metre of vegetation supplies between 30 – 31 litres of oxygen per day. Scientists need to know the best growing conditions in which plants can thrive on the long journey to Mars, to ensure that the rate of oxygen production is maintained or improved.

Which plants will produce most food and oxygen given the same amount of light and heat?

The optimum temperature and carbon dioxide levels also need to be studied for the maximum growth of the plants to be obtained.

It is likely that only a limited range of plants can be grown on board, and this could cause problems for the crew. The prospect of having only three or four different vegetables to eat as your entire diet for more than two years would be very unappealing. This is why food scientists are trying to develop- a machine that can produce food products with a range of textures, colours and flavours. This should help to prevent the diet from being too monotonous.

Since only a small range of plants can be grown, it is important to strike a balance between efficient photosynthesisers and ones which produce useful amounts of protein. What range of plants do you think would be needed?

Conclusion

The problem of providing a self-sustaining environment for a crew of up to eight astronauts on a mission to Mars and back are formidable. The alternative is to send up two year's (or more) supplies of concentrated food and bottle oxygen with the space craft. The cost of doing this would be enormous. It is extremely important that a study is carried out into the most efficient way of cultivating plants that will be needed on the flight. A self-sustaining ecosystem will make it possible for human beings to visit the only other planet in our solar system that we could land on safely.

APPENDIX 5: TOOLKIT ON SCIENTIFIC INQUIRY-TRANSLATING THE PRB IN CLASSROOM PRACTICE, AN EXAMPLE OF A LESSON PLAN

The following section presents a lesson plan for the first lesson on this PRB. Yelon's framework (1996) is used to inform the design of the activities within the plan. Yelon's framework invokes the notion that each lesson should contain the following types of activities: motivation, orientation, information, application and evaluation. While this may seem like a lot, one activity could potentially serve as a motivation or information activity, for example.

Day: Friday
Time: 13:30
No. of Students: 25
Length of Lesson: 80 Minutes
Subject: Science
Topic: Plants in Space:
Scientific Concept Covered (1) Photosynthesis: why it is important for life and what does it produce.
Lesson Number: 1

Previous Knowledge and Experience

Students should be aware of some of the basic ideas behind photosynthesis. They should be aware of the word equation for photosynthesis and understand that it is a process that plants do. The Students have previous experience of the Jigsaw Methodology and also in designing their own investigators.

Subject Matter

This introductory lesson will focus on testing out ways of showing that photosynthesis has taken place as well as factors which influence whether photosynthesis takes place. It will also begin to consider how a life sustaining system could be designed in space.

Table 5 shows the lesson aim, objectives, the organisation of learning experiences and the assessment strategies planned for in the lesson.

Resources

Activity Sheets (1x group- 5 groups), Apparatus for Investigation (suggest providing students with some red herrings also for them to be able to decide what apparatus should be used accordingly): light source, *Elodea Canadensis* (destarched), lit taper, starch solution (these will be needed for the next lesson), water bath, tin foil, funnel, beaker, flask, clamps, carbonated water, potassium hydroxide solution, cotton wool.

Self-Evaluation

I will think about the following after my class:

Table 5. The lesson aim, objective, organisation of learning experience and assessment strategies

Lesson Aim	Lesson Objectives	Organisation of Learning Experience	Assessment (of Student learning)
To develop an understanding of the biological factors which influence the creation of a life sustaining system in space.	1. ***Cognitive:*** Students will be able to describe the biology behind photosynthesis in terms of the production of oxygen and food. 2. ***Psychomotor:*** Students will design an investigation using the knowledge they have generated from their prior research. 3. ***Affective:*** Students will work collaboratively in groups. 4. ***Cognitive:*** Students will be able to understand the factors which influence the rate of photosynthesis.	1.Small group activity using the Jigsaw method of teaching 2. Group investigation 3. Group investigation 4. Group investigation	1. Higher order and lower order questioning 2. Self and peer- assessment 3. Teacher observation 4. Discussion during investigation and investigative report.

- Classroom management
- Time management
- Questioning skills
- Board work
- Observation/circulation
- Assessment of/for learning
- Use of positive reinforcement
- Enthusiasm
- Resources and how effective they were.

Table 6 shows the student and teacher activity within the lesson.

REFERENCES

Yelon, S. L. (1996). *Powerful Principles of Instruction*. Lancing, ML Longman Publishers.

Table 6. Showing the student and teacher activity

	Student Activity	Teacher Activity
Time		
3 minutes 5 minutes 20 minutes 40 minutes 12 minutes	**Motivation Activity** Students will enter the room promptly and will take out their science notebook and pens. **Orientation Activity** Students individually brainstorm ideas of what factors influence whether life can be sustained in space. Students contribute to a group discussion on this. **Information Activity** Students move into their home groups. Each group assigns a leader, a scribe and a time keeper. The leader of each group is briefed by the teacher. The leader of each group reads out the initial brief to their group (email from Daphne Jefferson) and assigns items to the different group members (pages 8, 9, 10 and 11 in the brief). Each group member goes to their expert group and within this group they read and analyse the information on their respective pages to draw conclusions. The leaders circulate around the classroom helping the other students to work out the information on their pages. **Application Activity** The members of the expert group then return to their home groups and share the information from their page with the rest of the group. From this the members of the home group develop an investigation plan where they design an investigation showing how photosynthesis produces food and oxygen but requires light and carbon dioxide. The students look at the resources laid out at the top of the class and are aware that some resources are "red herrings" are will not be needed in the investigation. The students should also consider a control. In their group they decide the procedure for the investigation and share this with the class. The students write the agreed procedure in their science notebooks. The students set up the investigation which the will complete in the next lesson. Together with the information from their pages and the experience of doing the investigation so far, the students begin to draw up an investigation report showing how they would design a life sustaining system in space. This report will be added to over the next few lessons. Students raise their hands and answer the question when directed to by the teacher. **Evaluation Activity** Students share their report with the entire class and students provide feedback to each other on their proposed ideas. Together the class decide on what they believe a life sustaining system could look like. The students write their task in their homework journals.	The teacher will direct the students to take their seats promptly. The teacher will inform the students of the key activities that will take place in today's lesson by writing the following on the whiteboard: *"You will act like a real scientist in today's lesson".* The teacher asks students to write down five key words associated with how life would be sustained in space. The teacher leads a group discussion and puts the ideas on the board. The teacher ends this discussion by saying that in today's lesson they will begin to test their hypothesis. Teacher then advises that they will be working in groups for today's lesson and assigns the students into groups of 5 which will be their home groups. Teacher advises students that the jigsaw methodology will be used to help them collect data and asks Students to assign the roles within the group. The teacher asks the leader to read out page 7 on the student activity sheets and then assigns the other pages to the group members. The teacher will circulate as students work on their respective pages. The teacher will help direct students to what they intend to investigate first i.e. that plants produce food and oxygen but light and carbon dioxide are needed for photosynthesis to happen. The teacher shares that this is only one investigation that can be carried out and the remaining investigations will be carried out in the next number of classes. The teacher listens to the procedures and then directs the class to reach a consensus. The teacher puts the finalised procedure on an overhead projector and informs the class that the results will not come for at least 24 hrs. The teacher also asks students to think about the following: ***How could physical activity affect oxygen consumption and carbon dioxide production?*** The teacher directs this peer-feedback. The teacher then assigns students with the task of gaining some additional information to help them to consider additional investigations that have come to their attention through reading the pages. They can research on the internet for homework and asks students to consider a new research question for the next lesson.

APPENDIX 6: TOOLKIT ON SCIENTIFIC INQUIRY- COMMUNICATING IN SCIENCE- SHARING OUR KNOWLEDGE WITH THE PUBLIC

Some PRBs provide students with opportunities to present reports from group investigations to the rest of the class. These provide valuable active writing experience. Posters, leaflets and PowerPoint presentations, etc. can be made to help communicate ideas to others.

Examples of active writing can produce a stimulating display for a school laboratory/classroom, for public areas and/or for a conference. Students can share responsibility for preparing and monitoring displays, which will help to promote their own self-esteem and contribute to the class, department and school ethos.

Students need considerable encouragement and practice in preparing and making presentations and displays.

It is a great advantage if students can develop the skills and confidence for presentations while at school. It helps prepare for careers in later life.

Tips for Presenting

Know your Objectives

- Why am I giving this presentation?
- What is its point?
- What am I trying to tell people?
- What science is involved?
- What information do I need to find out?

Know the Situation

- Who is in the audience?
- What will the room be like?
- What resources are available?
- What do they already know?
- How long will the presentation last?

Know your Subject

- Have I fully researched the subject?
- What questions might be asked?
- Will I be able to answer them?
- Should I include a demonstration?

Write the Presentation

- What structure will the presentation take?

- Is the sequence logical?
- How will I introduce the topic?
- Have I written it clearly?
- Have I included all the important points?
- Is the purpose of the presentation clear?

Rules for Presenting

DO:

- Speak clearly and at a steady pace
- Try to get rid of 'ers' and 'ums'
- Control your mannerisms - don't fiddle with clothes, pens, etc.
- Vary the volume and pitch of your voice
- Be yourself (there is no need to act)
- Be enthusiastic and smile.

DON'T:

- Be tempted to race on (use pauses)
- Speak to the screen, whiteboard, poster, etc. (look at the audience)
- Read from your notes (speak naturally).

An Example of a Conference Paper Template

The below template provides a suggested template for how students could communicate their findings.

Team Members:
Area of Science:
Our Prior Knowledge
Our Research Question

What Source Did We Use to Answer our Research Question?

Researching Internet
Analysing Articles
Experiment
Another
Our Method in more detail:
Our Key Data:
Our Main Conclusions:
What we did that a Scientist would also do?

Checklist

1. Did every member of the group have a role?
2. Did we make predictions before beginning to collect and analyse our data?
3. Did we have a control?
4. Were we accurate with the data we presented?

Tips for Creating a Poster

Balance

Creating a scientific poster has to do with balance. On the one hand of course you want to show all of your results… but on the other hand the poster should not bring your audience to tears because it contains so much text that it would take two hours to fully grasp the message within the poster…!

Rule Number 1:

Keep your poster as simple as possible, yet make sure everything important is on it.

Does not sound too hard, does it? Well it is quite tricky to find the right **balance** and scientific conferences are filled with posters that are flooding their audience with too much information.

Must Haves

On scientific posters you are basically telling the story of your scientific endeavour in a condensed form. Fairy tales often start with "Once upon a time…" and often end in "They lived happily ever after…"

Like in those fairy tales there are parts that everybody expects to be on your poster to make it an authentic scientific poster. Make sure the following are there:

- Introduction
- Experimental setup
- Results
- Interpretation of results
- Conclusions

Let's create a poster together and see what's what, shall we…?

Figure 3. Example of a poster

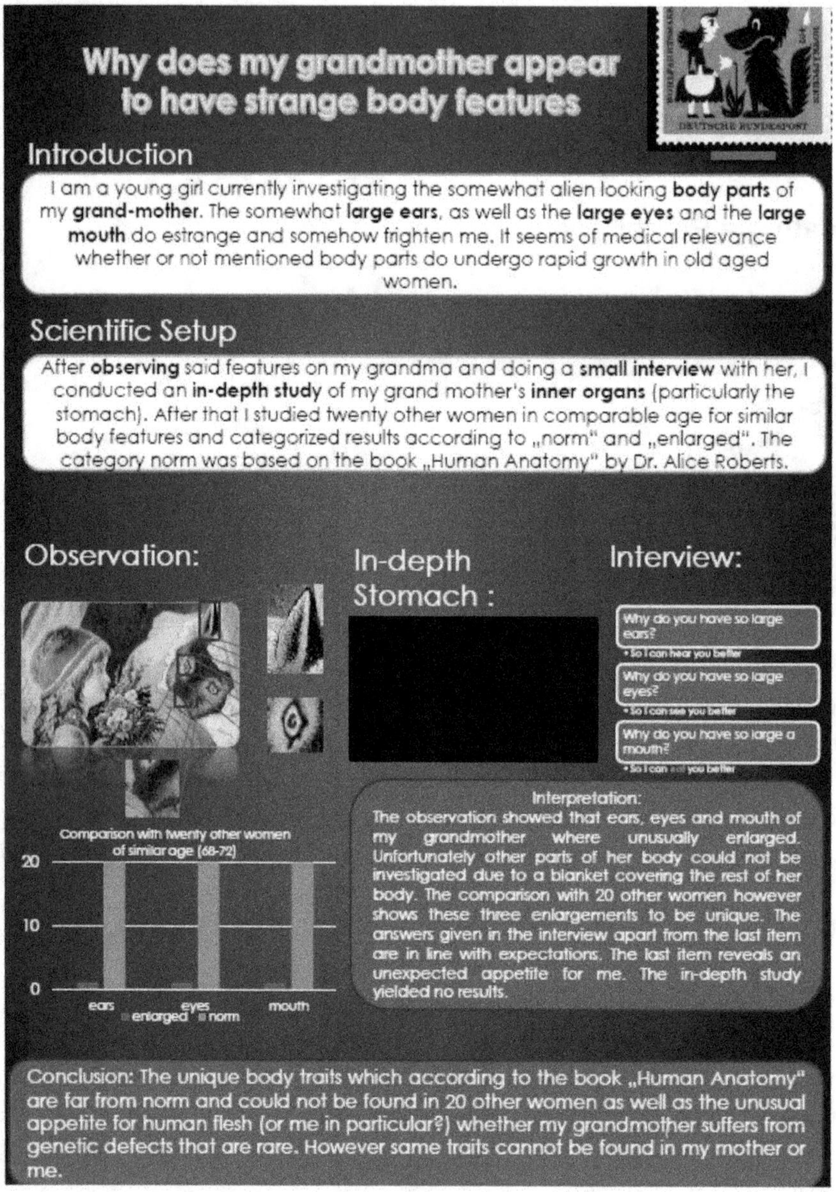

Creativity

When finding good ideas for a poster, do not hop on the first visual structure that comes to you. Look at the ideas and sketch them on paper to see if they visually look good. It is important to look for images which connect with your own ideas. When drawing, think of the format of the poster. Do as many sketches as you can – in the end you will use only some of these (or maybe only one) but creative processes include destruction also. When trying to decide the structure of your poster, do not use what is easiest – use what works best.

It is a great idea to write ideas on post-its but try to write no more than one idea per post-it. By doing so you can arrange and rearrange ideas.

You should take into account the following guidelines if you are going to use post-its:

- Do not judge your ideas, be open about them.
- It is a good idea to consider the "crazy ideas".
- The more ideas you think of the better.
- Continue with ideas that others had before you.

APPENDIX 7: TOOLKIT ON SCIENTIFIC INQUIRY-TOP TIPS FOR AN INQUIRY LESSON

Identify what your students need to learn and what they already know.

For example, in the brief 'Plants in Space', students need to know the following:

- Green plants photosynthesise when it is light.
- During photosynthesis light energy is absorbed by chlorophyll in chloroplasts in some plant cells.
- This light energy is used to convert carbon dioxide and water into sugar (glucose)
- Oxygen is released as a by-product.
- The rate of photosynthesis is limited by low temperatures or shortage of carbon dioxide and light.
- Plants need a range of mineral ions for healthy growth.
- Our body needs a balanced diet to remain healthy.
- Activity increases the need for energy, i.e. intake of carbohydrate and fat.

Find a real-life event or problem linked to the target knowledge to engage your students.

In the same brief, students take on the role of research scientists in a university department of bio-astronomy. As the potential of living on another planet excites people, students are asked to carry out some preliminary work to use in a competition to design a life-sustaining unit for use in a crewed flight to Mars. They are asked to carry out investigations into various aspects of photosynthesis, to work out what plants could best be used in the life-support system, and into human physiology to work out what needs a crew would have for oxygen and food on the journey.

Assign the project and student roles by sending an email to your students from a fictional university or research centre. Invite them to get engaged by solving a problem.

In groups students are invited to take part in a Space Agency competition, open to all institutions eligible for Research Council and Agency funding to design a Life- sustaining Unit for use on crewed space flights of over two years duration. A committee of international scientists will short-list 10 designs, which will form the main feature of a major public exhibition to be held in one year's time to celebrate the imagination, creativity and capability of the world's scientists and engineers. The winning design will be chosen and announced at a press conference during the final day of the exhibition.

After you ask students to take up the role of a scientist/researcher, ask them to consider and discuss in class what it takes to be a scientist. A Scientific Method diagram might come in handy.

Collate and simplify three of four articles on the target subject. Highlight terms in the articles for students to use as key terms during their literature research.

Give students briefing notes and invite them to work through an agenda, taking it in turns to give short presentations on each item in their articles. For example, in the 'Plants in Space' brief, students do not cover basic biology-related information on photosynthesis and tropism in plants (phototropism or geotropism). They can also run parallel searches on dietary analysis for eight people on a space mission, ways to evaporate and condense water and methods for recycling water, how physical activity affects oxygen consumption and CO2 production.

Invite your students to write research questions based on their first reading. After reading the above articles, students should come up with initial research questions.

For example:

1. What factors may affect the food and water needs of eight astronauts in space?
2. How can we recycle water in space?
3. How does physical activity affect CO2 production and oxygen consumption?

Raise students' awareness of source reliability, plagiarism and references.

Is this source of information reliable? Can I trust it?

Ask students to research those questions and present their findings in front of the whole class: Involve all students in class by assessing their contribution and coordinating their next step.

1. How are other students working toward the same problem/question?
2. How can we use what we've learned from others to inform our own inquiry?
3. How might we use what we've learned to help others in their learning?

When new questions or pathways emerge, divide students into groups so as to research different aspects of the same problem. ask them to present their action plans in class.

For example, your student teams could be investigating how hydroponics could be of use in their mission and what combination of nutrients are essential for a plant's healthy growth.

Coordinate your students' research without being prescriptive: Answer student questions with more questions.

1. What have we observed so far?
2. What are some of our initial understandings of the phenomenon being studied?
3. What are some of our initial theories?
4. What can we do to test our initial theories?
5. What are the questions that our whole group needs to work on to develop further understanding?

During experimentation, ask students to present their findings or challenges regularly or pose more questions.

Show students how to use graphs, tables and charts: units of measurements and legends.

Raise students' awareness of dependent and independent variables during experimentation.

Encourage peer assessment.

Ask students to evaluate their peers' reliability of measurement and critically analyse evidence from observations and experiments: You may assign them roles as in the six thinking hats:

1. Did we achieve our learning goal(s)?
2. How do we know?
3. What do we know now that we didn't know at the beginning?
4. What route did we take to get here?
5. Would we take the same route again?

Encourage students to investigate the possibility of visiting a University laboratory to use equipment for more accurate measurement.

Encourage students to investigate the possibility of a field trip, inviting or taking an interview from a scientist specialising in their research field.

Send students a Call for participation in a National Conference: ask them to put together a scientific poster and an oral presentation.

Raise students' awareness of presentation principles and scientific poster conventions by asking them to compare peers' and scholarly ones.

Tip: Anything your students ask is a potential research question; It is not a call for you to give them an answer. You are there to make sure they are on the right track.

APPENDIX 8: TOOLKIT ON SCIENTIFIC INQUIRY- BARRIERS TEACHERS MIGHT ENCOUNTER AND HOW TO OVERCOME THEM

The following section provides a summary of the possible barriers that teachers may encounter when doing the "Plants in Space" PRB and offers advice for overcoming these.

Terminology

Students often come across a number of new scientific and technical terms they are not familiar with. Terms that might need to be addressed in this research brief are: photosynthesis, tropism in plants, light, colour, energy, light intensity, oxygen consumption and dietary needs.

As well as active reading approaches, suggested in the general notes, teachers can encourage students to use simple experiments such as hand warming of different coloured lights can help relate colour to energy. Likewise, safely observing direct sun light during class time could be a starting point using, for example, transparent window glass, sunglasses and coloured plastic films. In order to understand photosynthesis we can start by asking students about the factors they think effect plant growth.

Requisite Student Prior Knowledge

1. Photosynthesis: Photosynthesis is the process used by plants, algae and certain bacteria to harness energy from sunlight and, together with some basic chemical elements, produce more complex compounds. Oxygenic photosynthesis is the most common and is seen in plants, algae and cyano-bacteria. During this process, light energy transfers electrons from water (H_2O) to carbon dioxide (CO_2), which in turn produces carbohydrates. In this transfer, the CO_2 is 'reduced' or receives electrons and the water becomes 'oxidized' or loses electrons. Ultimately, oxygen is produced along with carbohydrates.
2. Day and night in space: Light is needed to power the plants for the production of organic energy molecules. Light dependent reactions (light reactions) are initiated by coloured pigments, mainly green coloured chlorophylls. Light-independent reactions (dark reactions) make use of these organic energy molecules by providing the energy to fix the CO_2 (carbon dioxide) into carbohydrates. These facts must be taken into consideration when planning the experiment.
3. Tropism: Tropism is a biological phenomenon, indicating growth or turning movement of a biological organism, usually a plant, in response to an environmental stimulus. Phototropism is the movement or growth in response to lights or colours of light. Geotropism (or gravitropism) is the movement or growth in response to gravity.
4. Light spectrum and energy: Visualisation of light spectrum can be useful in understanding the differences between different colours. Light can be separated using a prism.
5. UV vs Visible vs IR wavelengths: The UV part of the spectrum carries more energy than that of the visible spectrum. If used though, it can damage the plants. It can also damage human skin. Plants must be chosen so that they can utilise safe areas of the light spectrum. The atmosphere of the earth absorbs most of the UV radiation. Outside the spaceship there is abundant UV radiation. This can be used for some processes that require a lot of energy.

6. Dietary needs of plants: Plants, like humans, require food. Plant food is different than that of humans. Plants use simple elements or compounds to build the complex food required by humans. As such, human waste can often be used to feed plants. Urine contains a lot of ammonia, while human excrements contain organic nutrients and other micronutrients the plants can reuse.
7. Dietary needs of humans: Another possible research question is: What are the minimum food requirements for the survival of people? Each living species requires different kinds and different amounts of food sources which facilitate or hinder energy conservation in space. In fact, the type of activity, the environment, size, and even gender play a big role. For instance, smaller humans usually spend less energy. Additionally, unnecessary activities can be minimised to save energy.
8. Selection of plants as potential food source: Humans require a variety of foods and nutrients. Different plants require different food sources, and produce different foods for humans. There is a great need for the best possible selection of plants to be used. Measurement of biomass might be an indication of that. Plants should provide as much food as possible for humans with minimum energy or light required.
9. Recycling in space: There are elements that can be recycled and others that cannot. What is the difference? How do you recycle? What are possible dangers or setbacks?

Students need to understand the difference between an element and a compound. Compounds can be built from elements. They require energy to be built; by plants for example. They release energy when used as food. They break down into simpler compounds or sometimes elements. Bond energy is turned into fuel for survival. Water can be recycled via the atmosphere of the spacecraft, human urine or wastewater. Water does not break down and not all types of water can be recycled. Some types are used for the synthesis of other compounds. Oxygen is used to oxidise food and release energy and some of it is turned into carbon dioxide. Carbon dioxide can be turned into oxygen again by photosynthesis, and thus it can be recycled. Organic compounds, such as sugars made by plants, cannot be recycled. They break down into simpler compounds (usually carbon dioxide and ammonia) and they release energy. Plants recycle CO_2 and NO_3^- to build new compounds with the help of the light.

Source Validity (In Search of Bibliographical Data)

When looking for information, students often stumble upon resources that are not always reliable or provide inadequate data. Teachers could design web quests in order to encourage students to cross-check data and raise their awareness of internet safety and criteria they can use in order to evaluate the reliability of sources available online. These criteria might be:

Authenticity

- Objectivity
- Timeliness
- Relevance
- Efficiency

Setting Up the Experiments

Students are often confused by the vast array of plants to choose from. As research time is limited, bibliographical data on plant nutritional value might be a good indication for selection. Alternatively, students might choose plants that are easily accessible to them.

Students should be encouraged to use everyday instruments and devices in order to make measurements of the plants' growth and height.

They might also take into consideration plant growth under natural and artificial light. Choice of light source (LED's or Fluorescent light bulbs) might take up more research time.

Temperature moderation is another factor that could be tested but available resources may not allow for successful monitoring.

Designing the Method

The idea of standardisation-bench mark is important and must be well understood. When a suggestion of making a measurement or a simple experiment is made, for instance, teachers should encourage it to be conducted by different students or groups so that they can compare their results in the following plenary session. What students need to notice and discuss is why their results might be similar or different, and what effect could a possible discrepancy have on the actual experiment. Students need to pay attention to hidden or inferred information. For example, if there are different lamps of the same colour, will they provide the plants with the same energy?

Understanding Dependent and Independent Variables

Before students set up an experiment, teachers should make sure that students are aware of scientific method and they can distinguish qualitative and quantitative data. They should be using tasks in class that facilitate students' writing of suitable hypothesis and identification of dependent and independent variables and control groups.

Given the scenario, the use of post-it notes can be very effective in order to break down the different variables. Also, making students aware that scientists should only tweak or change one of the variables every time is of great importance. At the same time, students should decide on the factors that should be kept constant during the experiment.

Light, water, soil, type of plant and temperature are key factors determining plant growth. So, for example, in order to measure the effect of light on plant growth, students should cultivate the same plant in the same substrate with the same amounts of water added and at the same temperature. They should also be made aware that they should be using the same instruments to conduct their measurements, i.e. thermometers.

At times, research projects like Plants in Space do not allow for simulation of variables such as gravity-free conditions. This might reduce the validity of the research. Students need to be prepared for a follow up question related to validity when presenting their work.

Students will often have to organise data in a table or graph. Students often need instructions on how to make different types of graphs. Common mistakes include the lack of legends and captions under figures, mislabelling the axes of a graph or missing units of measurements.

Keeping Track of Research Progress

As there should be multiple groups of students working on different research questions at the same time, teachers might feel disoriented and lacking control of what is going on. One way to keep track of progress is asking every team to present their experiments, data and conclusions in weekly plenary sessions. Working in small groups, invite students to share the information they have collected so far and to identify initial patterns. Also, making the most of blogs, students could post summaries of steps taken, experiments, data, setbacks, further questions and possible future actions. No matter what the inquiry path followed, it is important to bring the class together to "check in" and share and discuss the main ideas of the question at hand.

Overcoming an Urge to 'Spoon Feed' Your Students

One of the challenges for teachers is deciding if, how, when and where to intervene. Teachers should revisit concepts that students might have difficulty understanding and reflect on the ways students current understanding of the same concept differs now. They can also use guiding questions to help students draw up their plan of action, monitor their progress and consolidate their knowledge. A useful resource can be found here: http://www.edu.gov.on.ca/eng/literacynumeracy/inspire/research/CBS_InquiryBased.pdf

Compilation of References

(1989). How do i improve my practice? *Cambridge Journal of Education, 19*(1), 41–52.

American Association for the Advancement of Science. (1990). *Science for all Americans, Project 2061*. New York: Oxford University Press.

Anderson, L. W., Krathwohl, D. R., Airasian, P. W., Cruikshank, K. A., Mayer, R. E., Pintrich, P. R., ... Wittrock, M. C. (2001). *A Taxonomy for Learning, Teaching, and Assessing: A revision of Bloom's Taxonomy of Educational Objectives*. New York: Pearson, Allyn & Bacon.

Anderson, R. (2007). Inquiry as an organising theme for science curricula. In S. Abell & N. Lederman (Eds.), *Handbook of research on science education* (pp. 807–830). Mahwah, NJ: Lawrence Erlbaum Associates.

Anderson, R. D. (2002). Reforming science teaching: What research says about inquiry. *Journal of Science Teacher Education, 13*(1), 1–12. doi:10.1023/A:1015171124982

Atwater, M. M. (1994). Research on curricular diversity in the Classroom. In D. L. Gabel (Ed.), *Handbook of Research on Science Teaching and Learning*. New York: Macmillan.

Baird, J. R. (1990). Metacognition, purposeful inquiry and conceptual change. In E. Hegarty-Hazel (Ed.), *The Student Laboratory and the Science Curriculum*. London: Rutledge.

Ball, S. J. (2003). The teacher's soul and the terrors of performativity. *Journal of Education Policy, 18*(2), 215–228. doi:10.1080/0268093022000043065

Barron, B., & Darling-Hammond, L. (2010). Prospects and challenges for inquiry-based approaches to learning. *The nature of learning: Using research to inspire practice*, 199-225.

Barron, B., & Darling-Hammond, L. (2010). *Prospects and Challenges for inquiry based approaches to learning in The Nature of Learning - Using research to inspire practice*. Academic Press.

Bartholomew, H., Osborne, J., & Radcliffe, M. (2004). Teaching students "Ideas-about-science": Five dimensions of effective practice. *Science Education, 88*(5), 655–682. doi:10.1002ce.10136

Battikhi, A., Albashaireh, Z., & Nawash, O. (2016). Chain Reaction Develops Jordanian Student Inquiry Skills as Perceived by Participating. *International Journal of Education and Human Developments*, *2*(3), 21–29.

Baumert, J., Artelt, C., Klieme, E., Neubrand, M., Prenzel, M., Schiefeie, U., . . . Weiß, M. (Eds.). (2003). *PISA 2000 – Ein differenzierter Blick auf die Länder der Bundesrepublik Deutschland* [PISA 2000 – A critical look at the federal states in Germany]. Opladen, Germany: Leske + Budrich. Retrieved from https://link.springer.com/content/pdf/bfm%3A978-3-322-97590-4%2F1.pdf

Baumert, J., Bos, W., & Lehmann, R. (Eds.). (2000). TIMSS/III. Dritte Internationale Mathematik – und Naturwissenschaftsstudie. Mathematische und naturwissenschaftliche Bildung am Ende der Schullaufbahn, Band 1: Mathematische und naturwissenschaftliche Grundbildung am Ende der Pflichtschulzeit; Band 2: Mathematische und physikalische Kompetenzen am Ende der gymnasialen Oberstufe. Leske + Budrich.

Baumert, J., Lehmann, R., Lehrke, M., Schmitz, B., Clausen, M., Hosenfeld, I., . . . Neubrand, J. (1997). TIMSS – Mathematisch-naturwissenschaftlicher Unterricht im internationalen Vergleich. Deskriptive Befunde. Opladen, Germany: Leske + Budrich.

Bell, R.L. (2010). Teaching about scientific inquiry and the nature of science: Toward a more complete view of science. *The Journal of Mathematics and Science: Collaborative Explorations*, *13*, 5-25.

Bell, T., Urhahne, D., Schanze, S., & Ploetzner, R. (2010). Collaborative Inquiry Learning: Models, tools, and challenges. *International Journal of Science Education*, *32*(3), 349–377. doi:10.1080/09500690802582241

Berliner, D. (2011). Rational responses to high stakes testing: The case of curriculum narrowing and the harm that follows. *Cambridge Journal of Education*, *41*(3), 287–302. doi:10.1080/0305764X.2011.607151

Bevins, S. C., & Price, G. (2016). Reconceptualising inquiry in science education. *International Journal of Science Education*, *38*(1), 17–29. doi:10.1080/09500693.2015.1124300

Bianchini, J., & Colburn, A. (2000). Teaching the Nature of Science through Inquiry to Prospective Elementary Teachers: A Tale of Two Researchers. *Journal of Research in Science Teaching*, *37*(2), 177–209. doi:10.1002/(SICI)1098-2736(200002)37:2<177::AID-TEA6>3.0.CO;2-Y

Biesta, G., Pristley, M., & Robinson, S. (2015). The role of beliefs in teacher agency. *Teachers and Teaching*, *21*(6), 624–640. doi:10.1080/13540602.2015.1044325

Bissaker, K. (2014). Transforming STEM education in an innovative Australian school: The role of teachers' and academics' professional partnerships. *Theory into Practice*, *53*(1), 55–63. doi:10.1080/00405841.2014.862124

Blanchard, M. R., Southerland, S. A., & Granger, E. M. (2009). No silver bullet for inquiry: Making sense of teacher change following an inquiry-based research experience for teachers. *Science Education*, 321–360.

Blanchard, M. R., Southerland, S. A., & Granger, E. M. (2009). No silver bullet for inquiry: Making sense of teacher change following an inquiry-based research experience for teachers. *Science Education*, *93*(2), 322–360. doi:10.1002ce.20298

Blanchard, M. R., Southerland, S. A., Osborne, J. W., Sampson, V. D., Annetta, L. A., & Granger, E. M. (2010). Is inquiry possible in light of accountability?: A quantitative comparison of the relative effectiveness of guided inquiry and verification laboratory instruction. *Science Education*, *94*(4), 577–616. doi:10.1002ce.20390

Blatchford, P., Kutnick, P., Baines, E., & Galton, M. (2003). Toward a social pedagogy of classroom group work. *International Journal of Educational Research*, *39*(1), 153–172. doi:10.1016/S0883-0355(03)00078-8

Bloom, B. S. (Ed.). (1956). Taxonomy of Educational Objectives, Handbook I: The Cognitive Domain. New York: David McKay Co Inc.

Bolte, C., Hofstein, A., Holbrook, J., Mamlok-Naaman, R., Rannikmae, M., Rauch, F., & Streller, S. (2011). PROFILES: Professional Reflection-Oriented Focus on Inquiry-based Learning and Education through Science. *Proceedings of the Scientix Conference*.

Braun, V., & Clarke, V. (2013). *Successful Qualitative Research: A Practical Guide for Beginners*. Sage.

Brickman, P., Gormally, C., Armstrong, N., & Hallar, B. (2009). Effects of inquiry-based learning on students' science literacy skills and confidence. *International Journal for the Scholarship of Teaching and Learning*, *3*(2), 1–22.

Broggy, J., Boyle, R., & Forde, J. (2012). Current Views on the Practical Assessment of Science at Junior Cycle in Ireland. *Proceedings at the Science and Mathematics Education Conference*.

Brookfield, S. (1995). *Becoming a critically reflective teacher*. San Francisco: Jossey-Bass.

Bybee, R. (2000). Teaching science as inquiry. In J. Minstrel & E. H. Van Zee (Eds.), *Inquiring into Inquiry Learning and Teaching*. Washington, DC: AAAS.

Bybee, R. W. (2006a). *The BSCS 5E model Instructional Model Origins and Effectiveness*. Colorado Springs, CO: BSCS.

Campbell, T., Abd-Hamid, N., & Chapman, H. (2010). Development of instruments to assess teacher and student perceptions of inquiry experiences in science classrooms. *Journal of Science Teacher Education*, *21*(1), 13–30. doi:10.100710972-009-9151-x

Capps, K., & Crawford, B. A. (2012, August 26). Inquiry-based instruction and teaching about nature of science: Are they happening? *Journal of Science Teacher Education*.

Capps, D. K., Crawford, B. A., & Constas, M. A. (2012). A review of empirical literature on inquiry professional development: Alignment with best practices and a critique of the findings. *Journal of Science Teacher Education*, *23*(3), 291–318. doi:10.100710972-012-9275-2

Capps, D. K., Shemwell, J. T., & Young, A. M. (2016). Over reported and misunderstood? A study of teachers' reported enactment and knowledge of inquiry-based science teaching. *International Journal of Science Education*, *38*(6), 934–959. doi:10.1080/09500693.2016.1173261

Carleton, L. E., Fitch, J. C., & Krockover, G. H. (2008). An In-Service Teacher Education Program's Ef-fect on Teacher Efficacy and Attitudes. *The Educational Forum*, *72*(1), 46.

Centrum vedecko-technickych informacii Slovenskej republiky. (2012 – 2016). *Statistika prijimacieho konania na vysoke skoly SR*. Retrieved August 15, 2017 from http://www.cvtisr.sk/cvti-sr-vedecka-kniznica/informacie-o-skolstve/statistiky/statistika-prijimacieho-konania-na-vysoke-skoly-sr.html?page_id=9723

Cheevers, C., Eivers, E., & Shiel, G. (2006). *Implementing the revised Junior Certificate Science Syllabus: What teachers said*, Available online http://www.education.ie/en/Publications/Inspection-Reports-Publications/EvaluationReports-Guidelines/insp_jcssreview_pdf.pdf

Chinchaladze, I., & Burjanadze, T. (2016). *Experiments and Students Motivation*. Retrieved 8.05.17 from http://salis.iliauni.edu.ge/wp-content/uploads/2016/04/journal1.pdf

Cohen, J. (1996). Procedure and Substance in Deliberative Democracy. In S. Benhabib (Ed.), *Democracy and Difference: Contesting the Boundaries of the Political* (pp. 95–119). Princeton, NJ: Princeton University Press.

Colburn, A. (2000). An Inquiry Primer. *Science Scope*, *23*(6), 42–44.

Colburn, A. (2000b). An inquiry primer. *Science Scope*, *23*(6), 42–44.

Coolahan, J. (2002). *Teacher education and the teaching career in an era of lifelong learning*. Academic Press.

Cosgrove, J., Shiel, G., Oldham, E., & Sofroniou, N. (2004). A survey of mathematics teachers in Ireland. *Irish Journal of Education*, *35*, 20–44.

Crawford, B. A. (2000). Embracing the essence of inquiry: New roles for science teachers. *Journal of Research in Science Teaching*, *37*(9), 916–937. doi:10.1002/1098-2736(200011)37:9<916::AID-TEA4>3.0.CO;2-2

Crawford, B. A. (2007). Learning to teach science and inquiry in the rough and tumble of practice. *Journal of Research in Science Teaching*, *44*(4), 613–642. doi:10.1002/tea.20157

Crawford, B. A. (2007). Learning to Teach Science as Inquiry in the Rough and Tumble of Practice. *Journal of Research in Science Teaching*, *44*(4), 613–642.

Cullinane, A., & Liston, M. (2016). Review of the Leaving Certificate biology examination papers (1999–2008) using Bloom's taxonomy – an investigation of the cognitive demands of the examination. *Irish Educational Studies, 35*(3), 249–267. doi:10.1080/03323315.2016.1192480

das Graças, C., de Arruda Nascimento, M., & dos Reis, R.F. (2017). Teacher education: Perceptions of teachers who are entering the municipal education system in Rio de Janeiro. *Educação e Pesquisa, 43*(1), 49–64.

Day, C. (1987). Professional learning through collaborative in-service activity. In J. Smyth (Ed.), *Educating teachers: Changing the nature of pedagogical knowledge* (pp. 207–222). New York: Falmer Press.

Department for Children, Schools and Families. (2009). *National Curriculum*. QCDA.

Department for Education. (2006-2011). *Stem subject choice and careers: STEM Choices*. DfE.

Department for Education. (2011). *Maths and science education; the supply of high achievers at A level.* Research report DFE – RR079.

Department for Innovation, Universities and Skills. (2009). *Science so what? So everything.* Available online Retrieved on June 29th 2011 at: *English Secondary school students' perceptions of school science and science and engineering*. Author.

Department of Education & Science. (2003). *Junior Certificate Syllabus*. Dublin: Stationary Office.

Department of Education & Science. (2016). *School Self-Evaluation Guidelines*. Available online https://www.education.ie/en/Publications/Inspection-Reports-Publications/Evaluation-Reports-Guidelines/School-Self-Evaluation-Guidelines-2016-2020-Post-Primary.pdf

Department of Education and Skills. (2015). *A Framework for Junior Cycle*. STEM Education Policy Statement.

Dewey, J. (1991). Logic: The theory of inquiry. In J. A. Boydston (Ed.), *John Dewey: The Later Works, 1925–1953* (Vol. 12, pp. 1–5). Carbondale, IL: SIU Press. (Original work published 1938)

Dokme, I., & Aydinli, E. (2009). Turkish Primary School Students Performance on basic science process skills. *Procedia: Social and Behavioral Sciences, 1*(1), 544–548. doi:10.1016/j.sbspro.2009.01.098

Donnelly, D., McGarr, O., & O'Reilly, J. (2014). "Just be quiet and listen to exactly what he's saying": Conceptualizing power relations in inquiry-oriented classrooms. *International Journal of Science Education, 36*(12), 2029–2054. doi:10.1080/09500693.2014.889867

Dori, Y. D., & Herscovitz, O. (1999). Question-Posing capability as an alternative evaluation method: Analysis of an environmental case study. *Journal of Research in Science Teaching, 36*(4), 411–430. doi:10.1002/(SICI)1098-2736(199904)36:4<411::AID-TEA2>3.0.CO;2-E

Dori, Y. J., Sasson, I., Kaberman, Z., & Herscovitz, O. (2004). Integrating case-based computerized laboratories into high school chemistry. *Chemical Educator (Boise, Idaho)*, *9*, 4–8. Retrieved from http://chemeducator.org/bibs/0009001/910004yd.htm

Drechsel, B., & Prenzel, M. (2011). Internationale Vergleichsstudien [International comparative studies]. In E. Kiel & K. Zierer (Eds.), *Basiswissen Unterrichtsgestaltung – Unterrichtsgestaltung als Gegenstand der Wissenschaft* (Vol. 2, pp. 31–49). Hohengehren, Germany: Schneider Verlag.

Drew, V., Priestley, M., & Michael, M. K. (2016). Curriculum Development Through Critical Collaborative Professional Enquiry. *Journal of Professional Capital and Community*, *1*(1), 92–106. doi:10.1108/JPCC-09-2015-0006

Dumont, H., Istance, D., & Benavides, F. (2000). *Commission Working Document - Science, society and the citizen in Europe*. Retrieved from https://ec.europa.eu/research/swafs/pdf/pub_public_engagement/ss_en.pdf

Edelson, D. C. (2001). Learning-for-use: A framework for the design of technology-supported inquiry activities. *Journal of Research in Science Teaching*, *38*(3), 355–385. doi:10.1002/1098-2736(200103)38:3<355::AID-TEA1010>3.0.CO;2-M

Erduran, S., Simon, S., & Osborne, J. (2004). TAPping into argumentation: Developments in the application of Toulmin's argument pattern for studying science discourse. *Science Education*, *88*(6), 915–933. doi:10.1002ce.20012

Establish: European Science and Technology in Action: Building Links with Industry, Schools and Home. (2010). Retrieved from http://www.establish-fp7.eu

Europe 2020 - A European strategy for smart, sustainable and inclusive growth. (2010). Downloaded from: http://ec.europa.eu/eu2020/pdf/COMPLET%20EN%20BARROSO%20%20%20007%20-%20Europe%202020%20-%20EN%20version.pdf

European Commission Directorate-general for Enterprise. (2004). *Innovation Management and the Knowledge -Driven Economy*. Downloaded from https://publications.europa.eu/en/publication-detail/-/publication/dd46213f-89e1-4c20-ad21-f3adca0b0f7f/language-en/format-PDF/source-85259832

European Commission. (2000). *Commission Working Documents - Science, society and the citizen in Europe*. Downloaded from https://ec.europa.eu/research/swafs/pdf/pub_public_engagement/ss_en.pdf

European Commission. (2001). *Science and Society Action Plan*. Downloaded from: https://ec.europa.eu/research/swafs/pdf/pub_gender_equality/ss_ap_en.pdf

European Commission. (2004). *Europe needs more Scientists*. Brussels: European Commission.

European Commission. (2007). *Science Education now: A renewed Pedagogy for the future of Europe*. Brussels: Directorate General for Research, Science and Economy.

European Commission. (2010). *Communication from the Commission -Making a European Area of Lifelong Learning a Reality*. Downloaded from: http://aei.pitt.edu/42878/1/com2001_0678.pdf

European Commission. (2011). *Work Programme 2012- Capacities - Part 5 Science in Society*. Downloaded from http://ec.europa.eu/research/participants/data/ref/fp7/89382/s-wp-201201_en.pdf

European Commission. (2012). *Joint Report of the Council and the Commission on the implementation of the Strategic Framework for European cooperation in education and training (ET 2020) 'Education and Training in a smart, sustainable and inclusive Europe'.* Downloaded from http://eur-lex.europa.eu/legal-content/EN/TXT/PDF/?uri=CELEX:52012XG0308(01)&from=EN

Fibonacci: Disseminating Inquiry-based Science and Mathematics Education in Europe. (2010). Retrieved from http://www.fibonacci-project.eu

Flechsig, A., Knemeyer, J. P. & Marmé, N. (2017). *Schreibtraining - Wissenschaftliches Schreiben im naturwissenschaftlichen Unterricht* [A training on writing – scientific writing within science education contexts]. Doi:10.13140/RG.2.2.22283.69923

Fullan, M. (2007). *The New Meaning of Educational Change*. Routledge.

Gago, J. M., Ziman, J., Caro, P., Constantinou, C., Davies, G., Parchmannn, I., . . . Sjøberg, S. (2004). *Europe Needs More Scientists: Report by the High Level Group on Increasing Human Resources for Science and Technology*. Downloaded from: https://www.researchgate.net/publication/259705752_Europe_Needs_More_Scientists_Report_by_the_High_Level_Group_on_Increasing_Human_Resources_for_Science_and_Technology

Galano, S., Zappia, A., Smaldone, L., & Testa, I. (2015). Secondary students' views about scientific inquiry. *Il Nuovo Cimento*, *38*, C, 3, 94.

Geißler, K. A. (Ed.). (1985). Pädagogisch-psychologische Grundlagen für das Lernen in Gruppen: Lernen in Seminargruppen [Pedagogical-psychological basics for learning in groups: Learning in seminar groups]. Tübingen: DIFF.

Gess-Newsome, J., Southerland, S. A., Johnston, A., & Woodbury, S. (2003). Educational reform, personal practical theories, and dissatisfaction: The anatomy of change in college science teaching. *American Educational Research Journal*, *40*(3), 731–767. doi:10.3102/00028312040003731

Ginsburg-Block, M. D., Rohrbeck, C. A., & Fantuzzo, J. W. (2006). A meta-analytic review of social, self-concept, and behavioral outcomes of peer-assisted learning. *Journal of Educational Psychology*, *98*(4), 732–749. doi:10.1037/0022-0663.98.4.732

Gleeson, J. (2010). *Curriculum in Context: Partnership, Power and praxis in Ireland*. Bern: Peter Lang.

Gormally, C., Brickman, P., Hallar, B., & Armstrong, N. (2009). Effects of Inquiry-based Learning on Students' Science Literacy Skills and Confidence. *International Journal for the Scholarship of Teaching and Learning*, *3*(2), 16. doi:10.20429/ijsotl.2009.030216

Gräber, W. (1999). "Scientific Literacy" - Naturwissenschaftliche Bildung in der Diskussion ["Scientific literacy" – Science education in discussion]. In Qualitätsentwicklung im naturwissenschaftlichen Unterricht. Fachtagung am 15. Frankfurt am Main: GFPF.

Gunstone, R. F. (1991). Reconstructing theory from practical experience. In B. E. Woolnough (Ed.), *Practical science* (pp. 67–7). Milton Keynes, UK: Open University Press.

Guskey, T. R. (2002). Professional Development and Teacher Change. *Teachers and Teaching*, *8*(3/4), 2002.

Guskey, T. R. (2003). What makes professional development effective? *Phi Delta Kappan*, *84*(10), 748–750. doi:10.1177/003172170308401007

Hargreaves, A. (1989). *Curriculum and Assessment Reform*. Milton Keynes, UK: Open University Press.

Hargreaves, A., & Fullan, M. (2012). *Professional Capital: Transforming Teaching in Every School*. London: Routledge.

Hargreaves, A., & Fullan, M. (2012). *Professional capital: Transforming teaching in every school*. Teachers College Press.

Harker, R. (1984). On reproduction, habitus and education. *British Journal of Sociology of Education*, *5*(2), 117–127. doi:10.1080/0142569840050202

Harland, J., & Kinder, K. (1997). Teachers' continuing professional development: Framing a model of outcomes. *Journal of In-service Education*, *23*(1), 71–84. doi:10.1080/13674589700200005

Harwood, W. (2004). An activity model for scientific inquiry. *Science Teacher (Normal, Ill.)*, *71*(1), 44–46.

Hayes, S. (2011). *A Critical Examination and Evaluation of the Place of Science in the Irish Transition Year* (Unpublished thesis). University of Limerick.

Heinz, J., Enghag, M., Stuchlikova, I., Cakmakci, G., Peleg, R., & Baram-Tsabari, A. (2017). Impact of initiatives to implement science iquiry: A comparative study of the Turkish, Israeli, Swedish and Czech science education systems. *Cultural Studies of Science Education*, *12*(3), 677–708. doi:10.100711422-015-9704-6

Heise, M. (2009). *Informelles Lernen von Lehrkräften: ein Angebots-Nutzungs-Ansatz* [Informal learning of teachers: An offer-value-approach]. Münster: Waxmann.

Herron, M. D. (1971). The nature of scientific enquiry. *The School Review*, *79*(2), 171–212. doi:10.1086/442968

Hessisches Kultusministerium. (2017). Retrieved from https://la.hessen.de/irj/HKM_Internet?rid=HKM_15/HKM_Internet/nav/475/47570bc2-4148-5431-79cd-aae2389e4818%26_ic_uCon=f2640339-835f-5431-79cd-aae2389e4818.htm&uid=47570bc2-4148-5431-79cd-aae2389e4818

Hodson, D. (1993). Re-thinking old ways: Toward a more critical approach to practical work in school science. *Studies in Science Education*, *22*(1), 85–142. doi:10.1080/03057269308560022

Hofstein, A., & Walberg, H. J. (1995). Instructional strategies. In B. J. Fraser & H. J. Walberg (Eds.), Improving science education (pp. 1–20). Chicago: National Society for the Study of Education.

Hofstein, A., Eilks, I., & Bybee, R. (2011). Societal issues and their importance for contemporary science education—A pedagogical justification and the state-of-the-art in Israel, Germany, and the USA. *International Journal of Science and Mathematics Education*, *9*(6), 1459–1483. doi:10.100710763-010-9273-9

Hofstein, A., & Kind, P. M. (2012). Learning in and from science laboratories. In B. Fraser, K. Tobin, & C. McRobbie (Eds.), *Second international handbook of science education* (pp. 189–209). Dordrecht: Springer. doi:10.1007/978-1-4020-9041-7_15

Hofstein, A., & Lunetta, V. N. (1982). The role of the laboratory in science teaching: Neglected aspects of research. *Review of Educational Research*, *52*(2), 201–217. doi:10.3102/00346543052002201

Hofstein, A., & Lunetta, V. N. (2004). The laboratory in science education: Foundations for the twenty-first century. *Science Education*, *88*(1), 28–54. doi:10.1002ce.10106

Hofstein, A., Navon, O., Kipnis, M., & Mamlok-Naaman, R. (2005). Developing Students Ability to Ask More and Better Questions Resulting from Inquiry-Type Chemistry Laboratories. *Journal of Research in Science Teaching*, *42*(7), 791–806. doi:10.1002/tea.20072

Hofstein, A., Shore, R., & Kipnis, M. (2004). Providing high school chemistry students with opportunities to develop learning skills in an inquiry-type laboratory: A case study. *International Journal of Science Education*, *26*(1), 47–62. doi:10.1080/0950069032000070342

Hofstein, A., & Tuan, H. (2004). Inquiry in Science Education: International Perspectives. *Science Education*, *88*(3), 397–419. doi:10.1002ce.10118

Höttecke, D. (2010). Forschend-entdeckender Physikunterricht. Ein Überblick zu Hintergründen, Chancen und Umsetzungsmöglichkeiten entsprechender Unterrichtskonzeptionen [Inquiry based physics education. An overview on background, chances and possibilities for implementation of corresponding teaching concepts]. *Naturwissenschaften im Unterricht Physik*, *21*(119), 4–12.

Huberman, M. A. (1989). The professional life cycle of teachers. *Teachers College Record*, *91*(1), 31–57.

Huberman, M. A. (1993). *The Lives of Teachers*. New York: Teachers College Press.

IT Akademia – vzdelavanie pre 21. Storocie. (2017). Retrieved August 15, 2017, from http://itakademia.sk/sk/domov/

Janiuk, R. M., & Mazur, P. (2010). Basic Structure of the Educational System. *Teaching Chemistry Around the World*, 291.

Jannack, V., Seeberg, S., Knemeyer, J.-P., & Marmé, N. (2011). Windhösel Kraftwerke - ein Unterrichtsprojekt zum Thema Regenerative Energien [Windhösel powerplants – a science project for schools on sustainable energies]. *Naturwissenschaftliche Bildung als Beitrag zur Gestaltung partizipativer Demokratie. Jahrestagung der GDCP in Potsdam, 2010*, 312–314.

JeanPierre, B., Oberhauser, K., & Freeman, C. (2005). Characteristics of Professional Development That Effect Change in Secondary Science Teachers' Classroom Practices. *Journal of Research in Science Teaching*, (42), 668–690.

Jeskova, Z., Kires, M., McLoughlin, E., Finlayson, O., Ottander, Ch. & Ekborg, M. (2016). In-service and pre-service teacher education in IBSE: The ESTABLISH approach. *Il Nuovo Cimento C, 38*(3). doi:10.1393/ncc/i2015-15099-8

Jimenez-Aleixandre, M. P., & Erduran, S. (2008). Argumentation in Science Education: An Overview. In Argumentation in Science Education: Perspectives from Classroom-Based Research. Springer.

Johnson, W., Lustick, D., & Kim, M. (2011). Teacher Professional Learning as the Growth of Social Capital. Current Issues in Education, 14(3).

Johnson, D. W., & Johnson, R. T. (1989). *Cooperation and competition: Theory and research*. Edina, MN: Interaction Book Company.

Johnston, A. (2008). Demythologizing or dehumanizing? A response to Settlage and the ideals of open inquiry. *Journal of Science Teacher Education, 19*(1), 11–13. doi:10.100710972-007-9079-y

Johnstone, A. H., & Al-Shuaili, A. (2001). Learning in the laboratory; some thoughts from the literature. *University Chemistry Educ, 5*(2), 42–91.

Junior Cycle for Teachers. (2018). *JCT Development Strategy – Building on our Achievements.* Author.

Juntunen, M., & Aksela, M. (2013). Life-cycle analysis and inquiry based leaning in chemistry teaching. *Science Education International Journal., 24*(2), 150–166.

Kampanadze, M., & Eilks, I. (2014). Supporting reform in science education in central and Eastern Europe: Reflections and perspectives from the project TEMPUS-SALiS. *Eurasia Journal of Mathematics, Science & Technology Education, 10*(1), 47-58.

Kanter, D. E., & Konstantopoulos, S. (2010). *The impact of a project-based science curriculum on minority student achievement, attitudes, and careers: The effects of teacher content and pedagogical content knowledge and inquiry-based practices*. Retrieved on 24.04.2017 from http://onlinelibrary.wiley.com/doi/10.1002/sce.20391/full

Kapanadze, M., & Eilks, I. (2014). Supporting Reform in Science Education in Central and Eastern Europe - Reflections and Perspectives from the Project TEMPUS-SALiS. *Eurasia Journal of Mathematics, Science & Technology Education, 10*(1), 47-58.

Kapanadze, M., Janashia, S., & Eilks, I. (2010). From science education in the soviet time, via national reform initiatives, towards an international network to support inquiry-based science education – The case of Georgia and the project SALiS. In I. Eilks & B. Ralle (Eds.), Contemporary science education (pp. 237-242). Aachen: Shaker.

Katchevich, D., Hofstein, A., & Mamlok-Naaman, R. (2013). Argumentation in the chemistry laboratory: Inquiry and confirmatory experiments. *Research in Science Education*, *43*(1), 317–345. doi:10.100711165-011-9267-9

Katsampoxaki-Hodgetts, K., & Chaniotakis, N. (2015). IBSE Case Study: Student perceptions of science and inquiry after a two year IBSE project In a Greek secondary school. *Scientix National Conference, ScienceView Proceedings*.

Katsampoxaki-Hodgetts, K., Booth, J., & Chaniotakis, N. (2016). *Chain reaction outcomes on EU IBSE policy*. Retrieved from https://drive.google.com/file/d/0BwmqSk-8ziILdTA1TmJtRzlFU2M/view

Katsampoxaki-Hodgetts, K., Fouskaki, M., Siakavara, K., Moschochoritou, M., & Chaniotakis, N. (2015). Student and Teacher Perceptions of Inquiry Based Science Education in Secondary Education in Greece. *American Journal of Educational Research*, *3*(8), 968–976.

Kazempour, M., Amirshokoohi, A., & Harbour, W. (2012). Exploring Students' Perceptions of Science and Inquiry in a Reform-Based Undergraduate Biology Course. *Journal of College Science Teaching*, *42*(2), 38–42.

Kelly, G. J., Druker, S., & Chen, C. (1998). Students' reasoning about electricity: Combining performance assessment with argumentation analysis. *International Journal of Science Education*, *20*(7), 849–871. doi:10.1080/0950069980200707

Keys, C. W., & Bryan, L. A. (2001). Co-Constructing Inquiry-Based Science with Teachers: Essential Research for Lasting Reform. *Journal of Research in Science Teaching*, *38*(6), 631–645. doi:10.1002/tea.1023

Kind, P., Wilson, J., Hofstein, A., & Kind, V. (2011). Peer argumentation in school science laboratory – exploring effect of laboratory task feature. *International Journal of Science Education*, *33*(18), 2527–2558. doi:10.1080/09500693.2010.550952

King, D. (2017). *Junior Cycle Reform: Why Context Matters. A context-centric analysis of curriculum reform in lower Irish secondary education*. Doctoral Thesis.

King, D. (2017). *Junior Cycle Reform – Why Context Matters: A context-centric analysis of curriculum reform in lower Irish secondary education* (500th ed.). University of Sheffield. Available http://etheses.whiterose.ac.uk/18148/1/David%20King%20UoS%20Full%20Thesis%20FINAL%20Sept%2017%20Junior%20Cycle%20Reform%20Why%20Context%20Matters.pdf

Kires, M. (2016). *Badatelske aktivity v prirodovednom vzdelavani*. Bratislava: Statny pedagogicky ustav.

Kirk, D., & MacDonald, D. (2001). Teacher voice and ownership of curriculum change. *Journal of Curriculum Studies*, *33*(5), 551–567. doi:10.1080/00220270010016874

Kirschner, P. A., Sweller, J., & Clark, R. E. (2006). Why minimal guidance during instruction does not work: An analysis of the failure of constructivist, discovery, problem-based, experiential, and inquiry-based teaching. *Educational Psychologist, 41*(2), 75–86. doi:10.120715326985ep4102_1

Klahr, D., & Nigam, M. (2004). The equivalence of learning paths in early science instruction: Effects of direct instruction and discovery learning. *Psychological Science, 15*(10), 661–667. doi:10.1111/j.0956-7976.2004.00737.x PMID:15447636

Kleine, K., Brown, B., Harte, B., Hilson, A., Malone, D., & Moller, K. (2002). Examining inquiry. *Principal Leadership, 3*(3), 36–39.

Kleinpell, Ely, E. W., Williams, G., Liolios, A., Ward, N., & Tisherman, S. A. (2011). Web-based resources for critical care education. *Critical Care Medicine, 39*(3), 541–553. doi:10.1097/CCM.0b013e318206b5b5 PMID:21169819

Klusmann, U., Kunter, M., Trautwein, U., & Baumert, J. (2006). Lehrerbelastung und Unterrichtsqualität aus der Perspektive von Lehrenden und Lernenden [Teacher's burden and quality of teaching from a perspective of teachers and learners]. *Zeitschrift fur Padagogische Psychologie, 20*(3), 161–173. doi:10.1024/1010-0652.20.3.161

Krajcik, J., Mamlok, R., & Hug, B. (2001). Modern content and the enterprise of science: science education in the 20th century. In L. Corno (Ed.), Education Across A Century: The Centennial Volume (pp. 205-238). Chicago: National Society for the Study of Education (NSSE).

Kultusminister Konferenz. (2005a). Aktivitäten der Länder zur Weiterentwicklung des mathematisch-naturwissenschaftlichen Unterrichts [Activities of the German states concerning a further development of math and science education in schools]. *Beschluss der Kultusministerkonferenz vom 18.03.2005.* Retrieved from http://www.kmk.org/fileadmin/Dateien/pdf/PresseUndAktuelles/2000/nawi.pd

Kultusminister Konferenz. (2005b). *Beschlüsse der Kultusministerkonferenz. Bildungsstandards im Fach Physik für den Mittleren Schulabschluss (Jahrgangsstufe 10)* [Decisions of the Standing Conference of the Ministers of Education and Cultural Affairs of the Länder in the Federal Republic of Germany. Educational standards for physics education at the intermediate secondary level (grade 10)]. Retrieved from https://www.kmk.org/fileadmin/Dateien/veroeffentlichungen_beschluesse/2004/2004_12_16-Bildungsstandards-Physik-Mittleren-SA.pdf

Kultusminister Konferenz. (2017). *The Standing Conference of the Ministers of Education and Cultural Affairs of the Länder in the Federal Republic of Germany.* Retrieved from https://www.kmk.org/kmk/information-in-english.html

Lapitkova, V. (2008, February 5). Reforma vzdelavania – informacie, vychodiska a trendy pre fyzikalne vzdelavanie. Presentation given at the videoconference. *Reforma fyzikalneho vzdelavania.*

Lazarowitz, R., & Tamir, P. (1994). Research on using laboratory instruction in science. In D. L. Gabel (Ed.), Handbook of research on science teaching (pp. 94-127). New York: Macmillan Publishing Company.

Leat, D., Livingston, K., & Priestley, M. (2013). Curriculum deregulation in England and Scotland - Different directions of travel? In W. Kuiper & J. Berkvens (Eds.), *Balancing Curriculum Regulation and Freedom across Europe. CIDREE Yearbook, 2013* (pp. 229–248). Enschede, The Netherlands: SLO Netherlands Institute for Curriculum Development.

Lederman, N. G. (2006). Syntax Of Nature Of Science Within Inquiry And Science Instruction. In L.B. Flick & N.G. Lederman (Eds.), Scientific Inquiry and Nature of Science (pp. 301-317). Dordrecht: Springer.

Lederman, J. S., Lederman, N. G., Bartos, S. A., Bartels, S. L., Meyer, A. A., & Schwartz, R. S. (2014). Meaningful Assessment of Learners' Understandings About Scientific Inquiry—The Views About Scientific Inquiry (VASI) Questionnaire. *Journal of Research in Science Teaching*, *51*(1), 65–83. doi:10.1002/tea.21125

Leisen, J. (2008). Lesen ist schon schwer genug, dann auch noch Schreiben? [To read is difficult enough yet, then is writing to be added?]. *Naturwissenschaften im Unterricht. Physik*, *19*(104), 4–10.

Lemke, J. (1990). *Talking science: Language, learning, and values.* Norwood, NJ: Ablex.

Lewis, J., & Cowie, H. (1993). Cooperative group work: Promises and limitations. A study of teachers' values. *Education Section Review-British Psychological Society*, *17*, 77–77.

Linn, M. C., Bell, P., & Davis, E. A. (2004). *Internet environments for science education.* Lawrence Erlbaum Associates Publishers.

Lipowsky, F. (2009). Unterrichtsentwicklung durch Fort- und Weiterbildungsmaßnahmen für Lehrpersonen [Development of instruction through professional development measures for teachers]. *Beiträge zur Lehrerbildung,* (27), 346-360.

Lipowsky, F. (2013). Lehrerfortbildung neu und weiter denken. *AFB-Expertentagung zur Reform der Lehrerbildung in Hessen*. Retrieved from https://assets05.hessenspd.net/docs/doc_45511_2013731128.pdf

Lipowsky, F. (2010). Lernen im Beruf. Empirische Befunde zur Wirksamkeit von Lehrerfortbildung [In-service teacher trainings. Empirical results on the efficacy of teacher education programmes]. In F. H. Müller (Ed.), *Lehrerinnen und Lehrer lernen. Konzepte und Befunde zur Lehrerfortbildung* (pp. 51–70). Münster, Germany: Waxmann.

Lortie, D. C. (2002). *Schoolteacher: a sociological study*. Chicago: University of Chicago Press.

Luft, J. A., & Pizzini, E. L. (1998). The demonstration classroom in-service: Changes in the classroom. *Science Education*, *82*(2), 147–162. doi:10.1002/(SICI)1098-237X(199804)82:2<147::AID-SCE2>3.0.CO;2-E

Lunetta, V. N., Hofstein, A., & Clough, M. (2007). Learning and teaching in the school science laboratory: an analysis of research, theory, and practice. In N. Lederman & S. Abel (Eds.), Handbook of research on science education (pp. 393-441). Mahwah, NJ: Lawrence Erlbaum.

Lunetta, V. N. (1998). The school science laboratory: Historical perspectives and context for contemporary teaching. In B. Fraser & K. Tobin (Eds.), *International Handbook of Science Education* (pp. 349–264). Dordrecht: Kluwer Academic Publishers. doi:10.1007/978-94-011-4940-2_16

Lunetta, V. N., Hofstein, A., & Clough, M. P. (2007). Learning and teaching in the school science laboratory: An analysis of research, theory and practice. In S. K. Abell & N. G. Lederman (Eds.), *International handbook of science education* (pp. 393–441). Mahwah, NJ: Lawrence Erlbaum.

Mamlok-Naaman, R., Hofstein, A., & Penick, J. (2007). Involving Teachers in the STS Curricular Process: A Long-Term Intensive Support Framework for Science Teachers. *Journal of Science Teacher Education, 18*(4), 497–524. doi:10.100710972-007-9046-7

Mao, S., & Chang, C. (1998). Impacts of an Inquiry Teaching Method on Earth Science Students' Learning Outcomes and Attitudes at the Secondary School Level. *Proc Natl Sci Counc, 8*(3), 93–101.

Marmé, N., Glaser, J., Graf, S., & Welzel-Breuer, M. (2017). Schüler-Forschungsthemen im EU-Projekt Chain Reaction [The EUPRBs of the European project Chain Reaction]. In C. Maurer (Ed.), *Implementation fachdidaktischer Innovation im Spiegel von Forschung und Praxis. Gesellschaft für Didaktik der Chemie und Physik, Jahrestagung in Zürich 2016* (p. 540). Universität Regensburg.

Marmé, N., & Knemeyer, J. P. (2015). Kosmetikprojekt Marmétics verbindet außerschulisches Lernen mit dem regulären Schulunterricht [Marmétics combines informal learning and formal learning in the subject Sachunterricht]. *Naturwissenschaften im Unterricht. Chemie, 26*(147), 42–43.

Marshall, J., & Smart, J. (2013). Teachers' transformation to inquiry-based instructional practice. *Creative Education, 4*(2), 132–142. doi:10.4236/ce.2013.42019

McGuire, P. (2016, February 22). Where now for junior cycle reform? *The Irish Times*. Retrieved from https://www.irishtimes.com/news/education/where-now-for-junior-cycle-reform-1.2539986

McNeill, K. L., Marco-Bujosa, L. M., González-Howard, M., & Loper, S. (2018). Teachers' enactments of curriculum: Fidelity to Procedure versus Fidelity to Goal for scientific argumentation. *International Journal of Science Education, 40*(12), 1455–1475. doi:10.1080/09500693.2018.1482508

Méheut, M., & Psillos, D. (2004). Teaching-Learning sequences: Aims and tools for science educa-tion research. *International Journal of Science Education, 26*(5), 515–535. doi:10.1080/09500690310001614762

Menrath, M. (2014). *Schülerbefragung zu Inquiry Based Science Education im Rahmen eines EU-Projektes* [Student survey on inquiry based science education within the Chain Reaction project] (Unpublished thesis). Ruprecht-Karls-Universität Heidelberg, Germany.

Minner, D. D., Levy, A. J., & Century, J. (2010). Inquiry-based science instruction—What is it and does it matter? Results from a research synthesis years 1984 to 2002. *Journal of Research in Science Teaching, 47*(4), 474–496. doi:10.1002/tea.20347

Minstrell, J., & van Zee, E. H. (Eds.). (2000). *Inquiry in to Inquiry Learning and Teaching in Science*. Washington, DC: American Association for the Advancement of Science.

Moon, J. A. (2013). *Reflection in learning and professional development: Theory and practice*. Routledge. doi:10.4324/9780203822296

Moschochoritou, R., Siakavara, K., & Katsampoxaki-Hodgetts, K. (2014). *How sciences are taught in Greece*. University of Crete. Retrieved from http://www.chreact.gr

National Council for Curriculum and Assessment (2016). *Curriculum Specification for Junior Cycle Science*. Author.

National Council for Curriculum and Assessment. (2003). *Junior Certificate Science Syllabus*. Ordinary Level and Higher Level.

National Council for Curriculum and Assessment. (2006) *Junior Certificate Science, Draft Guidelines for Teachers*. Available online http://www.curriculumonline.ie/uploadedfiles/JC/JCScienceTGlin.pdf

National Council for Curriculum and Assessment. (2016). *Guidelines the Classroom Based Assessments and Assessment Task (Assessment Guidelines)* (1st ed.). Science.

National Research Council. (1996). *The National Science Education Standards*. Washington, DC: The National Academy Press.

National Research Council. (2012). *A Framework for K-12 Science Education: Practices, Crosscutting Concepts, and Core Ideas. Committee on a Conceptual Framework for New K-12 Science Education Standards. Board on Science Education, Division of Behavioural and Social Sciences and Education*. Washington, DC: The National Academies Press.

NCCA. (2012). *Towards a Framework for the Junior Cycle: Innovation and Identity*. Dublin: NCCA.

NCCA. (2013). *Background Paper and Brief for the Review of Junior Cycle Science*. Dublin: NCCA.

NCCA. (2016). *Junior Cycle Science Specification*. Available online https://www.curriculumonline.ie/getmedia/153bc83f-9848-49f0-ad87 0a0d6b9b596c/Specification-for-Jr-Cycle-Science-EV_20160126-(1).pdf

Nelson, T. H. (2001). *A Science teacher's wisdom of practice in teaching inquiry-based oceanography*. Paper presented at the annual meeting of the American Educational Research Association, Seattle, WA.

NRC, National Research Council. (1996). *National Science Education Standards*. Washington, DC: National Academy Press.

NRC, National Research Council. (2005). *National Science Education Standards*. Retrieved May 29, 2006, from: http://www.nap.edu/readingroom/books/nses/html/index.html

Nuthall, G. (2005). The cultural myths and realities of classroom teaching and learning: A personal journey. *Teachers College Record*, *107*(5), 895–934. doi:10.1111/j.1467-9620.2005.00498.x

OECD. (2014). *TALIS - The OECD Teaching and Learning International Survey*. Retrieved August 12, 2014, from http://www.oecd.org/edu/school/talis.htm

Omar, N. A. (2015). The Use Of Collaborative Learning to Improve Critical Thinking of General Certificate Of Education Advanced Level Psychology Students: An Experiment, Research. *Journal of Applied Sciences (Faisalabad)*, *10*(10), 594–597.

Organisation for Economic Co-operation and Development Global Science Forum. (2006). *Evolution of Student Interest in Science and Technology Studies Policy report*. Downloaded from: http://www.oecd.org/science/sci-tech/36645825.pdf

Organisation for Economic Co-operation and Development. (2006). *Evolution of Student Interest in Science and Technology Studies*. Retrieved August 15, 2017, from http://www.oecd.org/science/sci-tech/36645825.pdf

Osborne, J., & Dillon, J. (2008). *Science education in Europe: Critical reflections*. London: Nuffield Foundation.

Osborne, J., Erduran, S., & Simon, S. (2004). Enhancing the quality of argumentation in school science. *Journal of Research in Science Teaching*, *41*(10), 994–1020. doi:10.1002/tea.20035

Owen, R., Macnaghten, P., & Stilgoe, J. (2012). Responsible research and innovation: From science in society to science for society, with society. *Science & Public Policy*, *39*(6), 751–760. doi:10.1093cipolcs093

Palmer, D. H. (2009). Student interest generated during an inquiry skills lesson. *Journal of Research in Science Teaching*, *46*(2), 147–165. doi:10.1002/tea.20263

Penick, J. E., Crow, L. W., & Bonnsteter, R. J. (1996). Questions are the answers. *Science Teacher (Normal, Ill.)*, *63*, 26–29.

Perkins, R., Shiel, G., Merriman, B., Cosgrove, J., & Moran, G. (2012). *Learning for life: the achievement levels of 15 year olds in Ireland on mathematics, reading literacy and science in PISA 2012*. Education Research Centre.

Pintò, R., Ametller, J., Couso, D., Sassi, E., Monroy, G., Testa, I., & Lombardi, S. (2003). Some problems encountered in the introduction of innovations in secondary school science education and suggestions for overcoming them. *Mediterranean Journal of Educational Studies*, *8*(1), 113–134.

PISA. (2015). *Excellence and Equity in Education Volume 1*. Available from https://read.oecd-ilibrary.org/education/pisa-2015-results-volume-i_9789264266490-en#

Plummer, G., & Dudley, P. (1993). *Assessing Children Learning Collaboratively*. Chelmsford, UK: Essex Development Advisory Service.

Polakova, Z. (2016). *Badatelsky orientovane vyucovanie prirodovednych predmetov v podmienkach Gymnazia Andreja Sladkovica*. A talk given at the seminar "Inquiry-based education in science subjects". Banská Bystrica.

Porter, A. C., Smithson, J., Blank, R., & Zeider, T. (2007). Alignment as a teacher variable. *Applied Measurement in Education*, *20*(1), 27–51. doi:10.1080/08957340709336729

Priestley, M. (2017). *Approaches to specifying curriculum areas of learning*. Stirling: The Stirling Network for Curriculum Studies.

Priestley, M., Biesta, G., & Robinson, S. (2015). *Teacher Agency: An Ecological Approach*. London: Bloomsbury. doi:10.4324/9781315678573-15

Priestley, M., & Humes, W. (2010). The development of Scotland's Curriculum for Excellence: Amnesia and déjà vu. *Oxford Review of Education*, *36*(3), 345–361. doi:10.1080/03054980903518951

Primas: Promoting Inquiry in Mathematics and Science Education Across Europe. (2010). Retrieved from http://www.primas-project.eu

Qablan, A. M and DeBaz, T. (2014). Facilitating elementary science teachers' implementation of inquiry-based science teaching. *Teacher Development: An International Journal of Teachers' Professional Development*, 3-21. . doi:10.1080/13664530.2014.959552

Reaction, C. (2013). Retrieved 24/04/2017 from http://www.chreact.eu

Rees, C., Pardo, R., & Parker, J. (2013). Steps to opening scientific inquiry: Pre-service teachers' practicum experiences with a new support framework. *Journal of Science Teacher Education*, *24*(3), 475–496. doi:10.100710972-012-9315-y

Rickey, D., & Stacy, A. M. (2000). The role of Metacognition in learning chemistry. *Journal of Research in Science Teaching*, *77*, 915–920.

Rieser, S., Stahns, R., Walzebug, A., & Wendt, H. (2016). Einblicke in die Gestaltung des Mathematik- und Sachunterrichts. In TIMSS 2015. Mathematische und naturwissenschaftliche Kompetenzen von Grundschulkindern in Deutschland im internationalen Vergleich. Münster: Waxmann.

Rietschel, E. T., Arnold, E., Čenys, A., Dearing, A., Feller, I., Joussaume, S., . . . Stame, N. (2009). *Evaluation of the Sixth Framework Programmes for Research and Technological Development 2002-2006*. Downloaded from: https://ec.europa.eu/research/evaluations/pdf/archive/other_reports_studies_and_documents/fp6_ex-post_evaluation_expert_group_report.pdf#view=fit&pagemode=none

Roberts, G. G. (2002). *SET for success: the supply of people with science, technology, engineering and mathematics skills: The report of Sir Gareth Roberts' review*. Academic Press.

Rocard, M. (2007). *Science Education NOW: A renewed Pedagogy for the Future of Europe*. Brussels: European Commission. Retrieved from https://ec.europa.eu/research/science-society/document_library/pdf_06/report-rocard-on-science-education_en.pdf

Rocard, M., Csermely, P., Jorde, D., Lenzen, D., Walberg-Henriksson, H., & Hemmo, C. (2007). *Science Education Now: A Renewed Pedagogy for the Future of Europe*. Available: http://ec.europa.eu/research/science-society/document_library/pdf_06/report-rocard-on-science-education_en.pdf

Rocard, M., Csermely, P., Jorde, D., Lenzen, D., Walberg-Henriksson, H., & Hemmo, V. (2007). *Science education NOW: a renewed pedagogy for the future of Europe*. Downloaded from: http://ec.europa.eu/research/science-society/document_library/pdf_06/report-rocard-on-science-education_en.pdf

Rocard, M., Cesrmley, P., Jorde, D., Lenzen, D., Walberg-Herniksson, H., & Hemmo, V. (2007). *Science education NOW: A Renewed Pedagogy for the Future of Europe*. Brussels, Belgium: Office for Official Publications of the European Communities. Retrieved from http://ec.europa.eu/research/science-society/document_library/pdf_06/report-rocard-onscience-education_en.pdf

Ryan, R. M., & Deci, E. L. (2017). *Self-determination theory: basic psychological needs in motivation, development, and wellness*. New York: Guilford Press.

SAILS – Strategies for Assessment of Inquiry Learning in Science. (2015). Retrieved from http://www.sails-project.eu/

SALiS. (2013). *Science Education Research centre*. Retrieved from https://salis.iliauni.edu.ge/

Saltiel, E. (2005). *Inquiry-Based Science Education: Applying it in the Classroom*. Methodological Guide.

Sassi, E., Monroy, G., & Testa, I. (2005). Teacher Training about Real-Time Approaches: Research-based guidelines and materials. *Science Education*, *89*(1), 28–37. doi:10.1002ce.20041

Schwartz, R. S., & Crawford, B. A. (2006). Authentic Scientific Inquiry As Context For Teaching Nature Of Science: Identifying Critical Element. In L.B. Flick & N.G. Lederman (Eds.), Scientific Inquiry and Nature of Science (pp. 331-355). Dordrecht: Springer.

Seeberg, S., Jannack, V., Knemeyer, J. P., & Marmé, N. (2011). Marmétics - ein interdisziplinäres Kosmetikprojekt in der virtuellen Lernstadt Lucycity [Marmétics – an interdisciplinary project oft he virtual learningspace Lucycity]. *Naturwissenschaftliche Bildung als Beitrag zur Gestaltung partizipativer Demokratie. Jahrestagung der GDCP in Potsdam, 2010*, 306–308.

Sere, G. M. (2002). Towards renewed research questions from outcomes of the European lab-work in science education. *Science Education*, *86*(5), 624–644. doi:10.1002ce.10040

Shields, C. (2014). Aristotle. *The Stanford Encyclopedia of Philosophy*. Available from http://plato.stanford.edu/cgi-bin/encyclopedia/archinfo.cgi?entry=aristotle

Shiels, G., Perkins, R., & Gilleece, L. (2009). *OECD Teaching and Learning International Study (TALIS): Summary Report for Ireland*. Education Research Centre.

Shope, F., & McComas, W. (2015). Modeling scientific inquiry to guide students in the practices of science: The ED3U teaching model of conceptual change in action. In Inquiry-Based Learning for Science, Technology, Engineering, and Math (STEM) Programs: A Conceptual and Practical Resource for Educators. Academic Press. doi:10.1108/S2055-364120150000004013

Singer, J., Marx, R. W., Krajcik, J. S., & Chambers, J. C. (2000). Constructing extended inquiry projects: Curriculum materials for science education reform. *Educational Psychologist*, *35*(3), 165–178. doi:10.1207/S15326985EP3503_3

Slovinsky, E. (2012). SALiS and educational policy in Georgia. *Proceedings of the Student Active learning in Science Final Conference*.

Smith, E., & White, P. (2018). Where Do All the STEM Graduates Go? Higher Education, the Labour Market and Career Trajectories in the UK. *Journal of Science Education and Technology*, 1–15.

Smith, W. S., & Erb, T. O. (1986). Effect of women science career role models on early adolescents' attitudes toward scientists and women in science. *Journal of Research in Science Teaching*, *23*(8), 667–676. doi:10.1002/tea.3660230802

Smolleck, L. A., & Yoder, E. P. (2008). Further development and validation of the teaching science as inquiry (TSI) instrument. *School Science and Mathematics*, *108*(7), 291–297. doi:10.1111/j.1949-8594.2008.tb17842.x

State Exam Commission. (2010). *Junior Certificate Examination Science, Ordinary Level Chief Examiner's Report. Higher Level Chief Examiner's Report*. Athlone: SEC.

Statny pedagogicky ustav. (2015a). *Fyzika – nizsie stredne vzdelavanie*. Retrieved August 15, 2017, from http://www.statpedu.sk/files/articles/dokumenty/inovovany-statny-vzdelavaci-program/fyzika_nsv_2014-12-03.pdf

Statny pedagogicky ustav. (2015b). *Statny vzdelavaci program pre nizsie stredne vzdelavanie – 2. stupen ZS*. Retrieved August 15, 2017, from http://www.statpedu.sk/files/articles/dokumenty/inovovany-statny-vzdelavaci-program/svp_nsv_6_2_2015.pdf

Sutaz Misia Mars pozna svojich vitazov. (2016). Retrieved August 15, 2017 from https://www.seas.sk/clanok/sutaz-misia-mars-pozna-svojich-vitazov/273

Taitelbaum, D., Mamlok-Naaman, R., Carmeli, M., & Hofstein, A. (2008). Evidence-based Continuous Professional Development (CPD) in the Inquiry Chemistry Laboratory (ICL). *International Journal of Science Education*, *30*(5), 593–617. doi:10.1080/09500690701854840

Timperley, H., Wilson, A., Barrar, H., & Fung, I. (2008). Teacher professional learning and development. Academic Press.

Tint, S., & Nyunt, E. (2015). Collaborative Learning With Think -Pair - Share Technique. *Computer Applications: An International Journal*, *2*(1), 4.

Tobin, K. G. (1990). Research on science laboratory activities: In pursuit of better questions and answers to improve learning. *School Science and Mathematics*, *90*(5), 403–418. doi:10.1111/j.1949-8594.1990.tb17229.x

TPDC. (2012). Retrieved 24/04/2017 http://www.tpdc.ge/uploads/pdf_documents/sabunebismetyvelo.pdf (in Georgian)

Trna, J., & Trnova, E. (2016). Inquiry-based Science Education Modules and Their Effects on Teacher Education. In Education Applications & Developments II (31-46). Lisboa, Portugal: Science Press.

Trna, J., Trnova, E., & Sibor, J. (2012). Implementation of inquiry-based science education in science teacher training. *Journal of Instructional and Educational Studies in the World, 2*(4).

Trna, J. (2012). Implentation of inquiry based science education in science teaching training. *Journal of Educational and Instructional Studies in the World* , *2* (4), 2146–7463.

van den Akker, J. (1998). The science curriculum: Between ideals and outcomes. In *International handbook of science education* (pp. 421–447). Dordrecht: Kluwer. doi:10.1007/978-94-011-4940-2_25

Van Driel, J. H., & Berry, A. (2012). Teacher professional development focusing on pedagogical content knowledge. *Educational Researcher*, *41*(1), 26–28. doi:10.3102/0013189X11431010

Van Kampen, P. (2015). *Summary of the Junior Cycle Science Specification Process.* Available online http://www.juniorcycle.ie/Curriculum/Subjects/Science/Science

Varley, J. P., Murphy, C., & Veale, O. (2013). At the crossroads: The impact of new Irish science curricula on first year post-primary students. *Research in Science Education*, *43*(1), 275–298. doi:10.100711165-011-9252-3

Vygotsky, L. S. (1978). *Mind in society.* Cambridge, MA: Harvard University Press.

Wahl, D. (2013). *Lernumgebungen erfolgreich gestalten: vom trägen Wissen zum kompetenten Handeln* [Successfol realization of learning contexts – from inert knowledge to competent action]. Bad Heilbrunn: Verlag Julius Klinkhardt.

Wallace, C., & Priestley, M. (2016). Secondary Science Teachers as Curriculum Makers. *Mapping and Designing Scotland*, 1.

Walshe, A. (2015). *From Curriculum to Classroom in Upper Second Level Science.* Doctoral Thesis.

Webb, N. (1999). *Research monograph No. 18. Alignment of science and mathematics standards and assessments in four states*. Washington, DC: Council of Chief State School.

Welzel-Breuer, M., Marmé, N., Graf, S., Glaser, J., & Krebs, A.-K. (2017). Chain Reaction: ein FP 7-Projekt und seine Ergebnisse - Die Herausforderung: forschendes Lernen in den Unterricht integrieren [Chain Reaction: an FP7 project and its results – the challenge of integrating inquiry-based learning in science lessons]. In C. Maurer (Ed.), *Implementation fachdidaktischer Innovation im Spiegel von Forschung und Praxis. Gesellschaft für Didaktik der Chemie und Physik, Jahrestagung in Zürich 2016* (p. 536). Universität Regensburg.

Wenger, E. (1998). *Communities of practice: Learning, meaning and identity*. Cambridge, UK: Cambridge University Press. doi:10.1017/CBO9780511803932

Whitehead, J. (1989). Curriculum for Excellence. *Journal of Research in Science Teaching*.

White, R., & Gunstone, R. (1992). *Probing Understanding*. London: The Falmer Press.

Wilkinson, C., Franke, J., & Stroyan, J. (2016). *Ex-post Evaluation of Science in Society in FP7*. Retrieved from http://bookshop.europa.eu/en/ex-post-evaluation-of-science-in-society-in-fp7-pbKI0216492/

Wilson, C. D., Taylor, J. A., Kowalski, S. M., & Carlson, J. (2010). The relative effects and equity of inquiry-based and commonplace science teaching on students' knowledge, reasoning, and argumentation. *Journal of Research in Science Education*, *47*(3), 26–30.1.

Windschitl, M. (2003). Inquiry projects in science teacher education: What can Investigative experiences reveal about teacher thinking and eventual classroom practice? *Science Education*, *87*(1), 112–143. doi:10.1002ce.10044

Wong, J. L. N. (2010). What makes a professional learning community possible? *Asia Pacific Education Review*, *11*(2), 131–139. doi:10.100712564-010-9080-6

Zappia, A. (2018). *How secondary school science teachers implement scientific inquiry in classroom practice* (PhD thesis). Camerino, Italy: Università degli Studi di Camerino, School of Advanced Studies.

Zappia, A., Galano, S., Smaldone, L., & Testa, I. (2015). Science Teachers' Transformations When Implementing Inquiry-Based Teaching-Learning Sequences. In C. Fazio & R. M. Sperandeo (Eds.), *E-Book Proceedings of the GIREP-MPTL 2014 International Conference: Teaching/Learning Physics: Integrating Research into Practice* (pp. 897 - 906). Palermo, Italy: Università degli Studi di Palermo.

Zelina, M. (2007, December 5). A talk at the meeting of the Curricular Council, Casta, Pila.

Ziebell, N., Ong, A., & Clarke, D. (2017). Aligning curriculum, instruction and assessment. In T. Bentley & G. Savage (Eds.), *Educating Australia: Challenges for the decade ahead* (pp. 255–274). Melbourne: Melbourne University Publishing.

Related References

To continue our tradition of advancing academic research, we have compiled a list of recommended IGI Global readings. These references will provide additional information and guidance to further enrich your knowledge and assist you with your own research and future publications.

Aburezeq, I. M., & Dweikat, F. F. (2017). Cloud Applications in Language Teaching: Examining Pre-Service Teachers' Expertise, Perceptions and Integration. *International Journal of Distance Education Technologies, 15*(4), 39–60. doi:10.4018/IJDET.2017100103

Adera, B. (2017). Supporting Language and Literacy Development for English Language Learners. In J. Keengwe (Ed.), *Handbook of Research on Promoting Cross-Cultural Competence and Social Justice in Teacher Education* (pp. 339–354). Hershey, PA: IGI Global. doi:10.4018/978-1-5225-0897-7.ch018

Ahamer, G. (2011). How Technologies Can Localize Learners in Multicultural Space: A Newly Developed "Global Studies" Curriculum. *International Journal of Technology and Educational Marketing, 1*(2), 1–24. doi:10.4018/ijtem.2011070101

Ahamer, G. (2015). Conclusions from Social Dynamics in Collaborative Environmental Didactics. *International Journal of Technology and Educational Marketing, 5*(2), 68–92. doi:10.4018/IJTEM.2015070105

Ahamer, G. (2015). Designing and Analyzing Social Dynamics for Collaborative: Environmental Didactics. *International Journal of Technology and Educational Marketing, 5*(2), 46–67. doi:10.4018/IJTEM.2015070104

Ahamer, G. (2017). Quality Assurance for a Developmental "Global Studies" (GS) Curriculum. In I. Management Association (Ed.), Educational Leadership and Administration: Concepts, Methodologies, Tools, and Applications (pp. 438-477). Hershey, PA: IGI Global. doi:10.4018/978-1-5225-1624-8.ch023

Ahamer, G. (2017). Quality Assurance for a Developmental "Global Studies" (GS) Curriculum. In I. Management Association (Ed.), Educational Leadership and Administration: Concepts, Methodologies, Tools, and Applications (pp. 438-477). Hershey, PA: IGI Global. doi:10.4018/978-1-5225-1624-8.ch023

Alegre de la Rosa, O. M., & Angulo, L. M. (2017). Social Inclusion and Intercultural Values in a School of Education. In S. Mukerji & P. Tripathi (Eds.), *Handbook of Research on Administration, Policy, and Leadership in Higher Education* (pp. 518–531). Hershey, PA: IGI Global. doi:10.4018/978-1-5225-0672-0.ch020

Ambikairajah, E., Sethu, V., Eaton, R., & Sheng, M. (2014). Evolving Use of Educational Technologies: Enhancing Lectures. In F. Alam (Ed.), *Using Technology Tools to Innovate Assessment, Reporting, and Teaching Practices in Engineering Education* (pp. 241–258). Hershey, PA: IGI Global. doi:10.4018/978-1-4666-5011-4.ch018

Anderson, K. M. (2017). Preparing Teachers in the Age of Equity and Inclusion. In I. Management Association (Ed.), Medical Education and Ethics: Concepts, Methodologies, Tools, and Applications (pp. 1532-1554). Hershey, PA: IGI Global. doi:10.4018/978-1-5225-0978-3.ch069

Awdziej, M. (2017). Case Study as a Teaching Method in Marketing. In D. Latusek (Ed.), *Case Studies as a Teaching Tool in Management Education* (pp. 244–263). Hershey, PA: IGI Global. doi:10.4018/978-1-5225-0770-3.ch013

Bain, B. (2014). Exploring Assessment of Critical Thinking Learning Outcomes in Online Higher Education. In V. Wang (Ed.), *Handbook of Research on Education and Technology in a Changing Society* (pp. 1191–1202). Hershey, PA: IGI Global. doi:10.4018/978-1-4666-6046-5.ch089

Banas, J. R., & York, C. S. (2017). Pre-Service Teachers' Motivation to Use Technology and the Impact of Authentic Learning Exercises. In L. Tomei (Ed.), *Exploring the New Era of Technology-Infused Education* (pp. 121–140). Hershey, PA: IGI Global. doi:10.4018/978-1-5225-1709-2.ch008

Bariso, E. U. (2015). Educational Policy Analysis Debates and New Learning Technologies in England. In M. Khosrow-Pour (Ed.), *Encyclopedia of Information Science and Technology* (3rd ed.; pp. 2371–2378). Hershey, PA: IGI Global. doi:10.4018/978-1-4666-5888-2.ch230

Beycioglu, K., & Wildy, H. (2015). Principal Preparation: The Case of Novice Principals in Turkey. In K. Beycioglu & P. Pashiardis (Eds.), *Multidimensional Perspectives on Principal Leadership Effectiveness* (pp. 1–17). Hershey, PA: IGI Global. doi:10.4018/978-1-4666-6591-0.ch001

Beycioglu, K., & Wildy, H. (2017). Principal Preparation: The Case of Novice Principals in Turkey. In I. Management Association (Ed.), Educational Leadership and Administration: Concepts, Methodologies, Tools, and Applications (pp. 1152-1169). Hershey, PA: IGI Global. doi:10.4018/978-1-5225-1624-8.ch054

Bharwani, S., & Musunuri, D. (2018). Reflection as a Process From Theory to Practice. In M. Khosrow-Pour, D.B.A. (Ed.), Encyclopedia of Information Science and Technology, Fourth Edition (pp. 1529-1539). Hershey, PA: IGI Global. doi:10.4018/978-1-5225-2255-3.ch132

Bisschoff, T., & Rhodes, C. (2011). Transformation through Marketing: A Case of a Secondary School in South Africa. In P. Tripathi & S. Mukerji (Eds.), *Cases on Innovations in Educational Marketing: Transnational and Technological Strategies* (pp. 263–272). Hershey, PA: IGI Global. doi:10.4018/978-1-60960-599-5.ch016

Bodomo, A. B. (2010). Educational Technologies (WebCT): Creating Constructivist and Interactive Learning Communities. In A. Bodomo (Ed.), *Computer-Mediated Communication for Linguistics and Literacy: Technology and Natural Language Education* (pp. 252–290). Hershey, PA: IGI Global. doi:10.4018/978-1-60566-868-0.ch010

Bohjanen, S. L., Cameron-Standerford, A., & Meidl, T. D. (2018). Capacity Building Pedagogy for Diverse Learners. In J. Keengwe (Ed.), *Handbook of Research on Pedagogical Models for Next-Generation Teaching and Learning* (pp. 195–212). Hershey, PA: IGI Global. doi:10.4018/978-1-5225-3873-8.ch011

Brewer, J. C. (2018). Measuring Text Readability Using Reading Level. In M. Khosrow-Pour, D.B.A. (Ed.), Encyclopedia of Information Science and Technology, Fourth Edition (pp. 1499-1507). Hershey, PA: IGI Global. doi:10.4018/978-1-5225-2255-3.ch129

Brown, S. L. (2017). A Case Study of Strategic Leadership and Research in Practice: Principal Preparation Programs that Work – An Educational Administration Perspective of Best Practices for Master's Degree Programs for Principal Preparation. In V. Wang (Ed.), *Encyclopedia of Strategic Leadership and Management* (pp. 1226–1244). Hershey, PA: IGI Global. doi:10.4018/978-1-5225-1049-9.ch086

Brzozowski, M., & Ferster, I. (2017). Educational Management Leadership: High School Principal's Management Style and Parental Involvement in School Management in Israel. In V. Potocan, M. Üngan, & Z. Nedelko (Eds.), *Handbook of Research on Managerial Solutions in Non-Profit Organizations* (pp. 55–74). Hershey, PA: IGI Global. doi:10.4018/978-1-5225-0731-4.ch003

Cannaday, J. (2017). The Masking Effect: Hidden Gifts and Disabilities of 2e Students. In P. Dickenson, P. Keough, & J. Courduff (Eds.), *Preparing Pre-Service Teachers for the Inclusive Classroom* (pp. 220–231). Hershey, PA: IGI Global. doi:10.4018/978-1-5225-1753-5.ch011

Capobianco, B. M., & Lehman, J. D. (2010). Fostering Educational Technology Integration in Science Teacher Education: Issues of Teacher Identity Development. In J. Yamamoto, J. Kush, R. Lombard, & C. Hertzog (Eds.), *Technology Implementation and Teacher Education: Reflective Models* (pp. 245–257). Hershey, PA: IGI Global. doi:10.4018/978-1-61520-897-5.ch014

Chao, G. H., Hsu, M. K., & Scovotti, C. (2013). Predicting Donations from a Cohort Group of Donors to Charities: A Direct Marketing Case Study. In J. Wang (Ed.), *Optimizing, Innovating, and Capitalizing on Information Systems for Operations* (pp. 196–214). Hershey, PA: IGI Global. doi:10.4018/978-1-4666-2925-7.ch010

Chauhan, A. (2015). Beyond the Phenomenon: Assessment in Massive Open Online Courses (MOOCs). In E. McKay & J. Lenarcic (Eds.), *Macro-Level Learning through Massive Open Online Courses (MOOCs): Strategies and Predictions for the Future* (pp. 119–140). Hershey, PA: IGI Global. doi:10.4018/978-1-4666-8324-2.ch007

Coffman, T. L., & Klinger, M. B. (2013). Managing Quality in Online Education. In G. Kurubacak & T. Yuzer (Eds.), *Project Management Approaches for Online Learning Design* (pp. 220–233). Hershey, PA: IGI Global. doi:10.4018/978-1-4666-2830-4.ch011

Contreras, E. C., & Contreras, I. I. (2018). Development of Communication Skills through Auditory Training Software in Special Education. In M. Khosrow-Pour, D.B.A. (Ed.), Encyclopedia of Information Science and Technology, Fourth Edition (pp. 2431-2441). Hershey, PA: IGI Global. doi:10.4018/978-1-5225-2255-3.ch212

Cook, R. G. (2011). Educational Marketing: Coming Down from the Cloud Using Landing Gear. In U. Demiray & S. Sever (Eds.), *Marketing Online Education Programs: Frameworks for Promotion and Communication* (pp. 26–31). Hershey, PA: IGI Global. doi:10.4018/978-1-60960-074-7.ch003

Cook, R. G., & Ley, K. (2015). Past, Future and Presents: Meeting New Online Challenges with Primal Marketing Solutions. *International Journal of Technology and Educational Marketing*, *5*(2), 19–33. doi:10.4018/IJTEM.2015070102

Cooley, D., & Whitten, E. (2017). Special Education Leadership and the Implementation of Response to Intervention. In F. Topor (Ed.), *Handbook of Research on Individualism and Identity in the Globalized Digital Age* (pp. 265–286). Hershey, PA: IGI Global. doi:10.4018/978-1-5225-0522-8.ch012

Cosner, S., Tozer, S., & Zavitkovsky, P. (2017). Enacting a Cycle of Inquiry Capstone Research Project in Doctoral-Level Leadership Preparation. In I. Management Association (Ed.), Educational Leadership and Administration: Concepts, Methodologies, Tools, and Applications (pp. 1460-1481). Hershey, PA: IGI Global. doi:10.4018/978-1-5225-1624-8.ch067

Crawford, C. M. (2018). Instructional Real World Community Engagement. In M. Khosrow-Pour, D.B.A. (Ed.), Encyclopedia of Information Science and Technology, Fourth Edition (pp. 1474-1486). Hershey, PA: IGI Global. doi:10.4018/978-1-5225-2255-3.ch127

Crosby-Cooper, T., & Pacis, D. (2017). Implementing Effective Student Support Teams. In P. Dickenson, P. Keough, & J. Courduff (Eds.), *Preparing Pre-Service Teachers for the Inclusive Classroom* (pp. 248–262). Hershey, PA: IGI Global. doi:10.4018/978-1-5225-1753-5.ch013

Curran, C. M., & Hawbaker, B. W. (2017). Cultivating Communities of Inclusive Practice: Professional Development for Educators – Research and Practice. In C. Curran & A. Petersen (Eds.), *Handbook of Research on Classroom Diversity and Inclusive Education Practice* (pp. 120–153). Hershey, PA: IGI Global. doi:10.4018/978-1-5225-2520-2.ch006

Dass, S., & Dabbagh, N. (2018). Faculty Adoption of 3D Avatar-Based Virtual World Learning Environments: An Exploratory Case Study. In I. Management Association (Ed.), Technology Adoption and Social Issues: Concepts, Methodologies, Tools, and Applications (pp. 1000-1033). Hershey, PA: IGI Global. doi:10.4018/978-1-5225-5201-7.ch045

Davison, A. M., & Scholl, K. G. (2017). Inclusive Recreation as Part of the IEP Process. In C. Curran & A. Petersen (Eds.), *Handbook of Research on Classroom Diversity and Inclusive Education Practice* (pp. 311–330). Hershey, PA: IGI Global. doi:10.4018/978-1-5225-2520-2.ch013

DeCoito, I. (2018). Addressing Digital Competencies, Curriculum Development, and Instructional Design in Science Teacher Education. In M. Khosrow-Pour, D.B.A. (Ed.), Encyclopedia of Information Science and Technology, Fourth Edition (pp. 1420-1431). Hershey, PA: IGI Global. doi:10.4018/978-1-5225-2255-3.ch122

DeCoito, I., & Richardson, T. (2017). Beyond Angry Birds™: Using Web-Based Tools to Engage Learners and Promote Inquiry in STEM Learning. In I. Levin & D. Tsybulsky (Eds.), *Digital Tools and Solutions for Inquiry-Based STEM Learning* (pp. 166–196). Hershey, PA: IGI Global. doi:10.4018/978-1-5225-2525-7.ch007

Delmas, P. M. (2017). Research-Based Leadership for Next-Generation Leaders. In R. Styron Jr & J. Styron (Eds.), *Comprehensive Problem-Solving and Skill Development for Next-Generation Leaders* (pp. 1–39). Hershey, PA: IGI Global. doi:10.4018/978-1-5225-1968-3.ch001

Demiray, U., & Ekren, G. (2018). Administrative-Related Evaluation for Distance Education Institutions in Turkey. In K. Buyuk, S. Kocdar, & A. Bozkurt (Eds.), *Administrative Leadership in Open and Distance Learning Programs* (pp. 263–288). Hershey, PA: IGI Global. doi:10.4018/978-1-5225-2645-2.ch011

Dickenson, P. (2017). What do we Know and Where Can We Grow?: Teachers Preparation for the Inclusive Classroom. In P. Dickenson, P. Keough, & J. Courduff (Eds.), *Preparing Pre-Service Teachers for the Inclusive Classroom* (pp. 1–22). Hershey, PA: IGI Global. doi:10.4018/978-1-5225-1753-5.ch001

Dickerson, J., & Coleman, H. V. (2012). Technology, E-Leadership and Educational Administration in Schools: Integrating Standards with Context and Guiding Questions. In V. Wang (Ed.), *Encyclopedia of E-Leadership, Counseling and Training* (pp. 408–422). Hershey, PA: IGI Global. doi:10.4018/978-1-61350-068-2.ch030

Dickerson, J., Coleman, H. V., & Geer, G. (2012). Thinking like a School Technology Leader. In V. Wang (Ed.), *Technology and Its Impact on Educational Leadership: Innovation and Change* (pp. 53–63). Hershey, PA: IGI Global. doi:10.4018/978-1-4666-0062-1.ch005

Donne, V., & Hansen, M. (2017). Teachers' Use of Assistive Technologies in Education. In L. Tomei (Ed.), *Exploring the New Era of Technology-Infused Education* (pp. 86–101). Hershey, PA: IGI Global. doi:10.4018/978-1-5225-1709-2.ch006

Donne, V., & Hansen, M. A. (2018). Business and Technology Educators: Practices for Inclusion. In I. Management Association (Ed.), Business Education and Ethics: Concepts, Methodologies, Tools, and Applications (pp. 471-484). Hershey, PA: IGI Global. doi:10.4018/978-1-5225-3153-1.ch026

Dreon, O., Shettel, J., & Bower, K. M. (2017). Preparing Next Generation Elementary Teachers for the Tools of Tomorrow. In M. Grassetti & S. Brookby (Eds.), *Advancing Next-Generation Teacher Education through Digital Tools and Applications* (pp. 143–159). Hershey, PA: IGI Global. doi:10.4018/978-1-5225-0965-3.ch008

Drinka, D., Voge, K., & Yen, M. Y. (2005). From Principles to Practice: Analyzing a Student Learning Outcomes Assessment System. *Journal of Cases on Information Technology*, *7*(3), 37–56. doi:10.4018/jcit.2005070103

Durak, H. Y., & Güyer, T. (2018). Design and Development of an Instructional Program for Teaching Programming Processes to Gifted Students Using Scratch. In J. Cannaday (Ed.), *Curriculum Development for Gifted Education Programs* (pp. 61–99). Hershey, PA: IGI Global. doi:10.4018/978-1-5225-3041-1.ch004

Egorkina, E., Ivanov, M., & Valyavskiy, A. Y. (2018). Students' Research Competence Formation of the Quality of Open and Distance Learning. In V. Mkrttchian & L. Belyanina (Eds.), *Handbook of Research on Students' Research Competence in Modern Educational Contexts* (pp. 364–384). Hershey, PA: IGI Global. doi:10.4018/978-1-5225-3485-3.ch019

Ekren, G., Karataş, S., & Demiray, U. (2017). Understanding of Leadership in Distance Education Management. In I. Management Association (Ed.), Educational Leadership and Administration: Concepts, Methodologies, Tools, and Applications (pp. 34-50). Hershey, PA: IGI Global. doi:10.4018/978-1-5225-1624-8.ch003

Elmore, W. M., Young, J. K., Harris, S., & Mason, D. (2017). The Relationship between Individual Student Attributes and Online Course Completion. In K. Shelton & K. Pedersen (Eds.), *Handbook of Research on Building, Growing, and Sustaining Quality E-Learning Programs* (pp. 151–173). Hershey, PA: IGI Global. doi:10.4018/978-1-5225-0877-9.ch008

Ercegovac, I. R., Alfirević, N., & Koludrović, M. (2017). School Principals' Communication and Co-Operation Assessment: The Croatian Experience. In I. Management Association (Ed.), Educational Leadership and Administration: Concepts, Methodologies, Tools, and Applications (pp. 1568-1589). Hershey, PA: IGI Global. doi:10.4018/978-1-5225-1624-8.ch072

Everhart, D., & Seymour, D. M. (2017). Challenges and Opportunities in the Currency of Higher Education. In K. Rasmussen, P. Northrup, & R. Colson (Eds.), *Handbook of Research on Competency-Based Education in University Settings* (pp. 41–65). Hershey, PA: IGI Global. doi:10.4018/978-1-5225-0932-5.ch003

Farmer, L. S. (2017). Managing Portable Technologies for Special Education. In V. Wang (Ed.), *Encyclopedia of Strategic Leadership and Management* (pp. 977–987). Hershey, PA: IGI Global. doi:10.4018/978-1-5225-1049-9.ch068

Farmer, L. S. (2018). Optimizing OERs for Optimal ICT Literacy in Higher Education. In J. Keengwe (Ed.), *Handbook of Research on Mobile Technology, Constructivism, and Meaningful Learning* (pp. 366–390). Hershey, PA: IGI Global. doi:10.4018/978-1-5225-3949-0.ch020

Fındık, L. Y. (2017). Self-Assessment of Principals Based on Leadership in Complexity. In I. Management Association (Ed.), Educational Leadership and Administration: Concepts, Methodologies, Tools, and Applications (pp. 978-991). Hershey, PA: IGI Global. doi:10.4018/978-1-5225-1624-8.ch047

Flor, A. G., & Gonzalez-Flor, B. (2018). Dysfunctional Digital Demeanors: Tales From (and Policy Implications of) eLearning's Dark Side. In I. Management Association (Ed.), The Dark Web: Breakthroughs in Research and Practice (pp. 37-50). Hershey, PA: IGI Global. doi:10.4018/978-1-5225-3163-0.ch003

Floyd, K. K., & Shambaugh, N. (2017). Instructional Design for Simulations in Special Education Virtual Learning Spaces. In T. Kidd & L. Morris Jr., (Eds.), *Handbook of Research on Instructional Systems and Educational Technology* (pp. 202–215). Hershey, PA: IGI Global. doi:10.4018/978-1-5225-2399-4.ch018

Giovannini, J. M. (2017). Technology Integration in Preservice Teacher Education Programs: Research-based Recommendations. In M. Grassetti & S. Brookby (Eds.), *Advancing Next-Generation Teacher Education through Digital Tools and Applications* (pp. 82–102). Hershey, PA: IGI Global. doi:10.4018/978-1-5225-0965-3.ch005

Good, S., & Clarke, V. B. (2017). An Integral Analysis of One Urban School System's Efforts to Support Student-Centered Teaching. In J. Keengwe & G. Onchwari (Eds.), *Handbook of Research on Learner-Centered Pedagogy in Teacher Education and Professional Development* (pp. 45–68). Hershey, PA: IGI Global. doi:10.4018/978-1-5225-0892-2.ch003

Grobler, B. (2015). The Relationship between Emotional Competence and Instructional Leadership and Their Association with Learner Achievement. In K. Beycioglu & P. Pashiardis (Eds.), *Multidimensional Perspectives on Principal Leadership Effectiveness* (pp. 373–407). Hershey, PA: IGI Global. doi:10.4018/978-1-4666-6591-0.ch017

Hamidi, F., Owuor, P. M., Hynie, M., Baljko, M., & McGrath, S. (2017). Potentials of Digital Assistive Technology and Special Education in Kenya. In C. Ayo & V. Mbarika (Eds.), *Sustainable ICT Adoption and Integration for Socio-Economic Development* (pp. 125–151). Hershey, PA: IGI Global. doi:10.4018/978-1-5225-2565-3.ch006

Heavin, C., & Neville, K. (2015). Addressing the Learning Needs of Future IS Security Professionals through Social Media Technology. In M. Khosrow-Pour (Ed.), *Encyclopedia of Information Science and Technology* (3rd ed.; pp. 4766–4775). Hershey, PA: IGI Global. doi:10.4018/978-1-4666-5888-2.ch468

Henderson, L. K. (2017). Meltdown at Fukushima: Global Catastrophic Events, Visual Literacy, and Art Education. In R. Shin (Ed.), *Convergence of Contemporary Art, Visual Culture, and Global Civic Engagement* (pp. 80–99). Hershey, PA: IGI Global. doi:10.4018/978-1-5225-1665-1.ch005

Hismanoglu, M. (2012). Important Issues in Online Education: E-Pedagogy and Marketing. In I. Management Association (Ed.), E-Marketing: Concepts, Methodologies, Tools, and Applications (pp. 676-701). Hershey, PA: IGI Global. doi:10.4018/978-1-4666-1598-4.ch041

Howard, B. C. (2008). Common Features and Design Principles Found in Exemplary Educational Technologies. *International Journal of Information and Communication Technology Education*, *4*(4), 31–52. doi:10.4018/jicte.2008100104

Howard, B. C., & Tomei, L. A. (2008). The Classroom of the Future and Emerging Educational Technologies: Introduction to the Special Issue. *International Journal of Information and Communication Technology Education*, *4*(4), 1–8. doi:10.4018/jicte.2008100101

Hudgins, T., & Holland, J. L. (2018). Digital Badges: Tracking Knowledge Acquisition Within an Innovation Framework. In I. Management Association (Ed.), Wearable Technologies: Concepts, Methodologies, Tools, and Applications (pp. 1118-1132). Hershey, PA: IGI Global. doi:10.4018/978-1-5225-5484-4.ch051

Ion, G., Tomàs, M., Castro, D., & Salat, E. (2015). Analysis of the Tasks of School Principals in Secondary Education in Catalonia: Case Study. In K. Beycioglu & P. Pashiardis (Eds.), *Multidimensional Perspectives on Principal Leadership Effectiveness* (pp. 39–58). Hershey, PA: IGI Global. doi:10.4018/978-1-4666-6591-0.ch003

Janus, M., & Siddiqua, A. (2018). Challenges for Children With Special Health Needs at the Time of Transition to School. In I. Management Association (Ed.), Autism Spectrum Disorders: Breakthroughs in Research and Practice (pp. 339-371). Hershey, PA: IGI Global. doi:10.4018/978-1-5225-3827-1.ch018

Jesus, R. A. (2018). Screencasts and Learning Styles. In M. Khosrow-Pour, D.B.A. (Ed.), Encyclopedia of Information Science and Technology, Fourth Edition (pp. 1548-1558). Hershey, PA: IGI Global. doi:10.4018/978-1-5225-2255-3.ch134

Kaplan-Rakowski, R., & Rakowski, D. (2011). Educational Technologies for the Neomillennial Generation. In E. Dunkels, G. Franberg, & C. Hallgren (Eds.), *Interactive Media Use and Youth: Learning, Knowledge Exchange and Behavior* (pp. 12–31). Hershey, PA: IGI Global. doi:10.4018/978-1-60960-206-2.ch002

Karpinski, A. C., D'Agostino, J. V., Williams, A. K., Highland, S. A., & Mellott, J. A. (2018). The Relationship Between Online Formative Assessment and State Test Scores Using Multilevel Modeling. In M. Khosrow-Pour, D.B.A. (Ed.), Encyclopedia of Information Science and Technology, Fourth Edition (pp. 5183-5192). Hershey, PA: IGI Global. doi:10.4018/978-1-5225-2255-3.ch450

Kats, Y. (2017). Educational Leadership and Integrated Support for Students with Autism Spectrum Disorders. In I. Management Association (Ed.), Educational Leadership and Administration: Concepts, Methodologies, Tools, and Applications (pp. 101-114). Hershey, PA: IGI Global. doi:10.4018/978-1-5225-1624-8.ch007

Kaya, G., & Altun, A. (2018). Educational Ontology Development. In M. Khosrow-Pour, D.B.A. (Ed.), Encyclopedia of Information Science and Technology, Fourth Edition (pp. 1441-1450). Hershey, PA: IGI Global. doi:10.4018/978-1-5225-2255-3.ch124

Keough, P. D., & Pacis, D. (2017). Best Practices Implementing Special Education Curriculum and Common Core State Standards using UDL. In P. Dickenson, P. Keough, & J. Courduff (Eds.), *Preparing Pre-Service Teachers for the Inclusive Classroom* (pp. 107–123). Hershey, PA: IGI Global. doi:10.4018/978-1-5225-1753-5.ch006

Kilburn, M., Henckell, M., & Starrett, D. (2018). Factors Contributing to the Effectiveness of Online Students and Instructors. In M. Khosrow-Pour, D.B.A. (Ed.), Encyclopedia of Information Science and Technology, Fourth Edition (pp. 1451-1462). Hershey, PA: IGI Global. doi:10.4018/978-1-5225-2255-3.ch125

Konecny, L. T. (2017). Hybrid, Online, and Flipped Classrooms in Health Science: Enhanced Learning Environments. In I. Management Association (Ed.), Flipped Instruction: Breakthroughs in Research and Practice (pp. 355-370). Hershey, PA: IGI Global. doi:10.4018/978-1-5225-1803-7.ch020

Kowch, E. G. (2013). Towards Leading Diverse, Smarter and More Adaptable Organizations that Learn. In J. Lewis, A. Green, & D. Surry (Eds.), *Technology as a Tool for Diversity Leadership: Implementation and Future Implications* (pp. 11–34). Hershey, PA: IGI Global. doi:10.4018/978-1-4666-2668-3.ch002

Krezmien, M., Powell, W., Bosch, C., Hall, T., & Nieswandt, M. (2017). The Use of Tablet Technology to Support Inquiry Science for Students Incarcerated in Juvenile Justice Settings. In I. Levin & D. Tsybulsky (Eds.), *Optimizing STEM Education With Advanced ICTs and Simulations* (pp. 267–295). Hershey, PA: IGI Global. doi:10.4018/978-1-5225-2528-8.ch011

Leach, L. F., Winn, P., Erwin, S., & Benedict, L. P. (2015). What 21st Century Students Want: Factors that Influence Student Selection of Educational Leadership Graduate Programs. *International Journal of Technology and Educational Marketing*, *5*(1), 15–28. doi:10.4018/ijtem.2015010102

Leng, H. K. (2014). An Update on the Use of Facebook as a Marketing Tool by Private Educational Institutions in Singapore. In I. Lee (Ed.), *Trends in E-Business, E-Services, and E-Commerce: Impact of Technology on Goods, Services, and Business Transactions* (pp. 191–205). Hershey, PA: IGI Global. doi:10.4018/978-1-4666-4510-3.ch011

Leone, S. (2018). An Open Learning Format for Lifelong Learners' Empowerment. In M. Khosrow-Pour, D.B.A. (Ed.), Encyclopedia of Information Science and Technology, Fourth Edition (pp. 1517-1528). Hershey, PA: IGI Global. doi:10.4018/978-1-5225-2255-3.ch131

Ley, K., & Gannon-Cook, R. (2010). Marketing a Blended University Program: An Action Research Case Study. In S. Mukerji & P. Tripathi (Eds.), *Cases on Technology Enhanced Learning through Collaborative Opportunities* (pp. 73–90). Hershey, PA: IGI Global. doi:10.4018/978-1-61520-751-0.ch005

Loose, W., & Marcos, T. (2016). Instructional Design for Millennials: Instructor Efficiency in Streamlining Content, Assignments, and Assessments. In P. Dickenson & J. Jaurez (Eds.), *Increasing Productivity and Efficiency in Online Teaching* (pp. 1–25). Hershey, PA: IGI Global. doi:10.4018/978-1-5225-0347-7.ch001

Lovell, K. L. (2017). Development and Evaluation of Neuroscience Computer-Based Modules for Medical Students: Instructional Design Principles and Effectiveness. In J. Stefaniak (Ed.), *Advancing Medical Education Through Strategic Instructional Design* (pp. 262–276). Hershey, PA: IGI Global. doi:10.4018/978-1-5225-2098-6.ch013

Manuel, N. N. (2016). Angolan Higher Education, Policy, and Leadership: Towards Transformative Leadership for Social Justice. In N. Ololube (Ed.), *Handbook of Research on Organizational Justice and Culture in Higher Education Institutions* (pp. 164–188). Hershey, PA: IGI Global. doi:10.4018/978-1-4666-9850-5.ch007

Marouchou, D. V. (2015). The Impact of Academic Beliefs on Student Learning. In M. Khosrow-Pour (Ed.), *Encyclopedia of Information Science and Technology* (3rd ed.; pp. 4796–4804). Hershey, PA: IGI Global. doi:10.4018/978-1-4666-5888-2.ch471

McCormack, V. F., Stauffer, M., Fishley, K., Hohenbrink, J., Mascazine, J. R., & Zigler, T. (2018). Designing a Dual Licensure Path for Middle Childhood and Special Education Teacher Candidates. In D. Polly, M. Putman, T. Petty, & A. Good (Eds.), *Innovative Practices in Teacher Preparation and Graduate-Level Teacher Education Programs* (pp. 21–36). Hershey, PA: IGI Global. doi:10.4018/978-1-5225-3068-8.ch002

McDaniel, R. (2017). Strategic Leadership in Instructional Design: Applying the Principles of Instructional Design through the Lens of Strategic Leadership to Distance Education. In V. Wang (Ed.), *Encyclopedia of Strategic Leadership and Management* (pp. 1570–1584). Hershey, PA: IGI Global. doi:10.4018/978-1-5225-1049-9.ch109

Memon, R. N., Ahmad, R., & Salim, S. S. (2018). Critical Issues in Requirements Engineering Education. In I. Management Association (Ed.), Computer Systems and Software Engineering: Concepts, Methodologies, Tools, and Applications (pp. 1953-1976). Hershey, PA: IGI Global. doi:10.4018/978-1-5225-3923-0.ch081

Mendenhall, R. (2017). Western Governors University: CBE Innovator and National Model. In K. Rasmussen, P. Northrup, & R. Colson (Eds.), *Handbook of Research on Competency-Based Education in University Settings* (pp. 379–400). Hershey, PA: IGI Global. doi:10.4018/978-1-5225-0932-5.ch019

Mense, E. G., Griggs, D. M., & Shanks, J. N. (2018). School Leaders in a Time of Accountability and Data Use: Preparing Our Future School Leaders in Leadership Preparation Programs. In E. Mense & M. Crain-Dorough (Eds.), *Data Leadership for K-12 Schools in a Time of Accountability* (pp. 235–259). Hershey, PA: IGI Global. doi:10.4018/978-1-5225-3188-3.ch012

Mense, E. G., Griggs, D. M., & Shanks, J. N. (2018). School Leaders in a Time of Accountability and Data Use: Preparing Our Future School Leaders in Leadership Preparation Programs. In E. Mense & M. Crain-Dorough (Eds.), *Data Leadership for K-12 Schools in a Time of Accountability* (pp. 235–259). Hershey, PA: IGI Global. doi:10.4018/978-1-5225-3188-3.ch012

Mestry, R., & Naicker, S. R. (2017). Exploring Distributive Leadership in South African Public Primary Schools in the Soweto Region. In I. Management Association (Ed.), Educational Leadership and Administration: Concepts, Methodologies, Tools, and Applications (pp. 1041-1064). Hershey, PA: IGI Global. doi:10.4018/978-1-5225-1624-8.ch050

Monaghan, C. H., & Boboc, M. (2017). (Re) Defining Leadership in Higher Education in the U.S. In V. Wang (Ed.), *Encyclopedia of Strategic Leadership and Management* (pp. 567–579). Hershey, PA: IGI Global. doi:10.4018/978-1-5225-1049-9.ch040

Muthee, J. M., & Murungi, C. G. (2018). Relationship Among Intelligence, Achievement Motivation, Type of School, and Academic Performance of Kenyan Urban Primary School Pupils. In M. Khosrow-Pour, D.B.A. (Ed.), Encyclopedia of Information Science and Technology, Fourth Edition (pp. 1540-1547). Hershey, PA: IGI Global. doi:10.4018/978-1-5225-2255-3.ch133

Naranjo, J. (2018). Meeting the Need for Inclusive Educators Online: Teacher Education in Inclusive Special Education and Dual-Certification. In D. Polly, M. Putman, T. Petty, & A. Good (Eds.), *Innovative Practices in Teacher Preparation and Graduate-Level Teacher Education Programs* (pp. 106–122). Hershey, PA: IGI Global. doi:10.4018/978-1-5225-3068-8.ch007

Nkabinde, Z. P. (2017). Multiculturalism in Special Education: Perspectives of Minority Children in Urban Schools. In J. Keengwe (Ed.), *Handbook of Research on Promoting Cross-Cultural Competence and Social Justice in Teacher Education* (pp. 382–397). Hershey, PA: IGI Global. doi:10.4018/978-1-5225-0897-7.ch020

Nkabinde, Z. P. (2018). Online Instruction: Is the Quality the Same as Face-to-Face Instruction? In J. Keengwe (Ed.), *Handbook of Research on Digital Content, Mobile Learning, and Technology Integration Models in Teacher Education* (pp. 300–314). Hershey, PA: IGI Global. doi:10.4018/978-1-5225-2953-8.ch016

O'Connor, J. R. Jr, & Jackson, K. N. (2017). The Use of iPad® Devices and "Apps" for ASD Students in Special Education and Speech Therapy. In Y. Kats (Ed.), *Supporting the Education of Children with Autism Spectrum Disorders* (pp. 267–283). Hershey, PA: IGI Global. doi:10.4018/978-1-5225-0816-8.ch014

Okolie, U. C., & Yasin, A. M. (2017). TVET in Developing Nations and Human Development. In U. Okolie & A. Yasin (Eds.), *Technical Education and Vocational Training in Developing Nations* (pp. 1–25). Hershey, PA: IGI Global. doi:10.4018/978-1-5225-1811-2.ch001

Paciga, K. A., & Hoffman, J. L. (2015). Realizing the Potential of e-Books in Early Education. In M. Khosrow-Pour (Ed.), *Encyclopedia of Information Science and Technology* (3rd ed.; pp. 4787–4795). Hershey, PA: IGI Global. doi:10.4018/978-1-4666-5888-2.ch470

Paulson, E. N. (2017). Adapting and Advocating for an Online EdD Program in Changing Times and "Sacred" Cultures. In I. Management Association (Ed.), Educational Leadership and Administration: Concepts, Methodologies, Tools, and Applications (pp. 1849-1876). Hershey, PA: IGI Global. doi:10.4018/978-1-5225-1624-8.ch085

Petersen, A. J., Elser, C. F., Al Nassir, M. N., Stakey, J., & Everson, K. (2017). The Year of Teaching Inclusively: Building an Elementary Classroom for All Students. In C. Curran & A. Petersen (Eds.), *Handbook of Research on Classroom Diversity and Inclusive Education Practice* (pp. 332–348). Hershey, PA: IGI Global. doi:10.4018/978-1-5225-2520-2.ch014

Pfannenstiel, K. H., & Sanders, J. (2017). Characteristics and Instructional Strategies for Students With Mathematical Difficulties: In the Inclusive Classroom. In C. Curran & A. Petersen (Eds.), *Handbook of Research on Classroom Diversity and Inclusive Education Practice* (pp. 250–281). Hershey, PA: IGI Global. doi:10.4018/978-1-5225-2520-2.ch011

Preast, J. L., Bowman, N., & Rose, C. A. (2017). Creating Inclusive Classroom Communities Through Social and Emotional Learning to Reduce Social Marginalization Among Students. In C. Curran & A. Petersen (Eds.), *Handbook of Research on Classroom Diversity and Inclusive Education Practice* (pp. 183–200). Hershey, PA: IGI Global. doi:10.4018/978-1-5225-2520-2.ch008

Randolph, K. M., & Brady, M. P. (2018). Evolution of Covert Coaching as an Evidence-Based Practice in Professional Development and Preparation of Teachers. In V. Bryan, A. Musgrove, & J. Powers (Eds.), *Handbook of Research on Human Development in the Digital Age* (pp. 281–299). Hershey, PA: IGI Global. doi:10.4018/978-1-5225-2838-8.ch013

Rawlins, P., & Kehrwald, B. (2010). Education Technology in Teacher Education: Overcoming Challenges, Realizing Opportunities. In R. Luppicini & A. Haghi (Eds.), *Cases on Digital Technologies in Higher Education: Issues and Challenges* (pp. 50–63). Hershey, PA: IGI Global. doi:10.4018/978-1-61520-869-2.ch004

Rell, A. B., Puig, R. A., Roll, F., Valles, V., Espinoza, M., & Duque, A. L. (2017). Addressing Cultural Diversity and Global Competence: The Dual Language Framework. In L. Leavitt, S. Wisdom, & K. Leavitt (Eds.), *Cultural Awareness and Competency Development in Higher Education* (pp. 111–131). Hershey, PA: IGI Global. doi:10.4018/978-1-5225-2145-7.ch007

Riel, J., Lawless, K. A., & Brown, S. W. (2017). Defining and Designing Responsive Online Professional Development (ROPD): A Framework to Support Curriculum Implementation. In T. Kidd & L. Morris Jr., (Eds.), *Handbook of Research on Instructional Systems and Educational Technology* (pp. 104–115). Hershey, PA: IGI Global. doi:10.4018/978-1-5225-2399-4.ch010

Roberts, C. (2017). Advancing Women Leaders in Academe: Creating a Culture of Inclusion. In S. Mukerji & P. Tripathi (Eds.), *Handbook of Research on Administration, Policy, and Leadership in Higher Education* (pp. 256–273). Hershey, PA: IGI Global. doi:10.4018/978-1-5225-0672-0.ch012

Rodgers, W. J., Kennedy, M. J., Alves, K. D., & Romig, J. E. (2017). A Multimedia Tool for Teacher Education and Professional Development. In C. Martin & D. Polly (Eds.), *Handbook of Research on Teacher Education and Professional Development* (pp. 285–296). Hershey, PA: IGI Global. doi:10.4018/978-1-5225-1067-3.ch015

Romanowski, M. H. (2017). Qatar's Educational Reform: Critical Issues Facing Principals. In I. Management Association (Ed.), Educational Leadership and Administration: Concepts, Methodologies, Tools, and Applications (pp. 1758-1773). Hershey, PA: IGI Global. doi:10.4018/978-1-5225-1624-8.ch080

Ruffin, T. R., Hawkins, D. P., & Lee, D. I. (2018). Increasing Student Engagement and Participation Through Course Methodology. In M. Khosrow-Pour, D.B.A. (Ed.), Encyclopedia of Information Science and Technology, Fourth Edition (pp. 1463-1473). Hershey, PA: IGI Global. doi:10.4018/978-1-5225-2255-3.ch126

Rutaisire, J. (2011). Innovations in Technology for Educational Marketing: Stakeholder Perceptions and Implications for Examinations System in Rwanda. In P. Tripathi & S. Mukerji (Eds.), *Cases on Innovations in Educational Marketing: Transnational and Technological Strategies* (pp. 214–233). Hershey, PA: IGI Global. doi:10.4018/978-1-60960-599-5.ch013

Sabina, L. L., Curry, K. A., Harris, E. L., Krumm, B. L., & Vencill, V. (2017). Assessing the Performance of a Cohort-Based Model Using Domestic and International Practices. In I. Management Association (Ed.), Educational Leadership and Administration: Concepts, Methodologies, Tools, and Applications(pp. 913-929). Hershey, PA: IGI Global. doi:10.4018/978-1-5225-1624-8.ch044

Santamaría, A. P., Webber, M., & Santamaría, L. J. (2017). Effective School Leadership for Māori Achievement: Building Capacity through Indigenous, National, and International Cross-Cultural Collaboration. In I. Management Association (Ed.), Educational Leadership and Administration: Concepts, Methodologies, Tools, and Applications (pp. 1547-1567). Hershey, PA: IGI Global. doi:10.4018/978-1-5225-1624-8.ch071

Santamaría, L. J. (2017). Culturally Responsive Educational Leadership in Cross-Cultural International Contexts. In I. Management Association (Ed.), Educational Leadership and Administration: Concepts, Methodologies, Tools, and Applications (pp. 1380-1400). Hershey, PA: IGI Global. doi:10.4018/978-1-5225-1624-8.ch064

Sarafidou, J., & Xafakos, E. (2015). Transformational Leadership and Principals' Innovativeness: Are They the "Keys" for the Research and Innovation Oriented School? In K. Beycioglu & P. Pashiardis (Eds.), *Multidimensional Perspectives on Principal Leadership Effectiveness* (pp. 324–348). Hershey, PA: IGI Global. doi:10.4018/978-1-4666-6591-0.ch015

Segredo, M. R., Cistone, P. J., & Reio, T. G. (2017). Relationships Between Emotional Intelligence, Leadership Style, and School Culture. *International Journal of Adult Vocational Education and Technology*, *8*(3), 25–43. doi:10.4018/IJAVET.2017070103

Shaik, N., & Ritter, S. (2012). Social Media Based Relationship Marketing. In I. Management Association (Ed.), E-Marketing: Concepts, Methodologies, Tools, and Applications (pp. 88-110). Hershey, PA: IGI Global. doi:10.4018/978-1-4666-1598-4.ch006

Shalev, N. (2017). Empathy and Leadership From the Organizational Perspective. In Z. Nedelko & M. Brzozowski (Eds.), *Exploring the Influence of Personal Values and Cultures in the Workplace* (pp. 348–363). Hershey, PA: IGI Global. doi:10.4018/978-1-5225-2480-9.ch018

Siamak, M., Fathi, S., & Isfandyari-Moghaddam, A. (2018). Assessment and Measurement of Education Programs of Information Literacy. In R. Bhardwaj (Ed.), *Digitizing the Modern Library and the Transition From Print to Electronic* (pp. 164–192). Hershey, PA: IGI Global. doi:10.4018/978-1-5225-2119-8.ch007

Siozos, P. D., & Palaigeorgiou, G. E. (2008). Educational Technologies and the Emergence of E-Learning 2.0. In D. Politis (Ed.), *E-Learning Methodologies and Computer Applications in Archaeology* (pp. 1–17). Hershey, PA: IGI Global. doi:10.4018/978-1-59904-759-1.ch001

Siu, K. W., & García, G. J. (2017). Disruptive Technologies and Education: Is There Any Disruption After All? In I. Management Association (Ed.), Educational Leadership and Administration: Concepts, Methodologies, Tools, and Applications (pp. 757-778). Hershey, PA: IGI Global. doi:10.4018/978-1-5225-1624-8.ch037

Skibba, K., Moore, D., & Herman, J. H. (2013). Pedagogical and Technological Considerations Designing Collaborative Learning Using Educational Technologies. In J. Keengwe (Ed.), *Research Perspectives and Best Practices in Educational Technology Integration* (pp. 1–27). Hershey, PA: IGI Global. doi:10.4018/978-1-4666-2988-2.ch001

Slagter van Tryon, P. J. (2017). The Nurse Educator's Role in Designing Instruction and Instructional Strategies for Academic and Clinical Settings. In J. Stefaniak (Ed.), *Advancing Medical Education Through Strategic Instructional Design* (pp. 133–149). Hershey, PA: IGI Global. doi:10.4018/978-1-5225-2098-6.ch006

Slattery, C. A. (2018). Literacy Intervention and the Differentiated Plan of Instruction. In *Developing Effective Literacy Intervention Strategies: Emerging Research and Opportunities* (pp. 41–62). Hershey, PA: IGI Global. doi:10.4018/978-1-5225-5007-5.ch003

Smith, A. R. (2017). Ensuring Quality: The Faculty Role in Online Higher Education. In K. Shelton & K. Pedersen (Eds.), *Handbook of Research on Building, Growing, and Sustaining Quality E-Learning Programs* (pp. 210–231). Hershey, PA: IGI Global. doi:10.4018/978-1-5225-0877-9.ch011

Souders, T. M. (2017). Understanding Your Learner: Conducting a Learner Analysis. In J. Stefaniak (Ed.), *Advancing Medical Education Through Strategic Instructional Design* (pp. 1–29). Hershey, PA: IGI Global. doi:10.4018/978-1-5225-2098-6.ch001

Spring, K. J., Graham, C. R., & Ikahihifo, T. B. (2018). Learner Engagement in Blended Learning. In M. Khosrow-Pour, D.B.A. (Ed.), Encyclopedia of Information Science and Technology, Fourth Edition (pp. 1487-1498). Hershey, PA: IGI Global. doi:10.4018/978-1-5225-2255-3.ch128

Stocklin, S. (2015). Building Capacity by Managing a Mission. In J. Feng, S. Stocklin, & W. Wang (Eds.), *Educational Strategies for the Next Generation Leaders in Hotel Management* (pp. 115–139). Hershey, PA: IGI Global. doi:10.4018/978-1-4666-8565-9.ch005

Storey, V. A., Anthony, A. K., & Wahid, P. (2017). Gender-Based Leadership Barriers: Advancement of Female Faculty to Leadership Positions in Higher Education. In V. Wang (Ed.), *Encyclopedia of Strategic Leadership and Management* (pp. 244–258). Hershey, PA: IGI Global. doi:10.4018/978-1-5225-1049-9.ch018

Stottlemyer, D. (2018). Develop a Teaching Model Plan for a Differentiated Learning Approach. In *Differentiated Instructional Design for Multicultural Environments: Emerging Research and Opportunities* (pp. 106–130). Hershey, PA: IGI Global. doi:10.4018/978-1-5225-5106-5.ch005

Stottlemyer, D. (2018). Developing a Multicultural Environment. In *Differentiated Instructional Design for Multicultural Environments: Emerging Research and Opportunities* (pp. 1–27). Hershey, PA: IGI Global. doi:10.4018/978-1-5225-5106-5.ch001

Swami, B. N., Gobona, T., & Tsimako, J. J. (2017). Academic Leadership: A Case Study of the University of Botswana. In N. Baporikar (Ed.), *Innovation and Shifting Perspectives in Management Education* (pp. 1–32). Hershey, PA: IGI Global. doi:10.4018/978-1-5225-1019-2.ch001

Swanson, K. W., & Collins, G. (2018). Designing Engaging Instruction for the Adult Learners. In M. Khosrow-Pour, D.B.A. (Ed.), Encyclopedia of Information Science and Technology, Fourth Edition (pp. 1432-1440). Hershey, PA: IGI Global. doi:10.4018/978-1-5225-2255-3.ch123

Swartz, B. A., Lynch, J. M., & Lynch, S. D. (2018). Embedding Elementary Teacher Education Coursework in Local Classrooms: Examples in Mathematics and Special Education. In D. Polly, M. Putman, T. Petty, & A. Good (Eds.), *Innovative Practices in Teacher Preparation and Graduate-Level Teacher Education Programs* (pp. 262–292). Hershey, PA: IGI Global. doi:10.4018/978-1-5225-3068-8.ch015

Taliadorou, N., & Pashiardis, P. (2015). Emotional Intelligence and Political Skill Really Matter in Educational Leadership. In K. Beycioglu & P. Pashiardis (Eds.), *Multidimensional Perspectives on Principal Leadership Effectiveness* (pp. 228–256). Hershey, PA: IGI Global. doi:10.4018/978-1-4666-6591-0.ch011

Taliadorou, N., & Pashiardis, P. (2017). Emotional Intelligence and Political Skill Really Matter in Educational Leadership. In I. Management Association (Ed.), Educational Leadership and Administration: Concepts, Methodologies, Tools, and Applications (pp. 1274-1303). Hershey, PA: IGI Global. doi:10.4018/978-1-5225-1624-8.ch060

Tam, F. W., & Kwan, P. Y. (2011). School Images, School Identity, and How Parents Select Schools for Their Children: The Case of Hong Kong. In P. Tripathi & S. Mukerji (Eds.), *Cases on Innovations in Educational Marketing: Transnational and Technological Strategies* (pp. 87–103). Hershey, PA: IGI Global. doi:10.4018/978-1-60960-599-5.ch005

Tandoh, K. A., & Ebe-Arthur, J. E. (2018). Effective Educational Leadership in the Digital Age: An Examination of Professional Qualities and Best Practices. In J. Keengwe (Ed.), *Handbook of Research on Digital Content, Mobile Learning, and Technology Integration Models in Teacher Education* (pp. 244–265). Hershey, PA: IGI Global. doi:10.4018/978-1-5225-2953-8.ch013

Tinoca, L., Pereira, A., & Oliveira, I. (2014). A Conceptual Framework for E-Assessment in Higher Education: Authenticity, Consistency, Transparency, and Practicability. In S. Mukerji & P. Tripathi (Eds.), *Handbook of Research on Transnational Higher Education* (pp. 652–673). Hershey, PA: IGI Global. doi:10.4018/978-1-4666-4458-8.ch033

Tobin, M. T. (2018). Multimodal Literacy. In M. Khosrow-Pour, D.B.A. (Ed.), Encyclopedia of Information Science and Technology, Fourth Edition (pp. 1508-1516). Hershey, PA: IGI Global. doi:10.4018/978-1-5225-2255-3.ch130

Torres, M. L., & Ramos, V. J. (2018). Music Therapy: A Pedagogical Alternative for ASD and ID Students in Regular Classrooms. In P. Epler (Ed.), *Instructional Strategies in General Education and Putting the Individuals With Disabilities Act (IDEA) Into Practice* (pp. 222–244). Hershey, PA: IGI Global. doi:10.4018/978-1-5225-3111-1.ch008

Toulassi, B. (2017). Educational Administration and Leadership in Francophone Africa: 5 Dynamics to Change Education. In S. Mukerji & P. Tripathi (Eds.), *Handbook of Research on Administration, Policy, and Leadership in Higher Education* (pp. 20–45). Hershey, PA: IGI Global. doi:10.4018/978-1-5225-0672-0.ch002

Umair, S., & Sharif, M. M. (2018). Predicting Students Grades Using Artificial Neural Networks and Support Vector Machine. In M. Khosrow-Pour, D.B.A. (Ed.), Encyclopedia of Information Science and Technology, Fourth Edition (pp. 5169-5182). Hershey, PA: IGI Global. doi:10.4018/978-1-5225-2255-3.ch449

Usman, L. M. (2011). Adult Education and Sustainable Learning Outcome of Rural Widows of Central Northern Nigeria. *International Journal of Adult Vocational Education and Technology*, *2*(2), 25–41. doi:10.4018/javet.2011040103

Vettraino, L., Castello, V., Guspini, M., & Guglielman, E. (2018). Self-Awareness and Motivation Contrasting ESL and NEET Using the SAVE System. In M. Khosrow-Pour, D.B.A. (Ed.), Encyclopedia of Information Science and Technology, Fourth Edition (pp. 1559-1568). Hershey, PA: IGI Global. doi:10.4018/978-1-5225-2255-3.ch135

Wang, V. C. (2013). Marketing Educational Programs through Technology and the Right Philosophies. In P. Tripathi & S. Mukerji (Eds.), *Marketing Strategies for Higher Education Institutions: Technological Considerations and Practices* (pp. 15–24). Hershey, PA: IGI Global. doi:10.4018/978-1-4666-4014-6.ch002

Wiemelt, J. (2017). Critical Bilingual Leadership for Emergent Bilingual Students. In I. Management Association (Ed.), Educational Leadership and Administration: Concepts, Methodologies, Tools, and Applications (pp. 1606-1631). Hershey, PA: IGI Global. doi:10.4018/978-1-5225-1624-8.ch074

Williams, D. D. (2006). Measurement and Assessment Supporting Evaluation in Online Settings. In D. Williams, M. Hricko, & S. Howell (Eds.), *Online Assessment, Measurement and Evaluation: Emerging Practices* (pp. 1–9). Hershey, PA: IGI Global. doi:10.4018/978-1-59140-747-8.ch001

Wolf, F., Seyfarth, F. C., & Pflaum, E. (2018). Scalable Capacity-Building for Geographically Dispersed Learners: Designing the MOOC "Sustainable Energy in Small Island Developing States (SIDS)". In U. Pandey & V. Indrakanti (Eds.), *Open and Distance Learning Initiatives for Sustainable Development* (pp. 58–83). Hershey, PA: IGI Global. doi:10.4018/978-1-5225-2621-6.ch003

Woodley, X. M., Mucundanyi, G., & Lockard, M. (2017). Designing Counter-Narratives: Constructing Culturally Responsive Curriculum Online. *International Journal of Online Pedagogy and Course Design*, *7*(1), 43–56. doi:10.4018/IJOPCD.2017010104

Woods, P. A., & Woods, G. J. (2011). Lighting the Fires of Entrepreneurialism?: Constructions of Meaning in an English Inner City Academy. *International Journal of Technology and Educational Marketing*, *1*(1), 1–24. doi:10.4018/ijtem.2011010101

Yell, M. L., & Christle, C. A. (2017). The Foundation of Inclusion in Federal Legislation and Litigation. In C. Curran & A. Petersen (Eds.), *Handbook of Research on Classroom Diversity and Inclusive Education Practice* (pp. 27–52). Hershey, PA: IGI Global. doi:10.4018/978-1-5225-2520-2.ch002

Zhao, J. (2011). China Special Education: The Perspective of Information Technologies. In P. Ordóñez de Pablos, J. Zhao, & R. Tennyson (Eds.), *Technology Enhanced Learning for People with Disabilities: Approaches and Applications* (pp. 34–43). Hershey, PA: IGI Global. doi:10.4018/978-1-61520-923-1.ch003

Zinger, D. (2016). Developing Instructional Leadership and Communication Skills through Online Professional Development: Focusing on Rural and Urban Principals. In A. Normore, L. Long, & M. Javidi (Eds.), *Handbook of Research on Effective Communication, Leadership, and Conflict Resolution* (pp. 354–370). Hershey, PA: IGI Global. doi:10.4018/978-1-4666-9970-0.ch019

Zutshi, A., Pogrebnaya, M., & Fermelis, J. (2014). Wellness Programs in Higher Education: An Australian Case. In N. Baporikar (Ed.), *Handbook of Research on Higher Education in the MENA Region: Policy and Practice* (pp. 391–419). Hershey, PA: IGI Global. doi:10.4018/978-1-4666-6198-1.ch017

About the Contributors

Stuart Bevins is a Principal Research Fellow and lead for science education research. He has designed and led both large- and small-scale research projects and written extensively about science teacher professional development and classroom inquiry in science education. Stuart was the Coordinator for the Chain Reaction project.

Louise Lehane is a Lecturer in Education at St Angela's College in Sligo, Ireland. She is a qualified science teacher and completed both her undergraduate studies and her PhD in the University of Limerick, Ireland. She now specializes in teaching general and science-specific pedagogics. Her main research interests include Nature of Science, teacher PCK development, curriculum policy and change, the sociology of education, development education through a science teaching lens and self-efficacy development through the use of professional learning communities. She worked on the Chain Reaction project from 2013-2016.

Josephine Booth has worked at Sheffield Hallam University since 2009, following the successful completion of her PhD in Biological Sciences at the University of Sheffield. She has worked in the private and third sector as well as in an academic environment. Josephine's research interests are wide ranging and include environmental and outdoor education, the impact of teacher education routes, SEND in schools and inquiry-based science education policy. She co-managed the Chain Reaction project throughout its lifetime.

* * *

Anwar Battikhi is President of JSSR and a Professor of Soil Physics in Soil, Water and Environment Department at the Jordan University.

Eleanor Byrne has worked at Sheffield Hallam University for over 10 years. An experienced project manager, her research interests are around cross-curricular learning, teacher leadership, teaching methods and an understanding of leadership change within the school context. She is the co-author of "Cross-curricular Teaching and Learning in the Secondary School: Science" published in 2012. She co-managed the Chain Reaction project throughout its lifetime.

Giuliana Capasso is a math and physics teacher. Her main research interest concerns the design of teaching learning sequences in modern physics.

Nikos Chaniotakis is a professor of Chemistry, and Dean of the School of Science and Engineering at the University of Crete. His scientific interests include basic and applied research in the area of Analytical Chemistry, as well as in higher education issues. Has participated in a variety of EU supported projects in the area of chemical sensors, biosensors and education.

Silvia Galano is fellowship owner at the Department of Physics "E. Pancini" of Naples. Her main research interests concern astronomy education and inquiry-based science education.

Jonathan Glaser is a teacher for chemistry, physics, and mathematics in secondary schools. He works in Baden-Württemberg, Germany. In 2013 he graduated at the University of Education Heidelberg. In 2016 he received a Master's Degree in Educational Sciences (study field: science teaching strategies) at the University of Education Heidelberg. From 2013 to 2016 he worked in the Chain Reaction project at the University of Education Heidelberg.

Sönke Graf is a teacher of Physics and English, a doctoral student in the field of physics education and scientific staff member in the physics department of the University of Education Heidelberg. He has been working in the field of university education for many years. Due to writing his dissertation in the field of teacher trainings his professional expertise lies in informal learning inside of formal trainings. He is familiar in creating environments that foster informal communication. Sönke Graf has been actively involved in the COMENIUS project CAT and the FP7 project Chain Reaction.

Liam Guilfoyle is a research officer in the University of Oxford. His research interests are in teacher education, the role of epistemic beliefs in science teacher education and the theory-practice divide in the teaching profession. Liam worked on the Chain Reaction project as a research assistant from 2015 to 2016.

Stanislav Holec is a head of the Physics Department at the Faculty of Natural Sciences, Matej Bel University. He has more than 30 years of experience teaching on physics teacher training programmes. He has published widely; his research work concentrates on didactics of physics and science, especially on real and model physics and science school experiments aided by the computer. He has a wide experience with coordinating projects focused on a development of science literacy of young people as well as a wider population, and a curricular and methodical reform and innovation of physics and science education.

Martin Hruska was born in 1974; MSc degree in 1997 in Teachers training for Physics and Chemistry subjects at the Faculty of Sciences of Masaryk University in Brno, Czech Republic; PhD degree in 2009 at the Department of Physics at the Faculty of Natural Sciences Banská Bystrica, in Education of Physics; Assistant Professor at the Department of Physics at the Faculty of Natural Sciences, Matej Bel University (since 1997). Research areas: didactics of physics and science - science school experiments aided by the computer; integration approaches in science education; new methods in physics teaching especially those using ICT; inquiry based teaching and learning methods; popularisation of physics through various activities.

Marika Kapanadze is professor of science education, a head of science education research centre SALiS and science teacher professional development programs at Ilia State University, Tbilisi, Georgia. She is engaged in development, implementation and evaluation of student-centred curricula and new curricular materials. Her research interests are teacher professional development, science curriculum development and investigation of teachers' attitudes and students' interest in science. She develops the courses about inquiry based learning and active learning methods in Physics and Science. She is a co-author of methodological books in education and in science education in Georgia, and regularly publishes her research in national and international journals.

Kallia Katsampoxaki-Hodgetts holds a Masters of Education in Teaching English as a Foreign Language (MED TEFL, University of Bristol, UK). She teaches in-sessional English for Academic Purposes programmes in the School of Science and Technology of the University of Crete (UOC) since 2001 and she is the author of ESAP textbooks such as "English For Chemistry EAP" (2017), "English for Biology EAP"(2018) and "English for Mathematics" (2018). She has also designed and run online open access courses (1. Academic English 2. Teaching approaches of Inquiry Based Science Education- IBSE), is the co-ordinator of the international pre-sessional summer course "Academic English for Chemistry Graduates" at UOC and she has given many lectures and workshops on technical writing in Europe. Kallia has been the leader of the technical board in the European programme "Chain Reaction" which aimed at the dissemination, training and sustainability practices of inquiry based education. Her research interests include collaborative writing in ESP, Peer-feedback as a learning tool, Content Based Instruction for ESP, Dictogloss, Genre-analysis as a learning tool, IBSE, and inquiry based learning.

David King is a Science teacher from West Waterford, Ireland. He holds a Masters in Education Management from Waterford Institute of Technology and a Professional Doctorate in Education from The University of Sheffield. At the time of writing, he was seconded as Team Leader for Science with the Junior Cycle for Teachers Support Service. David's research interests include Science education, curriculum, education policy, educational change, outcomes based education and teacher professional development.

Ann-Katrin Krebs studied at the University of Hildesheim Biology, German language and literature studies and Sociology and graduated in 2014 with a Bachelor‘s degree in Biology. Between 2015 and 2017, she was scientific staff member of the German "Chain Reaction" team and began her Master studies in Educational Sciences at the University of Education Heidelberg. She did her Master's thesis on the impact of the teacher briefings during "Chain Reaction" and graduated in June 2017.

Rachel Mamlok-Naaman got her BSc in chemistry from the Hebrew University in Jerusalem, her MSc and PhD (Suma cum Laude) in Science Education from Bar Ilan. After conducting her postdoctoral research at the University in Michigan (2000) she joined the Weizmann Institute of Science in the Department of Science Teaching, where she is the head of the National Center for Chemistry Teachers, and the coordinator of the chemistry group at the Department of Science Teaching. Rachel is engaged in development, implementation, and evaluation of new curricular materials; research on students' perceptions of chemistry concepts (especially bonding ans structure); inquiry-type skills and activities – teaching and learning in different cultures (e.g., argumentation, asking questions, hypothising); relevance in chemistry education; the nature of science; professional development of chemistry teachers (in-serice and pre-service). Thus, her publications focus on the above topics. In her work, she uses the experience which she gained as a chemistry teacher who taughed high school students for 26 years, and as a regional consultant for chemistry teachers. She is the coordinator of chemistry teachers' programs in the framework of the Rothschild-Weizmann MSc program for science teachers, and of projects in the framework of the European Union (the FP7 programs) in Israel - PROFILES (which ended on July 2015), and TEMI. In addition, she is an ACS member, the formal representative of the Israel Chemical Society to the various international organizations of Chemistry Education, including the IUPAC CCE and the EuCheMS DivCED; the ICS contact person for all matters of Chemistry Education and relevant activities, and membership in editorial and advisory boards of journals and organizations of science education.

Nicole Marmé has a diploma in Chemistry and a state examination for teaching Chemistry and Physics. She gained a PhD (2002) and a postdoctoral lecture qualification (2012) in Physical Chemistry. Recently, she achieved a second postdoctoral lecture qualification in didactics of science at the University of Education Heidelberg. Her scientific focus is the development and evaluation of problem based teaching units, scientific writing and working. Currently, Nicole is Professor at the University of Education in Heidelberg and Dean for Students of the Faculty for the Faculty for Sciences, Geography and Technology.

Oraib Nawash got her PhD in Plant Ecology from the University of Hoheneheim/Stuttgart-Germany in 2006. Dr. Nawash worked previously on an EU project that promotes for EU grants including FP7 and ENPI CBC MED as activity officer for all Jordanian Universities. She published many scientific papers and children's books and essays that adopt environment conservation and awareness messages.

John O'Reilly is a lecturer in Initial Teacher Education in the University of Limerick addressing areas such as professionalism, planning, curriculum development, active learning, classroom management and assessment for the Postgraduate Masters in Education programme. He is also course director for the undergraduate science education programmes. His research interests are all schools-based with strong partnerships in the areas of science education (IBSE and PCK in particular) and curriculum development with schools(Negotiated Integrated Curriculum), professional development agencies and the National Council for Curriculum and Assessment. He served as a member of the junior cycle science specification development group and has just completed leading the FP7 project "Chain Reaction" focused on professional development for IBSE in Ireland.

Janka Raganova graduated in 1988 from the Faculty of Mathematics and Physics of the Comenius University in Bratislava, where she studied mathematical physics, and gained a teacher qualification at this faculty as well. In 2006 she gained a PhD degree with a dissertation focused on pupils' investigative activities within physics education. Her habilitation theses (2014) have focused at development of methods and tools of active physics learning. Her research interests include problems of theory of physics education, new approaches to school physics based on pupils' inquiry activities and science popularization. She has been involved in several international and national curriculum and resources development projects, including national coordination of international ComLab-2, Chain Reaction and HOPE projects.

Luigi Smaldone is retired professor of Astronomy. His main research interests concern planetariums dissemination and astronomy education in general.

Miriam Spodniakova Pfefferova has graduated from Master's degree in the Education of Subjects – Informatics and Physics – at the Faculty of Natural Sciences of UMB in 2001. In 2006 she was awarded PhD with a dissertation focused on implementation of ICT into physics education. Besides this interest, her research is nowadays oriented on inquiry-based education with using ICT. She has gained some experiences in solving various national as well as international projects focused on the development of physics or science education.

Italo Testa is researcher in physics education. His main interests concern the design of innovative Teaching Learning Sequences and Teacher Training.

Manuela Welzel-Breuer accomplished her PhD study in 1994 at the University of Bremen, Germany, focussing individual learning processes of single students through analysing videotapes in order to investigate their interactions while learning physics. Since 1999, Manuela Welzel-Breuer is working as Professor for Physics and Physics Education at the University of Education Heidelberg. Here, she is educating pre-service science teachers of different school types. In parallel, she is involved in science education research and development in the fields of physics and natural science education for different age groups, science education for gifted children, science education as part of education for sustainable development, the use of computer aided learning environments and the design of effective teaching-learning environments in formal and informal contexts. She co-ordinated a European project on the use of computer aided teaching and learning materials (CAT) in science education, which resulted in an international teacher training course offer. For several years, Manuela Welzel-Breuer served as member of the executive board of the German Physical Society (DPG) and the European Science Education Research Association (ESERA). From 2011 to 2015 she was leading ESERA as President.

Alessandro Zappia is a physics teacher. His main research interests concerns teacher training and inquiry based science education.

Index

R

S

T

U

www.ingramcontent.com/pod-product-compliance
Lightning Source LLC
LaVergne TN
LVHW062145200926
848484LV00007B/1864